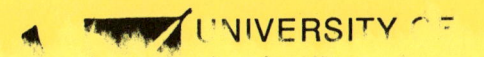

WP 0836846 5

HANDBOOK OF WEAR DEBRIS ANALYSIS AND PARTICLE DETECTION IN LIQUIDS

HANDBOOK OF WEAR DEBRIS ANALYSIS AND PARTICLE DETECTION IN LIQUIDS

Trevor M. Hunt
*Consulting Engineer, Westbury-on-Trym,
Bristol, UK*

ELSEVIER APPLIED SCIENCE
London and New York

ELSEVIER SCIENCE PUBLISHERS LTD
Crown House, Linton Road, Barking, Essex IG11 8JU, England

WITH 40 TABLES AND 250 ILLUSTRATIONS
© 1993

THIS ENGLISH EDITION
© 1993 ELSEVIER SCIENCE PUBLISHERS LTD

British Library CIP Data Applied for

ISBN 1-85166-962-0

Library of Congress CIP Data Applied for

Disclaimer of Warranty

Neither the Author nor Publisher makes any warranty, express or implied, that the calculation methods, descriptions, procedures and programmes included in this book are free from error. The application of these methods and procedures is at the user's own risk, and the author and publisher disclaim all liability for damages, whether direct, incidental or consequential, arising from such applications or from any other use of this book.

No responsibility is assumed by the Publisher or Author for any injury and/or damage to persons or property as a matter of products liability, negligence or otherwise, or from any use or operation of any methods, products, instructions or ideas contained in the material herein.

Special regulations for readers in the USA

This publication has been registered with the Copyright Clearance Center Inc. (CCC), Salem, Massachusetts. Information can be obtained from the CCC about conditions under which photocopies of parts of this publication may be made in the USA. All other copyright questions, including photocopying outside the USA, should be referred to the publisher.

All rights reserved. No part of this publication may be reproduced, stored in a retrieval system, or transmitted in any form or by any means, electronic, mechanical, photocopying, recording, or otherwise, without the prior written permission of the publisher.

Printed in Great Britain by Galliard (Printers) Ltd, Great Yarmouth

Preface

It has been said of one person, that if all the things he did were written down, "even the whole world would not have room for all the books" (Ref: John 85). I do not suggest that wear debris and particle detection is so great a subject, but there do seem to be endless papers progressing at an ever advancing rate on the topic. Each paper is also becoming narrower in its field, with a complication that is designed to keep off all would-be contenders.

This book is written by an engineer for engineers. By that, I mean practical engineers who want to achieve something better with the equipment they use, or want to find out what other ideas are around. I mean also, student engineers who want to learn the subject of Condition Monitoring and how wear debris analysis fits into the total package. Management, too, is not forgotten, for here is the evidence they need to get started and a directory which will be in constant use. It can also be valuable for improving the communication between physicist and engineers, as I attempt to redefine the almost sacrosanct terminology of the 'scientist'.

Wear debris monitoring starts with particle detection. There can be no wear debris analysis until the wear debris is located. That is why the two subjects are included in this book. However, particle detection is a much larger subject than just looking for debris; it concerns looking for contaminants in hydraulic fluids; it concerns determining the distribution of substances in process fluids; it concerns the control of super clean fluids used in microprocessing. These are only really touched on in passing, but the inclusion of the types of instruments they use was felt essential to provide a full awareness of the subject.

'Condition monitoring' sounds good to an engineer. He might even use the expression 'health monitoring' to bring it closer to home. But when a maintenance man is asked how he monitors his machines, the answer is not that encouraging. Most times one finds out that the monitoring has been passed on from previous generations — "We've always done it that way". It might be quite good. It might be useless.

Wear debris monitoring and analysis is no longer the poor relation to vibration monitoring. For many years it was felt that only the extension of the wheel tapper's art could provide sufficient information to an engineer to enable him to maintain his equipment. Vibration was the be-all and end-all which could not be challenged.

However, cracks began appearing in this theory, when it was realised that vibration had limitations both with slow running machinery which could not raise a sound, and, at the other extreme, the highly complex engines with their multiplicity of confusing signals. An emphasis towards the simpler magnetic chip plugs began to occur and achieve success; Ferrography in the early 1970's, hit the headlines; spectrometric analysis was being used by engineers as well as 'scientists' in the absurdly entitled SOAP. Or was it so absurd? after all, engines were beginning to be cleaned up and last longer!

But this book is not an historical review. It is designed to be both a text book and a reference book. Whilst written with sufficient coverage to provide a good basic grounding in the subject, the book is ideally suited as a check on possible practical methods for a particular application. The early chapters are there to set the scene; it is essential to read them first in order to appreciate the finer points which occur later. However, the arrangement of the chapters is such that specific features can be fully examined without recourse to other parts, although there is cross reference where this could provide helpful additional information.

The book is concerned with particles in oils, hydraulic fluids, lubricants, fuels and other liquids such as process chemicals and, even, water. The particles are normally called 'solid particles' (although they may be semi-porous or even hollow). But their origin can either be from wear, as designated by 'wear debris', or can be from a whole host of extraneous sources which are foreign to the process and hence are classed as 'contaminants'. It is important that both are understood otherwise the engineer can easily be misled by the observations he makes.

It should be noted, too, that both Wear Debris Analysis and Contamination Control (as it is called) are subjects of more than passing interest. They have become essential in critical applications — almost all the operators of British aircraft monitor wear debris (amongst other features) as a regular means of engine condition monitoring; and the Royal Navy, for another, examine the contaminant levels as a routine practice on every ship. However, the 'control' aspect of the contamination control is only briefly mentioned because that is a subject of its own.

The description of equipment covers many disciplines and, at first sight, the choice may appear daunting. The current use of the equipment in Wear Debris and the Contamination Control fields is explained in their respective chapters. However, it is quite possible that there will be cross fertilisation in coming years and it will be to the advantage of the user to be aware of the other instruments, which previously have been anathema because of their long complex sounding names! Particularly is this true of the instruments designed to detect and measure particles in the sub-micron range; here of necessity the book must needs simplify the processes involved, because the theory is only really understandable to a physicist and it is only the instrument maker that must be totally aware of the method.

Because the book is practical, there are the occasional examples of real use in the text itself. However, such examples are not only necessary to sustain interest, they are essential to prove the true effectiveness of such monitoring. It is considered invaluable for management to be aware of genuine examples where savings have been substantial. Establishments, who are pleased they took the step of installing such monitoring, because they now see substantial benefits, are, therefore, given a special chapter all to themselves. (Alas not all are willing to share their experience, because they realise how absurd had been the situation previously!)

I would like to sincerely thank the many contributors to the text and illustrations in this book. There are those referred to because they have written papers or books, but there are also several others who have generously given of their expertise, and filled in numerous gaps in my education! In their case, sadly, they are nameless; all that I have been able to do is to mention their companies.

The intention of the book is to improve mechanical systems by ensuring a greater reliability through wear debris analysis and contamination control. I would be greatly honoured if users who achieve just that, would kindly pass on their experiences to me for inclusion in later editions; so that their good fortune could encourage others also.

Trevor Hunt,
Bristol, April 1992

Contents

Preface		v
Chapter 1	Machine Condition Monitoring and Wear Debris Analysis	1
2	Fluid Condition	19
3	Classification of Debris	36
4	Contamination Control	54
5	Methods of Collecting Debris	79
6	Methods of Detecting Particles	106
7	Quantifying Debris	142
8	Counting Debris — by 'size'	179
9	Chemical Identification of Debris	204
10	Physical Identification of Debris	239
11	Multi-value Equipment	254
12	Levels and Standards	271
13	Examples of Successful Experience	298
14	Putting Wear Debris Analysis into Practice	315
15	Bibliography	330
Appendix 1	Listing of Instruments and Services	357
Appendix 2	Listing of Addresses & Functions	446
Subject Index		470
Author Index		486

CHAPTER 1

Machine Condition Monitoring and Wear Debris Analysis

Condition monitoring has one purpose — life.

That is why it is occasionally called 'health monitoring'. In technical terms, the health of a machine is really its reliability. And with reliability, of course, go a whole host of other connected benefits, such as confidence, assurance, continuity, output, security, safety, value and, providing it has not cost too much, good financial return. In other words, life.

Condition monitoring of machines can only provide a return of capital if it increases the reliability of the mechanism or system. In itself, it does nothing. It merely observes. But that very observation, at the most opportune time, triggers off action which brings the required reliability.

Reliability may have been an objective of the '70 and 80's, but it became a practical reality only when condition monitoring was accepted — accepted as the major feature of maintenance. True, there were certain financial short term benefits by using Time Based Maintenance with its regular check on the gauges; maybe even the Unplanned Maintenance was not too much of a disaster providing sufficient spares were in the stores and the fitters enjoyed getting up in the middle of the night (and were paid for it). But Preventive Maintenance involving condition monitoring was a bonus which lifted the whole concept of machine reliability and acceptability to a new plane.

Reliability and the maintenance routine will be discussed more fully in Chapter 14 as we see wear debris analysis put into practice. Before then we look at the practicalities. In this chapter we see what condition monitoring choices are open to the engineer

MACHINE CONDITION MONITORING

Condition monitoring may commence with quite small inroads into the maintenance schedule. This is for two reasons. Firstly cost, and secondly, lack of experience. Cost is understandable; no manager will commit himself or herself to spending money on a function which is 'unproductive' unless the long term benefits can be clearly seen. Experience is related here. The newly appointed maintenance officer may well just not know of sufficient case studies to be able to show that cost benefit; he leaves himself to plead that he might "have a try to see if it works", and most times it does not! The story ends there. (The reason it does not work is usually that no failure mode was experienced during the trial period and hence the monitor was neither successful nor unsuccessful.)

Experience is growing, though. This book will cite many examples from literature, or the author's experience, where just one type of monitoring, wear debris analysis, was successful. Other books on condition monitoring will no doubt add to the success story.

Just what is Condition Monitoring? As has already been said, it is observation; it is a science of detection. It is concerned, through its outcome, with achieving and maintaining satisfactory life of a function at an economical cost. So it is more than observation, it is there to prevent failure and be cost effective. It is this three-fold action which actually brings the acceptable reliability; in other words, the monitoring detects

    a fault which might otherwise eventually cause failure,

    it allows the system to be suitably changed to prevent failure

    and gives sufficient warning, so that the maintenance team can choose a slack time when the plant is shut down, in which to effect the repair or replacement with the minimum of financial loss.

These days medical teams are trying to encourage (at least for the old and the very young) the use of check-ups. They have realised that too much money is spent on corrective medicine and surgery, and too little on preventive medicine. Whereas the rewards in both money and human happiness are far greater in the preventive field, some people may be too set in their ways to take up this offer, but those who do understand long-term benefits will not be slow in coming forward. So it must also be in the mechanical realm.

Just as there are many types of medical monitoring of the body, so there are considerable variations in mechanical engineering monitoring. Below is listed a range of mechanical monitoring methods (and, for good measure, their equivalent medical counterparts in layman's terms).

They are separated into three groups — Steady State, Dynamic State and Fluid State.

Mechanical Monitoring	Medical Monitoring
Steady State	
Appearance	Appearance
Crack testing	X-Ray
Energy Consumption	Appetite
Flow	Ability to pass
Leakage	Loss of fluid
Performance	Work done
Pressure	Blood Pressure
Speed	Movement
Temperature	Temperature
Dynamic State	
Acoustic Emission	Heart beat
Noise	Breathing
Pressure Pulses	Pulse
Response Rate	Reflexes
Ultrasonics	Brain and Heart scan
Vibration	Shaking
Fluid State	
Content	Blood test
Particle Count	Output Fluid test

Immediately it will be noticed how close the monitoring styles are between the monitoring of the body and the monitoring of a machine. It must also be recognised that just as the medical monitoring has been so successful, the mechanical monitoring, if used, augurs well for the future. It offers considerable benefit for long term life and reliability.

FLUID MONITORING

This book, however, only looks at the fluid monitoring. But this is no great disadvantage. The monitoring of the blood in the body is an important technique, if not the major technique, for sensing the condition of a human. The monitoring of the fluid in the machine offers just as much promise.

The fluid, by definition, is a carrier. It carries might and matter from one place to another. It is a force channel, particularly in fluid power systems, conveying power from a prime mover to a practical machine. Or it is the bearer of the lubricant, to be forced between hostile surfaces. It also takes on board unwanted debris which is being shed and needs to be be removed; at the same time it might unfortunately do the opposite and carry solid particulate into critical components as a contaminant.

The evidence of what has happened, in the form of wear debris, and what may well happen, in the form of harmful particles, is all there within the fluid. This is explained in detail in Chapter 2; but suffice for the moment that the reader notes that there is more in the fluid than one might think at first sight. Wilmott 1990 gives an idea of how different techniques can cope with different applications; wear debris ('oil debris') is seen as a major contributor (Fig. 1.1). (I have added some ?? where I am aware of additional possibilities.)

Although oils may come to mind as the most likely fluids in a system, they are by no means the only fluid. Indeed there are important reasons, such as cost and fire resistance, which necessitate the use of other liquids. Some of the fire resistant fluids, such as the phosphate esters,

Technique / Application	Thermography	Vibration & shock pulse	Fluid film thickness	Ferrography	Oil debris Particle On line	Oil debris Particle Off line	Spectro & chemical	Torque measurement	Orifice testing
Rolling Element Bearings – Grease lubricated	x	x		?		?	?		
Rolling Element Bearings – Oil lubricated		x			x	x	x	x	
Plain Hydrodynamic – Oil filled bearings	x		x		x	x	x		
Electric Motors	x	x						x	
Hydraulic Systems – Super clean					x	x	x		x
Hydraulic Systems – General				x	x	x	x		x
Gear Systems – General		x		x			x		
Recirculatory Lube Systems (Turbo)	x	x			x	x	x		
Recirculatory Lube Systems – General		x					x		
Internal Combustion Engines							x		

Fig. 1.1 Technique possibilities for applications (Wilmott 1990)

are highly expensive; but conversely where water and water based fluids can be used, the costs are dramatically reduced. Because there is this range of fluids, this book deals with methods which may be applied to low viscosity liquids, such as water, as well as thick gear oils.

WEAR

'Wear' and 'old age' are words we try to avoid, but they happen, nevertheless. Machinery has a limited life unless parts are constantly renewed. 'Wear' is undesirable removal. 'Wear out' is when there is not enough material left to permit a reasonable function to continue any longer.

Faure 1991 has described in great detail the wear associated with gear meshing. He helpfully lists, first of all, the features which may influence wear:

"the operating conditions
the type of load applied
the relative speeds of surfaces in contact
the temperature
the lubrication
the surface hardness
the compatibility and nature of material present."

He then goes on to define 'wear' in the classical manner, ie. as being the outcome of two surfaces sliding together. This can be caused by a number of processes involving not only the gear teeth themselves, but also the intermediary action of unwelcome particulate in the oils ('contaminant' in the sense used in this book):

"Abrasive wear with two bodies
streaks and scoring
polishing
scuffing (cold or hot)
abrasive wear with three bodies
scratches and grooves
interference wear."

In a conventional spur gear design of involute form, only at the pitch circle is there rolling motion. Elsewhere on the tooth face there has to be sliding and hence wear. In a helical gear form the sliding is even more pronounced.

Faure also gives a chart (redrawn below as Fig. 1.2) indicating the progression from acceptable wear, to excessive wear, to complete breakage. Although it is interesting to be able to detect 'normal wear', it is highly desirable to be able to observe the 'moderate wear' in good time and thus prevent an untimely failure. (It should be pointed out that 'excessive wear' is too late, because at this stage inefficiencies in the system in the form of noise, additional friction and loss of precision will require possible repeat operations as well as extra power requirements.)

	Abrasive wear with 2 bodies		Adhesive wear				Wear with 3 bodies		Interference Wear
	Wear with 2 bodies	Scoring streaks	Polishing	Adhesions or metal pull-off	Hot scuffing	Cold scuffing	Scratches Grooves	Abrasive wear	
Normal wear - slow progress	○→○		○				○	○	
Moderate wear - watch progress			○	○			○	○	○
Excessive wear - gear limit		○		○	○		○	○	
				Breakage					

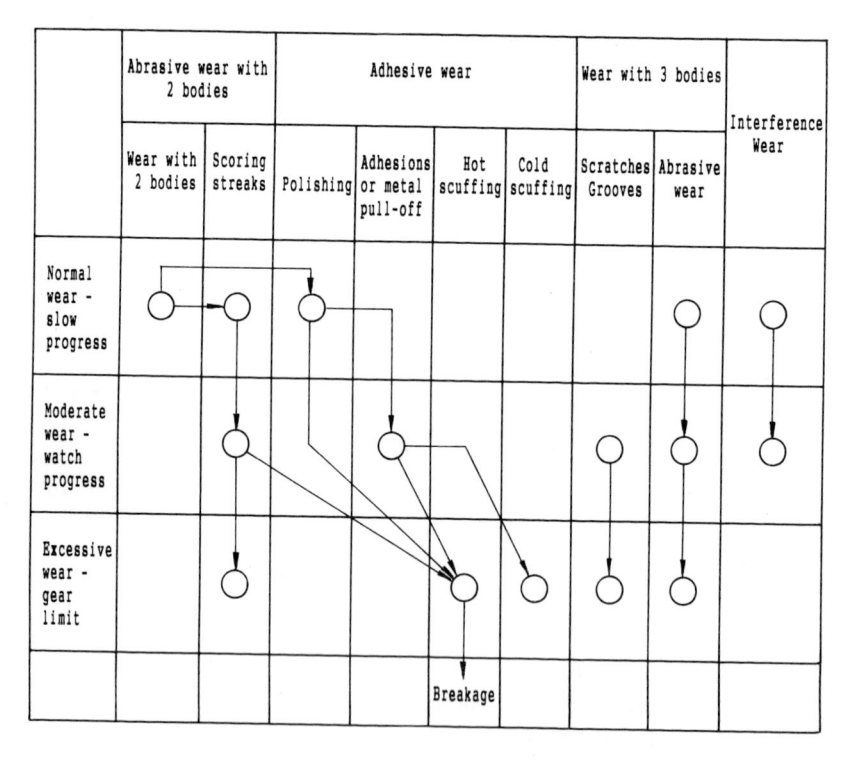

Fig. 1.2 *Progression of wear in gear meshing (after Faure 1991)*

Tweedale 1991 charts the wear in gear teeth more graphically. He looks at two of the features which can be the most drastic in the influence of wear — load and speed. (Other factors could also be the lubrication, temperature, environmental conditions including local vibration and, of course, ingested debris.) In looking at load and speed Tweedale highlights 'zones of distress' where he does include lubrication and debris contamination (Fig. 1.3).

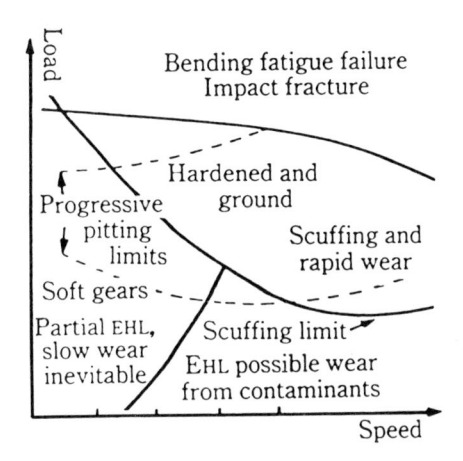

Fig. 1.3 Zones of distress (after Tweedale 1991)

Wear debris particles, in enormous numbers, can be generated by quite simple operations. Fitch 1979 reports on the particles produced by a single stainless steel threaded coupling when screwed into an anodised fitting –

Number of particles greater than	7 μm	57,000
	25 μm	17,000
	50 μm	2,000
	100 μm	17

Although that was unlubricated, an almost similar number occurred when an oil was applied. (Note a micron (μm) is one thousandth of a millimetre, ie. one English inch unit 'thou' is 25.4 μm.)

However, wear, in more general terminology, is not just the result of two surfaces rubbing together, although that is the major cause. Wear can also be considered as taking place when, because of some other motion or external influence, material is lost from the surface of a component. This is usually due to vibration or multiple impact and, hence, could be called 'fatigue wear', but corrosion can also be responsible. Excessive load causing breakage or plastic flow, may also have some influence in developing a kind of wear; at least material is lost from its rightful place and debris is shed.

Fatigue wear debris may either come from the surface in flakes, as is the case with flexing components, or as chunks or spheres from cracks or pits which develop under the surface. Again particles appear in large numbers. Corrosion and cavitation debris also come from a destruction commencing at the surface and moving inwards.

Harvey & Herraty 1989 suggest that bearing wear is not really necessary at all! If the user uses the bearings as the manufacturer indicates (in their case, SKF), then there should be an infinite life. The problem is that the bearings are misused due to such things as inadequate lubrication or excessive debris in the oil, as well as due to design errors concerned with load and speed.

Thus wear occurs. It may not be able to be stopped, but it can be anticipated to a certain extent, and action taken before wear-out. This is particularly true of excessive wear occurring because of a fault in the system. This is where condition monitoring helps in improving reliability, by identifying the presence of a fault.

FLUID MONITORING COMPARED WITH OTHER MONITORING METHODS

Is fluid monitoring really necessary? If one intends using the other types of monitoring mentioned above, perhaps one would be tempted to say that fluid monitoring is not necessary. However, there are two very important points which need to be appreciated —

The evidence in the fluid is to be found nowhere else.

The cost benefit ratio is better than most other techniques.

To be fair to other techniques, some are much more appropriate for certain machine operations than is fluid monitoring. Indeed in certain cases there is no fluid present! Refractory material which is deteriorating is admirably detected by remote thermography. Fatigue in structures may not emit particles into an oil but it does emit high frequency signals which can be detected by ultrasonics or acoustic emission sensors. Electrical machinery and switch gear is another non-fluid function which is well catered for by using power monitoring.

There are also some fluid situations where the analysis of the fluid is not only difficult but may not really be what is required. Consider the use of ram pumps underground. They are fluid power pumps using high water content (95%) fluids at some 290 L/min at 138 bar. The pumps are designed for long life, and would have that life except for one problem — the fluid. But it is not the debris content of the fluid which causes failure, it is the lack of fluid itself. Blockage of the inlet, leakage and drop in reservoir fluid level, can contribute to an exceptionally low inlet pressure which then destroys the pump. In this case the monitoring technique proposed by the author (Hunt 1988), and found to be the most successful, was simply a fluid inlet pressure monitor (with flow rate as an extra bonus).

Another difficult 'fluid' monitoring situation is that which involves sealed systems. A bearing may be 'sealed for life'. Here the withdrawal of the fluid for analysis is highly complex and would not normally be recommended because of the iatrogenic effect on the bearing system; after all, the whole objective is the achievement of a stated life without interference.

A semi-sealed mechanism is not as bad, where it uses grease instead of oil; a very slow flow occurs and by suitable dilution it is possible to analyse the resultant fluid with a measure of success. It is somewhat complex as Jones 1990 has found in his work at Swansea, but the evidence is there; it is the technique which requires further thought. Earlier work by Bowen, Bowen & Anderson 1978 had indicated that the debris was present in the grease; but that work had been restricted to the use of Ferrography which is biassed towards ferromagnetic particles only. More of this later in Chapter 5.

The real comparison between monitoring methods exists where there is a possibility of testing a variety of monitors. Most industrial concerns do not have the time nor the finance to initiate a complete survey of monitors in the hope of finding one which which might suit them best. One objective of this book is to illustrate successful applications of fluid monitoring (within the text and in Chapter 13) and these may well be close enough to show the relevance to a particular industrial example in which the reader is involved. But there could be a lingering disquietude or lack of confidence because of not knowing what other methods might give or not give. The idea of comparative tests and, to some degree, the results, are seen in tests which have already been undertaken; two are outlined below. (A word of caution, though. The results relate to the machinery tested; they may not be applicable to any other mechanical system.)

Example of bearing break-up (Stewart 1975)
This comparative analysis work was conducted at Southampton Institute of Sound and Vibration Research. Because it took place a number of years ago it is of particular significance as regards basic concepts. It looked at a simple loaded bearing rig; the different monitors were far from sophisticated; the 'failure' was a bearing spalling leading to the stalling of the motor. The comparative results are shown in the Fig. 1.4 opposite.

Basic vibration (accelerometer rms), temperature of the outer race, off-the-shelf vibration meter, iron content in the oil, magnetically collected chips and an 'advanced' vibration analysis were tried. Most of the monitors reacted far too late — only the 'advanced' vibration and the magnetically collected debris were in any way successful.

Such a test, as Stewart rightly points out, and as observed above, may indicate the type of result for that application, but it can also give a lead into other related machinery situations, with care. The main difficulty, in moving into a more complex mechanism, is that of cross correlation. What effect are adjacent components having on the monitors being used in the system under test? Vibration, be it solid borne or airborne, is notoriously difficult to isolate; temperature, too, is conducted, and even debris can be carried around the circuit to appear at some quite remote test point.

Stewart's work at that time certainly revealed the cost effectiveness of debris monitoring. There was absolutely no comparison between the cost of the chip-plug and the advanced vibration technique he was developing. However, he chose to investigate and further develop the vibration monitoring into a highly successful unit for aircraft use (Pipe 1987).

Before discussing the next example, a word of caution. It must be underlined that there is a need for diligence in deciding the most appropriate monitor. It is very easy to rush into choosing one which appears to work well, but it may not be the 'best'. Page 321 describes a simple check-list which will enable the prospective maintenance engineer, or his boss, to acquire the optimum monitor. The point is really, which monitor is the most suitable to be able to anticipate the failures which are expected to occur?

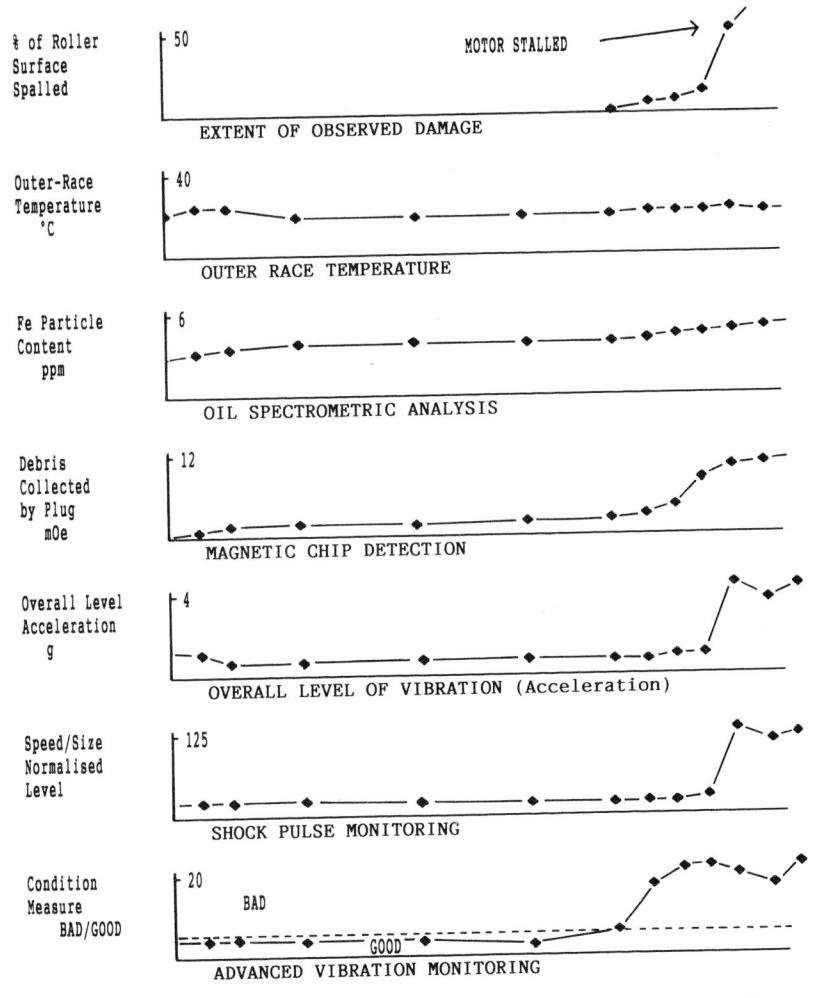

Fig. 1.4 Terminal fault monitoring for rolling-element bearings (from Stewart 1975)

Example of hydraulic pump test (Hunt 1982, Hunt & Kibble 1986)

This example concerns a high efficiency, high pressure, fluid power pump. Fluid is very much in evidence here and hence wear debris monitoring is a possibility. What are the possible failures? Some 14 possibilities (individual components) are described, none of which has a history of failure in its own right; however, four components had deteriorated due to misuse of one sort or another — a swash plate, a roller bearing, a plain bearing and a Glacier DX bearing pad.

Misuse is as important as an inherent weakness. If the result is failure, then monitoring is important, although it may be monitoring the poor operation rather than the deteriorating component direct. Various operators of this pump model had failed to follow the maker's instructions; one had run at below atmospheric inlet pressure causing radial cavitation, another had used a polyglycol fluid which was incompatible with the swash plate bearing material and caused seizure, another ran with inadequate lubrication with the inevitable spalling of the taper roller bearing.

In addition to the above failures, it was anticipated that wear could be caused by inadequate fluid filtration allowing excessive contaminant to get into the piston cylinders. When testing with a high contaminant level it was found that both the piston/cylinder combinations and the control spools wore away in a highly significant manner, producing a drop in efficiency of over 30%. Another failure feature was generated when using a slightly oversize cylinder bore (although this should not get past the manufacturer's inspection, but it might); in this case severe scraping of the piston occurred.

So an all-round analysis of the possible failures indicated the following components/failures:

Swash Plate	Cavitation
Taper Roller Bearing	Spalling
Plain Bearing	Grooving
Bearing Pad	Etching
Piston/Cylinders	Wear
Control Spools	Wear

Each one of these sheds debris. Hence this is an ideal example for the use of Wear Debris monitoring as the method of detection. However, there is a catch. Whilst the swash plate, bearing pad, piston/cylinders and control spools all operate with the hydraulic fluid, the other bearings are lubricated by a separate, self-contained supply. Thus two debris monitors would be needed if all features were to be covered.

A multi-monitor test was conducted on the pump, with a gradually increasing severity of wear taking place. (The wear was provided by means of injected silicaceous dust — AC Fine Test Dust — in the hydraulic fluid.) Although the chip detector worked in a few moments when particles were shed from the piston rubbing in the oversize cylinder

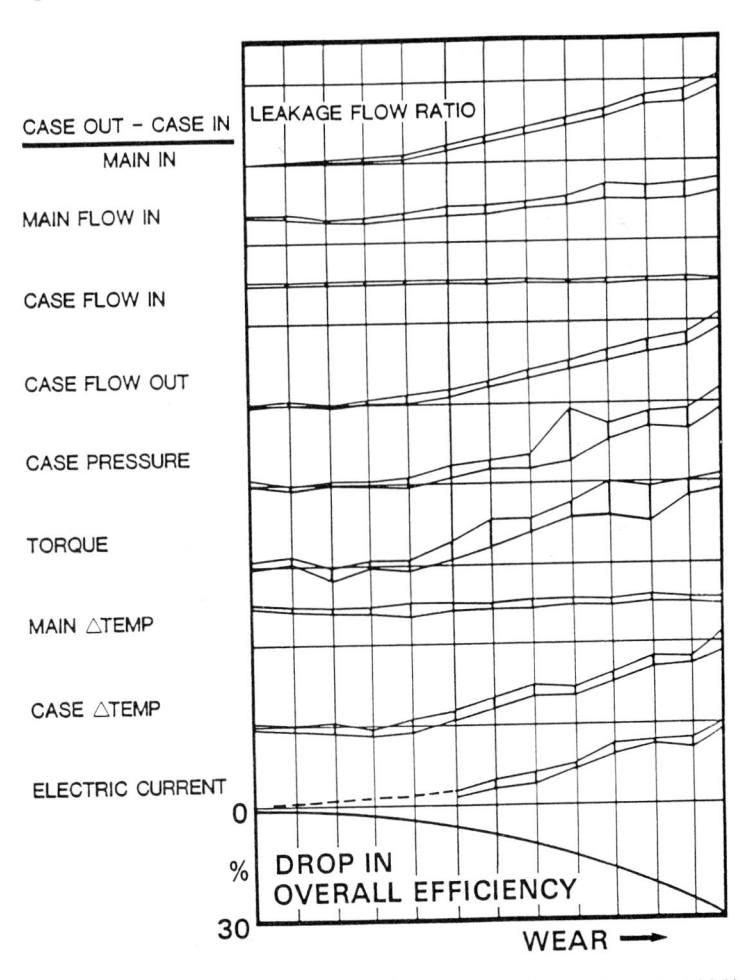

Fig. 1.5 Change in sensor output with pump wear (Hunt & Kibble 1986)

(and thus justified its position as a satisfactory monitor) the tests were primarily to check on other methods which might vie for supremacy. The chart, Fig. 1.5, shows the detectable change with drop in efficiency of the pump (through wear). Each parameter is scaled to the same vertical magnitude (except the one which showed no change).

Although most showed a change, not all were steady and not all were progressive. In the end, only the function 'Leakage Flow Ratio' was seen as a good alternative or addition to the chip plug. Because it only required flow meters in the fluid lines it was certainly easy to fit and reasonably inexpensive. It must be understood, though, that a wear debris monitor might still be the most cost effective; that is where the financial evaluation comes in, to determine the optimum choice.

INDUSTRIAL APPLICATIONS FOR FLUID MONITORING

So far we have identified a variety of components which are suitable for the use of wear debris monitoring. In each case the component likely to fail is 'washed' by a lubricant or liquid fluid in order for the evidence of wear to be transported to a sensor. Any machinery with fluids operating in this way should include wear debris monitoring in the short-list of monitors.

As a survey of possibilities, consider the following:
- Aircraft
- Cranes
- Machine tools
- Mining
- Quarrying
- Road Vehicles
- Robotics
- Ships
- Steel Mills
- Trains.

In fact anything which includes:
- Lubricated or hydraulic controls
- Engines
- Gear boxes.

An immediate thought will come to the experienced maintenance person

— surely vibration monitoring is suitable for all these. Yes, maybe. But which is the most cost effective for that early warning? The answer to this will vary, but the question needs to be asked and all possible monitoring methods considered.

What benefit will accrue with the monitoring? Remember, it is not the detection of complete failure which is required, it is the detection of the initial stages early enough to prevent failure. Or if failure is going to occur, that sufficient warning is given to allow replacement at a convenient time with the minimum of expense and down-time.

The underground mining industry is one where several operating strictures apply. Machinery is fitted in cramped conditions requiring considerable care and time to replace. The shift system means that maximum output is expected at every working shift, and no allowance is made for unexpected stoppage. It is a production line where if one machine stops all have to stop. Instrumentation needs to be to an Intrinsic Safety standard with the added bulk and expense attached. A monitor must not normally be allowed to stop a machine, it can only convey a signal that there is trouble, which the miner may disregard if he so wishes. All told, a very hazardous environment, but one where effective advance warning scores highly; one where wear debris monitoring can pay its way very quickly.

Machine shops are not usually bothered with the idiosyncrasies of the coal industry, except possibly the production requirements. Machine problems in machine tools or robots cause a deviation from the required finished shape. Sizes begin to waver, control is less precise and, if there is close inspection, scrap rate increases alarmingly. Some sort of absolute value needs to be agreed, a level above (or below) which a trend must not be allowed to go. Monitoring the size of the finished product is probably the most appropriate monitor in this case. However, it should be noted that any such deviation which does begin to occur is likely to be due to wear in some part of the system — wear in sliding parts, rotating shafts and gears and bearings — all of which is generating the evidence into the oil or hydraulic fluid.

Engines, be they internal combustion or steam or gas turbine, all require oil, primarily for lubrication but also for cooling. The bearings can be monitored with wear debris systems where there is a flow of oil rather than drip feed or sealed packs. All gear boxes are ideal candidates. The 'power sections' vary; pistons and cylinders are usually oil washed but the steam and gas turbines need the water or air to be analysed, rather than the oil, to assess what is happening to the blading. (Normally, of course, movement monitors including balance would be more appropriate for such components, eg. vibration and torque grids.)

Slow moving machinery has a special claim on wear debris analyses. The sliding motion undoubtedly produces wear material but virtually no vibration or heat generation. Swing bridges and cranes gradually silt up their oil sumps with the evidence but no one bothers to check because it is not easy to notice any change in performance. But it is happening and the evidence is in the oil.

In the field of high power hydraulics another kind of debris is present in the fluid. It is present in the lubricants, too, but because of the bigger clearances in non hydraulic machinery it has little effect. This is the built-in and absorbed particulate which causes problems with refined controls. This will be discussed more fully in Chapter 4, but suffice to say for now, this debris needs to be monitored just as much as the generated wear debris. Rather than being the evidence of what has already happened, it is the prediction of what may happen in the future.

An example of an industrial use where debris monitoring is essential is that of the flushing out of a new system. All the pipework, pumps, motors, actuators and controls have been put together but in the very operation of the connection of the components, although each may have been cleaned, a great deal of debris is generated. This will be from the screw threads, seals and disturbance of only partially cleaned features. This debris needs to be thoroughly removed in case it gets into a critical component and causes blockage or jamming and the inevitable failure.

As well as hydraulic systems such as hydraulic power packs, individual hydraulically powered machines need constant checking to ensure particle contamination is kept to an acceptable level. (This is discussed more fully in Chapter 12.)

THE EVIDENCE REMOVED

There is one great nuisance to successful wear debris monitoring — filtration. Undoubtedly it is essential for high precision equipment and hydraulic machinery to operate; they need refined fluids for long life.

But the feature which helps most in giving that purity of fluid is the same feature which removes the evidence of wear — filtration.

A filter is designed to remove particles above a certain size. True it will also stop a fair proportion of smaller particulate which agglomerates or sticks on the back of larger particles in the filter media. It also does let some larger particles through which may have an unusual shape or length/width ratio. But, basically, the debris is being stopped.

There are, however, two encouraging features. One is that sampling of the fluid can take place upstream of the filter. The other is that debris is kept and can be removed for later examination, particularly spectrometric analysis (Chapter 9). Another possibility is to have a filter upstream of an important component (not a bad idea, anyway) and then sample immediately downstream; in this way only debris from that particular component will be detected and separate identification will not be necessary. Positioning of sample points is discussed in Chapters 5 and 14.

THE WRONG EVIDENCE

Before leaving this introductory chapter it is important to realise that it is possible to get misleading evidence. If a wear monitor is attached to a system so that there is no possibility of external influence, then the evidence is likely to be reliable. This is called on-line or in-line monitoring as shown in Fig. 1.6 below.

If, however, sample points need to be opened and closed as instruments or sample collection bottles are attached and removed (off-line), the process in itself adds particulate to the test sample.

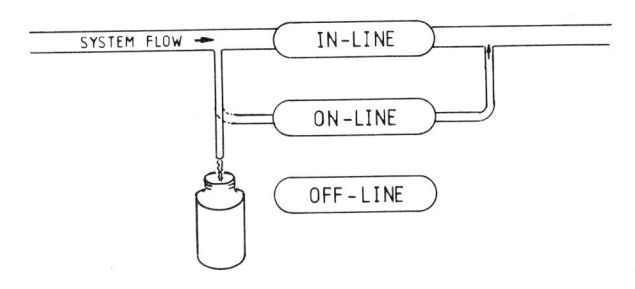

Fig. 1.6 Three types of monitoring position

Dirt in a bottle used for taking a sample will also give a faulty impression about the sample. Wiping sample test points or connections with any but the cleanest non-shedding material will also leave tell-tale evidence. This is a very important part of the wear debris analysis and is explained fully in Chapter 5. The three types of monitoring (shown in Fig. 1.6) are discussed on page 107.

CHAPTER 2

Fluid Condition

THE SECRET IS IN THE FLUID

The great advantage of monitoring a fluid is that the fluid is a ready made communication circuit between the various parts of the mechanical system. It is highly likely to carry the evidence of faults from a variety of positions to some point where a monitor can be fitted.

In some ways this may be a disadvantage. The precise location of a fault is not immediately identifiable; indeed, several faults may be present from different components and the mixture may be confusing. However, the thought that "if there is a fault at all, then it will be detected" far outweighs the disadvantages mentioned above. Here is an effective remote logging method; the information is not transmitted by cable or radio wave, but by the fluid which already exists in the system. What could be cheaper!

Before proceeding, it is only fair to point out, as was mentioned in the previous chapter, that not every fault is washed by the lubricating or hydraulic oil. There will indeed be the 'dry' components which will be failing whose 'cry' of distress will only be detected by other means. Also the grease packed and non-flowing oil arrangement hardly help in this way; but all is not lost in their case, as we shall see later (Chapter 5).

Although normally the fluid monitoring would see evidence from component distress through the whole fluid circuitry as the foreign substance is carried around, there are ways of identifying the origin. One such way is to have a number of sampling points around the system, both upstream and downstream of critical components and filters. Chapters 5 and 14 include some detail on possible positions. This requires the use of a sampling means to be fitted both to high pressure and low pressure lines. There will still be a certain amount of evidence from all the components, but the concentration will be highest immediately downstream of a particular problem.

There are three types of evidence in the fluid — liquid, gaseous and solid.

These are usually quite distinct. 'Debris', the solid one of the trio, is the subject of this book, and will be dealt with in full. For the moment, though, it is worth examining the other two types in order to get the complete spectrum, particularly as these two may have a significant influence on the third. In certain cases it is difficult to differentiate between them, for example in slurries, or where the gaseous content is partly absorbed by the other two substances. However, in all three the evidence is considered to be a 'foreign' substance, ie. it should not be present.

LIQUID FOREIGN SUBSTANCE

There are many chemical additives within an oil. They are present to encourage the fluid to last as long as possible and, at the same time, enable an extended life to be experienced by the hardware. Such additives include

Anti-wear
Anti-corrosion
Anti-foam
Anti-oxidation
Vapour phase inhibitor
Viscosity Index improver.

These are not 'foreign' bodies. They are meant to be there, to help in improving the life of the oil and the components, but they may present hazards to the wear debris analyst. It is possible to draw out the additives in a test (they are generally submicron) and they will be indicated in any submicron spectrometric analysis as specific elements. Secondly, the additive can react unfavourably with some components. The classic example is that of the anti-wear additive ZDDP (zinc dialkyl dithio phosphate) which reacts insidiously with silver plating. Although ZDDP is an excellent anti-wear additive, oils containing it must be restricted to systems which contain no silver or bronze; for those with these elements, the less efficient and older TCP (tri cresyl phosphate) or the more modern 'secret' additives are available. (Producers of all high quality oils always state whether ZDDP is included or not, as a warning to the user.)

The less favourable, foreign, aspects usually relate to compatibility. Compatibility of the oil or hydraulic fluid with components and seals is often a serious system problem. For those involved in any 'unusual' liquid, it is essential that not only is the liquid manufacturer contacted and asked for a compatibility list, but the maker of the system is challenged as to whether all his parts do conform. Each liquid has its own idiosyncrasies, and if they are not complied with, debris will be shed. Pieces of seal, as well as dissolved metal, can lead the analyst away from the true wear debris. Fig. 2.1 below gives a selection of common compatibility levels.

Then again, other unwanted foreign chemicals are present which will destroy both fluid and mechanics. They may have entered the system because of leakage through seals or worn pistons and actuators. They may have been formed because of excessive temperatures causing oxidation.

MATERIAL	MINERAL OIL	PHOSPHATE ESTER	WATER GLYCOL	WATER/OIL	WATER
Natural Rubber	X	X	?	X	O
Nitrile	O	X	O	O	O
Neoprene	X	X	?	X	?
Fluorocarbon (Viton)	O	O	O	O	O
Ethylene Propylene	X	O	O	X	O
Aluminium	O	O	?	O	?
Bronze	O	O	?	O	O
Copper/Brass	O	O	O	O	O
Ferrous	O	O	O	O	X
Lead	O	O	X	X	?
Acrylics	O	X	O	O	O
Nylon	O	O	O	O	O
(X - not compatible, ? - possible, O - OK) (Note the ? varies with surface treatment)					

Fig. 2.1 Compatibility of liquids with seals and metals

The chemicals could be acidic or alkaline, or just 'corrosive' in some other way, like water on non-treated aluminium parts.

Biological growth and gum precipitates could also be thought to be a liquid contaminant. Their presence is due to the atmospheric conditions, hence the need for anti-microbial additives. This type of problem is acute in dilute emulsions (oil-in-water). The presence of the micro-organisms in the water element causes loss of emulsion stability and lubricating properties, blockage of small control apertures including the filters and a general corrosiveness and unpleasant odour. Measurement for these biological features has usually to be undertaken by means of culture tests using pretreated dip-slides. (It should be mentioned that the possible presence of the micro-organisms is one of the main reasons for the 5% solcenic oil in a dilute emulsion. True it is important to have some lubricity and anti corrosion property; but the solcenic fluid is a special mix including anti-microbial content and, hence, there is some guarantee that microbial growth will be contained in the final working fluid.)

The Lantos team (see Appendix 2 — Asesoramiento Técnico en Lubricación Industrial) have come up with some interesting findings on diesel fuel oil. The effect of solid contaminant in the oil is mentioned in Chapter 4, but in this chapter it is of interest to describe a reverse feature, namely the deterioration of the fuel system because of the affect of the oil on the particles of debris. They found that filters became clogged, the fuel system damaged and combustion dropped in efficiency because of this fuel debris. From their experience they noted that only rarely was there a satisfactory elimination of contaminants, instead there could be present additional debris because of the mixing of fuels causing sludges, and the solid contaminants in the fuel oil becoming colloidally dispersed.

Water exists in all oils. New mineral oils may well have between 50 and 100 parts per million (ppm) of dissolved water content. But it is when that water content increases to a level where it begins to come out of solution and form a fine dispersion of water droplets that problems commence. In hydraulic oils this super saturated level must be avoided to prevent loss in performance and reduce oxidation problems; such oil should be replaced if the level exceeds 400 ppm. A reduction in temperature reduces the saturation limit.

Lubricating oils are capable of withstanding much higher levels of water content, but the presence of water is always undesirable, if only because

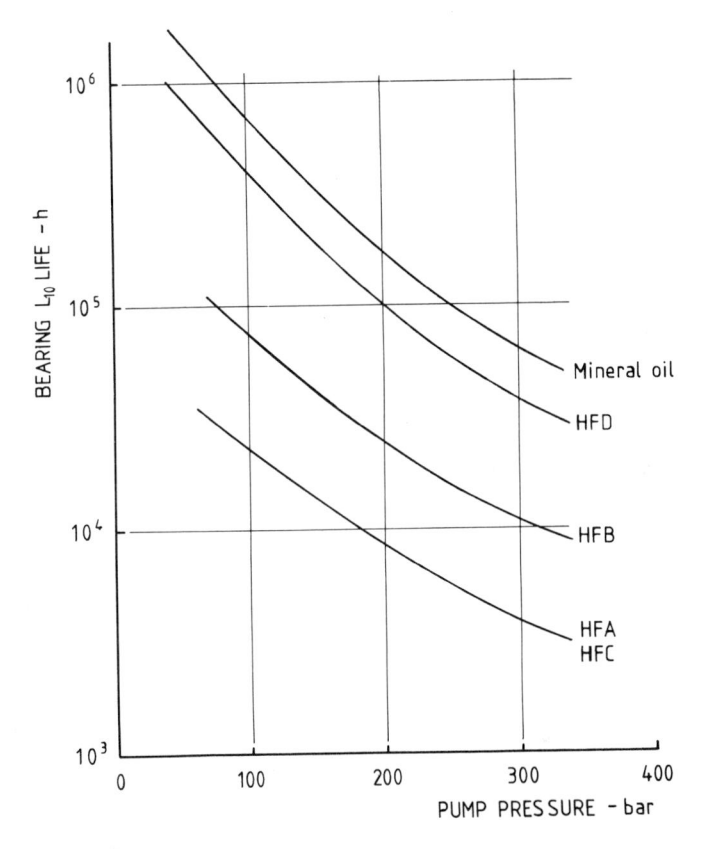

Fig. 2.2 L_{10} *life of bearings in a loaded piston pump*

of the corrosive effect on common metals. Ionnides & Jacobson 1989 go out of their way to emphasise that the presence of water on rolling element bearings is much more significant than many would imagine. Even 0·01% (100 ppm) has been found to drop the life by almost a half. It is not fully understood why this should be except, obviously, the lubricity will drop and also the film thickness with the lowering of viscosity. This drop in life is perhaps even more vividly displayed when normal bearings are run in fire resistant fluids, of varying water content, as shown in Fig. 2.2. (This is a private communication from a pump manufacturer.)

The classifications given in Fig. 2.2 have the following meaning:

HFA — Oil in water emulsion (eg. 5% oil, 95% water)
HFB — Water in oil emulsion (eg. 60% water, 40% oil)
HFC — Water glycol (eg. 40% water)
HFD — Synthetic liquid (eg. phosphate ester, 0% water)

The above example is not a situation of water being a foreign influence, but it serves to show the penalty which will be incurred if water is allowed to enter the system, whether in hydraulic fluid or in lubricating oil.

Fig. 2.3 shows a sodium sulphate crystal stemming from deposits from an exhaust manifold of a diesel engine fed with fuel contaminated with salt water. Such ingress of salt water can cause corrosion deposits on valve seats, exhaust manifolds, turbo-charger blades, etc.

The presence of coolant liquid contaminant can be a gauge of the condition of an engine, or even of the way it is being operated. Analysis of

Fig. 2.3   Salt crystal on exhaust manifold (Lantos)

oil samples for wear debris quite often also includes extra tests for water and chemicals, using tests such as TAN or water monitors, as indicated in Appendix 1, both in the main instrument section and the section on Services available.

> Definition: TAN — Total Acid Number — the quantity of base, expressed in milligrams of potassium hydroxide (KOH) that is required to titrate all acidic constituents present in 1 gram of sample.

The TAN value of a base oil tends to be around 0·2 and gradually rise with degradation. The change is caused by running at an excessive temperature with resulting oxidation; this is accelerated if there is a more than normal aeration (or oxygen presence). Fluids with additives such as ZDDP will start with a TAN around 1·0, or a little over, then begin to drop as they oxidise, and then commence rising with deterioration. This analysis is undertaken comparatively rather than by looking at an absolute value, due to the different commencing levels; in this, a change greater than 1·0 in TAN will indicate that the fluid needs replacing. (It should be pointed out that the viscosity is also likely to rise with high levels of oxidation, with additional corrosion and lubrication problems. The oil will usually darken in colour and the smell will be more obvious.)

TBN, the Total Base Number is the opposite of the TAN value. It conveys the alkalinity of a lubricant, and is particularly of importance where it is known that dilution by weak acids may occur, such as in crankcases. Normally such oils are supplied with a strong TBN in order to balance the dilution effect for some while. It would normally be recommended that an oil is replaced when the TBN has dropped to 50% of its original value, and in this way minimize the likelihood of corrosion.

Water content in the oil can be assessed in a variety of ways. In the laboratory, Dean & Stark glassware is common, or the more modern coulometric Karl Fischer titration. Simple tests like the 'crackle' test or 'visi-cross' give an idea. Even a visual test, in the case of liquids which are normally highly transparent, will show up perhaps 500 ppm or greater by a measure of turbidity or cloudiness indicating the presence of fine water droplets. There are also some quite sophisticated on-line monitors using microwaves, or steam expansion, or infrared, for example.

In diesel engines which burn high sulphur fuels, highly alkaline detergent type additives are used. The concentration of these additives, and hence the ability to neutralize acids, is given by the converse of the

TAN, namely the TBN (Total Base Number) or alkalinity value. The TBN level is mainly of value in smaller engines where the same oil is used for cylinder and crankcases, and hence the neutralisation ability is of importance.

GASEOUS FOREIGN SUBSTANCE

Air is the prime foreign gas. If other gases are present, they are there solely because of the unusual process for which the machinery is being used, although it is just possible that some chemical reaction has occurred resulting in an unwanted gaseous content usually obvious through an obnoxious smell.

Air is present to a certain degree in all liquids. There is a natural absorption, the level of which depends on the atmospheric pressure. What is a problem is when the air content causes bubbles to appear and the compressibility to change. Also the likelihood of corrosion is substantially increased.

If the air content is measured as a monitoring exercise and found to be excessive, then it is quite possible that all that is being indicated is that the liquid flow is turbid (probably due to the design of the system), rather than anything about the machine condition. However, one possibility is that air is getting into the system because the fluid level is low, or there is a leak, or the design of the tank or reservoir is poor. (The tank needs to have a fluid level above the return line of the fluid circuit, unless there is a likelihood of reverse flow causing a problem; this is discussed later on page 85 in relation to reservoir design. A settling region is also helpful; this may not be possible if the tank has to be small in comparison with the total flow being circulated.)

Air can be measured by determining the comparative density of the fluid or by measuring the change in volume when a vacuum is applied to a known quantity of oil. It can also been seen by a gradual cloudiness when the oil being used is relatively clean and clear.

Before examining the solid debris aspect, it is worth looking at the review given by Kohlhaas 1991 on reasons for oil change relating to a number of different parameters in the liquid/gaseous/solid categories:

Viscosity
Pour point
Compressibility
Air separation capability
Foaming characteristics
Demulsibility
Corrosion stability
Oxidation stability
Flash point
TBN
Solid particle contamination
Water content
Silicon content
Chlorine content.

This list in Kohlhaas 1991 is amplified to show what actually influences these particular parameters to cause them to deteriorate, and then goes into detail on the effect on the system because of these values. The overall effect is that of reducing the efficiency and reliability of the system. In other words, the system becomes very expensive to run. The cost of monitoring is much less than the financial losses incurred.

SOLID FOREIGN SUBSTANCE (DEBRIS)

'Debris' always implies a 'solid' foreign body. 'Solid' only in the sense that it is neither liquid nor gaseous; it may be heavily aerated, or with a pitted surface, but it is basically solid. In common parlance 'debris' is the left over from some other bigger system, such as a space rocket which sheds its no-longer-needed accessories. In the engineering field the 'debris' might be considered as having come from a mechanical part, but this is not necessarily true. It can also be considered as, and in this book it is considered as, other solid particles which are present in the liquid as foreign bodies.

An alternative word is 'contaminant' although it has to be taken in context as more common uses of that word are in reference to radioactive or chemical effects. The phrase 'wear debris' is more distinct, although, again, it can be misleading because it can refer to particles which have been shed by mechanical components without actually 'wearing', eg. fracture, fatigue, dislodged.

PARTICLE TYPE	DEFINED BY PROCESS	EASY TO DISLODGE	DIFFICULT TO DISLODGE	IMPOSSIBLE TO DISLODGE
IMPLANTED	Manufactur'g Handling Packaging Storing Transport Flushing Commission'g.	Light weight loose debris such as chips, slag, fibres, etc.	Heavy weight particles with some adherence including gummy items.	Sticky debris such as rust, lacquers, varnish, etc.
GENERATED	Wear Chemical Electrolysis.	Light weight small wear debris and rust.	Heavy weight large particles.	Lacquers, varnish, chemical products, etc.
INGESTED	Breathing Inadequate sealing.	Airborne particles, moisture, etc	None.	None.
INDUCED	Maintenance services.	Airborne particles, moisture, fibres, etc.	None.	None.
ESCAPED	Released by filter.	All debris initially captured by filter.	None.	None.

Fig. 2.4 Categories of contaminant (from Prakash & Gandhi 1991)

Debris, in the sense of contaminant, can be identified in a number of ways. Prakash & Gandhi 1991 suggest five distinct categories as shown in the Fig. 2.4 based on their work. Also included is their secondary feature relating to the ease by which particles could be dislodged within a system.

However, I prefer the simpler classification which uses just three classes as shown in Fig. 2.5.

Built-in debris
There are very few production operations which do not cause debris to be released from the core stock. Even deformation processes cause minute flaking to occur on the surface. Machining is even more obvious, because that is the purpose of the process. Blasting and peening again has this defined objective of removing surface, and not only is surface particulate generated but the very particles used to wear away the

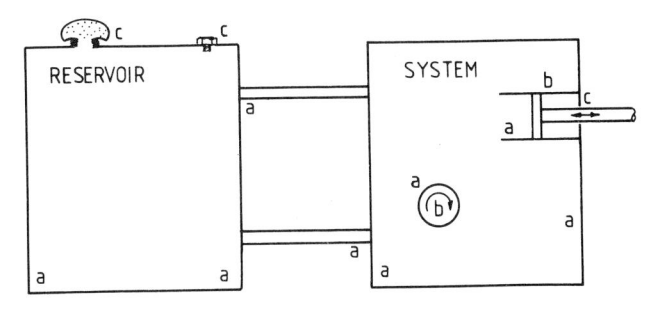

Fig. 2.5 Debris origin — a. Built-in, b. Generated, c. Ingested

surface can sometimes remain. Welding is a prime candidate; the tiny spheres seen like tracer bullets in the air, are very often destined for the minute cavities within a machine, even the oil circuit. (See the description of spheres on page 37.)

Any one or more of these particles may land on the component being prepared or on those close by. True, in a well made mechanism, careful cleaning will be required for each component before the build, but invariably a large quantity of this contaminant will be built into the machine at assembly. As will be seen later in the text, particles are sometimes, most times, invisible to the naked eye, which can normally only see 40 μm and larger.

Then there is the assembly itself. Twisting, turning, tightening, heating, pushing, bending, sliding, etc. are all features present in the build of machines; Chapter 1 gives the example of joining a stainless steel threaded coupling and the thousands of particles which are generated in that simple movement. Each movement is producing debris, most of which is entering the system and getting trapped in dead-ends, in screw threads, in hidden corners.

High quality hydraulic systems are flushed out during build and after the final completed assembly. This usually takes the form of passing a fluid of lower viscosity than the ultimate machine oil, through the system at a higher velocity than will occur later. The full flushing operation is described on page 68; however, it is important to say in this chapter, that flushing is never 100% effective, so debris does remain within the system to a certain degree.

The debris monitors described in later chapters, are now used extensively in these flushing operations. The old fashioned process was to either assume that given a certain time the process was complete, or to take a membrane sample every so often for analysis. Whilst the membrane sample did give the required approval, it took a considerable time if the sample had to be sent to a laboratory.

Generated debris

The 'generated' debris meant here is the typical 'wear debris', plus a few other related types. Despite the lubrication provided to sliding surfaces, high spots are knocked off and a gradual wear process occurs (this is the debris which, strictly speaking, is 'wear debris'). Slivers of metal are released, large quantities of wearing-in 'silt' particulate enter the fluid. This may not be too disastrous to the system, it may actually improve the running of the machinery — a running-in process — but the particles are there and need to be trapped and checked.

Another type of debris, much more of a concern, is that including the bigger chunks given off due to fatigue and over stress. These chunks are not that big, necessarily, but they are bigger than the 5 μm ('five micron') silt particles. They are distinct and need to be related to some specific point in the system.

Some will be obvious, particularly if they are even larger. Pieces of gear teeth are recognisable due to the hardened surface and score lines and sharp edges. Spalled chunks of bearing surfaces have a similar appearance. Non-hardened surfaces are less crisp at the edges. Colour will convey not only heat treatment but also the type of metal involved, such as brass or steel. (This is discussed more fully in Chapter 3.)

Some generated debris will be due to melting. Low melting point white metal plain bearings and other alloys may flow due to local hot spots and then reform as globules in the oil.

Still other debris produced by the machinery will be due to the process being undertaken. In fact this may well confuse any laboratory to which a sample of oil has been sent, unless full details have been given about the process. (The maintenance engineer on site is less likely to be fooled.) Examples of this are industries using specific materials like rubber, plastics, paint. It is important, nevertheless, to assess the presence of such substances if they are appearing in the oil.

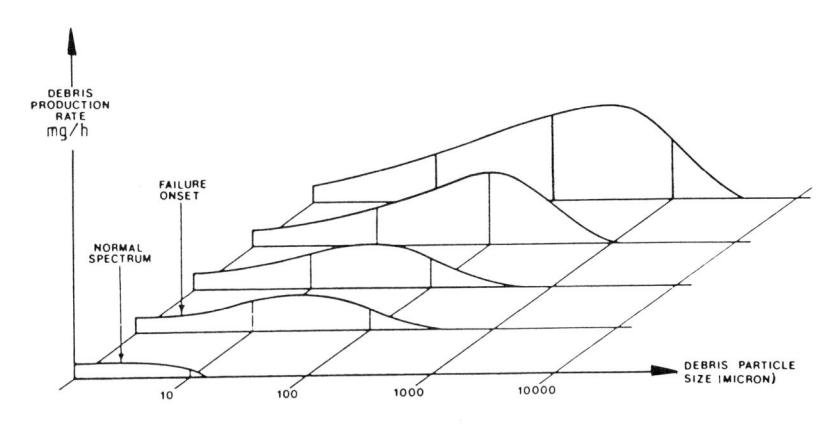

Fig. 2.6   Debris generation rate

Surface treatments can also gradually become eroded in use. One example known to the author was that of the special protective coating given to an underground tank to prevent corrosion from the water in the dilute emulsion being used. Because of heavy condensation and the decision to use ordinary mild steel containers, a highly water resistant coating had been applied. Regrettably the surface had not been prepared properly and large sheets of the coating came off after a while to be broken up and enter the system. This would have caused a greater problem than the corrosion if it had not been spotted in time.

Fig. 2.6 often appears in papers in slightly varying guises and it is not quite clear where it originated (eg. Day 1991, Newell 1991). It does serve as an excellent illustration of the change in debris generation rate which occurs during the life of a system.

Wear debris, the main subject of this book, can thus be seen as 'generated foreign debris'. Whilst it is important to recognise this, it is also essential to realise that there is other foreign particulate which is present and this may cause confusion when the analysis stages are reached.

Ingested debris
Ideally, debris in the atmosphere should be completely isolated from the system fluid circuit. However, this is by no means easy in a real system, and generally some allowance must be made for the likelihood of such ingression from the surrounding air.

There are methods to reduce any interrelationship between machine and atmosphere, but there are some features, such as actuators, which require a certain amount of open lubrication to operate. They could all be covered up, but this is quite often impracticable due to the movement requirements of equipment such as mobile machinery. Hydraulic seals are designed to be wet; the metal components are wiped by the seal rings, but not wiped dry; thus at each stroke, just like the oil exits, a small amount of external contaminant can enter. It is usually sucked in because of a pressure drop within. Having said this it should be pointed out that it is possible to have a forced flow of fluid outwards to prevent this happening; this, however, is wasteful and not normally employed.

Another common entry point is in the reservoir or tank. Quite frequently no attempt is made to seal the tank, indeed this is not desirable unless some separate expansion chamber is provided. Bolt and screw holes on the top surface are left open and debris enters with no trouble. Filler caps are another source on the same surface; not only is the trouble with the air 'breathing' in but the actual filling process passes much debris into the system. This is dealt with on page 76.

FILTRATION

Filters in the system are of interest to the oil user in two ways –

> How well does the filter remove debris?
> How badly does the oil block the filter?

It is necessary to repeat the warning given earlier with reference to filtration. The real purpose of filtration is to remove the debris. If it has been successful, then there is no point in taking samples of oil for debris analysis after the filter — the debris will not be there! One could examine the filter media, and this is done sometimes where a critical or serious failure has occurred and the investigators need to know something about the failure mode. But this is messy, and not normally recommended (see page 105).

Having given the above warning it should be stated that the effectiveness of filtration is rarely 100%, and hence there will be some, maybe just a little, evidence still present in the oil. If one cannot avoid sampling after a filter, then it is worth checking on the efficiency rating of the filter to give some loading effect to the debris counts which have been achieved downstream of it.

Filters are usually described in terms of the size of hole in a plate or mesh, or, for hydraulic filters, the minimum size of particle which generally is trapped. As only a tiny proportion of particles are regular in shape (ie. spheres or cubes), the stated filter rating may only relate to the minimum dimension of what can be a quite large particle which has a high aspect ratio. (This is discussed more fully in Chapter 3 and the efficiency of a hydraulic filter is explained in detail on page 73).

Another type of filter is one which includes a magnetic field. Sometimes this is included within a conventional filter, but mostly such filters are simply a magnetic plug which pulls the ferrous material from the oil and holds on to it. Care has to be taken to regularly examine such plugs, because if there has been a high collection of debris it is possible for this to be released as the flow drag gets larger than the magnetic attraction. Chip plugs are thus only acceptable features for collecting the wear debris, provided the debris is removed regularly — see Chapter 5 for more detail.

The second aspect of interest to the oil user is the ability of the filter to continue operating as a filter without getting blocked by the oil (rather than the solids) — the 'filterability'. It is not immediately obvious as to whether one oil is better than another. It is not that the viscosity of one liquid is higher than another — viscosity differences just mean it takes longer for the oil to pass through a filter — that is all right. Poor filterability, however, means that a deposit is formed on the filter media which has a more permanent blocking effect.

Oil manufacturers and suppliers are aware of the importance of filterability, and at least two have produced documentation explaining how it can be tested. BP Oil Ltd, for instance, undertook demonstrable comparative testing in the past. The information produced at the time, now no longer available, but still applicable, states:

"In broad terms, filterability is a measure of a fluid's ability to continue passing through a fine filter after a considerable time in service. This characteristic is influenced largely by the presence or absence of gelatinous substances formed when deterioration occurs in additive materials, since such substances can block filters to the extent that they hinder the removal of normal contaminating particles; eventually the fluid itself may not be able to pass through."

BP goes on to describe the ever and ever increasing precision and hence finer filtration required in machinery. This has necessitated a rethink as to what additives can be used effectively, especially those which can actually generate a soft contaminant. This can happen, for example, when the additives have an affinity to trace amounts of water (such as zinc-based oils). Other serious additives are those in motor oils included as detergent additives or corrosion inhibitors which produce soapy deposits on filters.

Although oil suppliers do not normally publicise their filterability values — they use such expressions as "good filterability" or "perform excellently" — there are tests which are recognised. The two most common are the Abex Dennison Method and the ASTM F52-65T 'Silting Index' Method, but neither is exactly consistent as there are variables which the analyst can alter. They are explained below.

> **THE ABEX DENNISON METHOD** 100 mL of fluid is timed through 1·2 μm membrane filter under a certain vacuum (the vacuum level depending on the viscosity). That is the 'dry' time. A 'wet' time is achieved by first adding 2% of distilled (and hopefully, prefiltered) water to the sample and shaking for 5 minutes before immediately testing through another 1·2 μm membrane.
>
> **THE ASTM F52-65T 'SILTING INDEX' METHOD** A large quantity of the fluid is forced under a certain fixed pressure (usually a fixed gravitational load on a piston) through an 0·8 μm membrane filter. The time to pass the fluid is constantly monitored and plotted.

Whilst it is recognised that these will, in the same laboratory, give comparative results, because of their variability it can be understood why the actual measured values are not quoted.

Both BP Oil and Shell Lubricants UK have developed their own techniques based on the Abex Dennison Method. BP Oil use a range of filters from 10 μm to 0·45 μm and add 0·1% of water which is allowed to mix for seven days (with agitation) to cause degradation. Shell contaminate the oil with small quantities of water and calcium, a combination known to cause filterability problems with many hydraulic oils and typical of contaminants found in industrial hydraulic oils in service. The contaminated oil is then stored to allow any reaction to take place

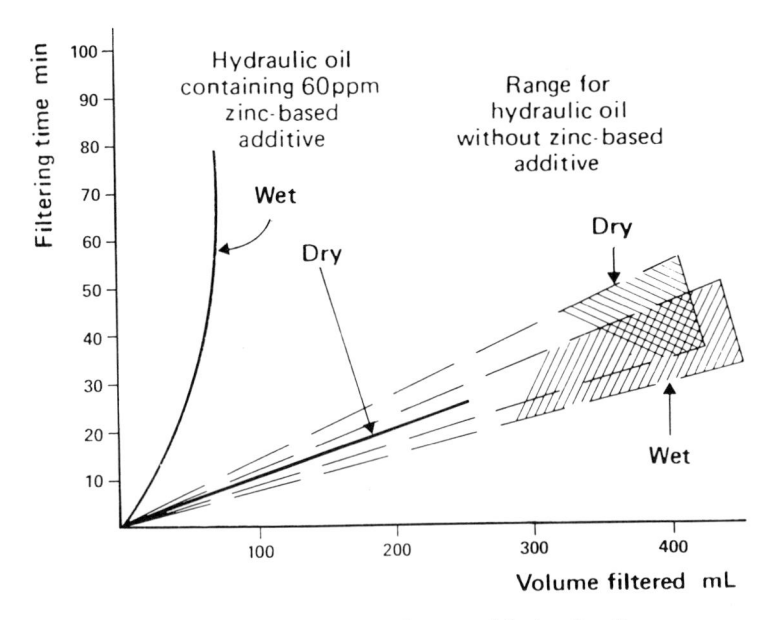

Fig. 2.7 *Filtering time for several hydraulic oils*

before being filtered through a 1·2 μm filter. They consider a satisfactory oil will pass cleanly through the filter within a specified time.

One result using the **BP Oil** method is shown in Fig. 2.7 using a nylon membrane of pore size 1·0 μm

CHAPTER 3

Classification of Debris

The origin of debris has been discussed in the previous chapter. Some mechanical systems may have a proliferation of debris from one kind of source more than another, but basically all three types of origin usually apply every time –

>built-in at the start,
>generated by the machinery and
>ingested from the atmosphere.

It is not surprising that the debris found in the oil is as varied as novelties in a Christmas cracker!

There is not always a clear-cut distinction between one debris particle shape and another; there will be those particles which defy classification because of secondary damage. However, by looking at real shapes, and knowing how they have been formed, it is possible to make a list of sensible descriptions to facilitate more rapid analysis later.

It should be mentioned that Wear Particle Atlases have many photographs from different applications. The photographs may be from optical microscopy or scanning electron microscopy (SEM). Many are given in colour and the particles are described in detail regarding their likely origin. The reader is recommended to purchase at least one such volume in conjunction with this book, if he or she is going to get involved with visual identification of particles. Three are mentioned in the biography; in order of size, smallest first, —

>British Coal Wear Particle Atlas (1984)
>Guide to Wear Particle Recognition (Swansea Tribology Centre)
>Wear Particle Atlas (1982/91) (Spectro Incorporated)

Seven different designations are described below, all based on the appearance of the particle, but also relevant as regards the likely source. The magnitude of the particles in general would vary between submicron and perhaps a millimetre or two (2000 μm). (Where a shape is only really related to a more restricted size range, this is mentioned.)

The seven are:

1. Spheres
2. Distorted smooth ovoids (pebbles)
3. Chunks and slabs
4. Platelets and flakes
5. Curls, spirals and slivers
6. Rolls
7. Strands and fibres.

SHAPE 1 — SPHERES

We start with a sphere. It is the particle we all like to analyse because it is three-dimensionally uniform. Particle counters of the optical type do very well with this shape. Its size is precise and its shape convenient. It adheres well to theory such as Stokes' Law.

The presence of spheres in oils is quite frequent — as are also the arguments as to where they originated! For that reason the description of spheres in this chapter is in some detail. There appear to be two major possibilities, perhaps three, and only the application and history of the machine will give a clear lead as to which is right. (Elemental analysis and micro-sectioning, as described below, are other ways to bring a positive proof, but they are expensive and not always available.)

Jin & Wang from the Tsinghua University of Beijing, published in 1989 a very extensive survey of spheres found during the running-in period of a diesel engine; that work is helpful in suggesting the identification of, and reason for, spheres in other industrial situations.

They were aware that sometimes spheres are found in the new lubricating oil as it comes from a container; the as-supplied oil was, therefore, examined but no such trace was found in their case. In order to compare the spheres generated in the diesel engine with other types, they used a Timken sliding wear machine to produce spheres with a non circulating oil supply. Another type was produced by electric arc sparking where the micro sparks are actually microdroplets of molten steel which assume a spherical shape as they rapidly cool under the action of surface tension. A further source was also noted in the carbon deposit on the surface of the top of the pistons; rough spherical particles of ferrous material were found mixed with metallic oxides and the carbon.

Jin & Wang 1989 used Ferrography to isolate the spheres from the oil, and optical microscopy and scanning electron microscopy to identify their shape and sizes. Etching and polishing were used to examine the interiors of the spheres. They found a variation in size from 1 μm to 20 μm, a surface texture from smooth to highly wrinkled, and both solid and hollow spheres, some with severe shrinkage holes. A selection of these has been drawn diagrammatically below in Fig. 3.1.

The smooth spheres, microspheres (1 μm–5 μm), were found in the first few minutes of the running-in period. A slightly smaller quantity of red oxide spheres were also found at the same time (probably due to the oxidation of some of the ferrous microspheres). The authors felt that all these were actually present before the start (due to previous machining [and welding ?]) and were then being flushed out as the system went to work. Although they continued to appear, particularly at surges in the running, their quantity reduced considerably.

The major wear function in running-in occurs during one and five hours after start. During this period a large number of rough surfaced spheres (5 μm–20 μm) were found; they looked very similar to the Timken machine generated spheres. With the high loads, more of the rough larger wear spheres were generated in the oil.

The outer surface of the spheres was examined by scanning electron microscope (SEM). It was found that the 'melting' spheres (approx 1 μm–12 μm) were almost smooth, but had a very fine surface tracking.

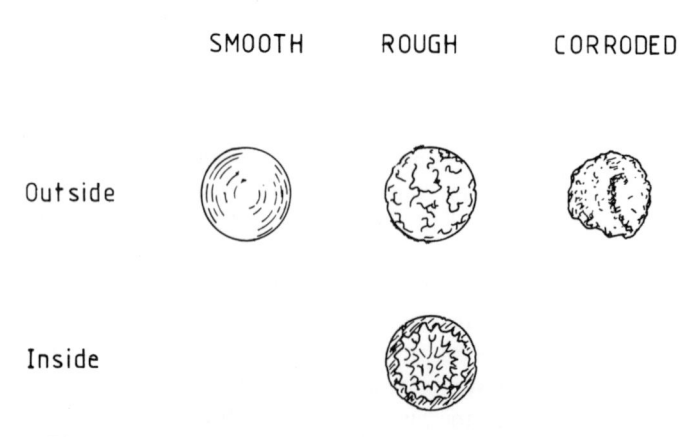

Fig. 3.1 Typical spheres found in oil (after Jin & Wang 1989)

The rough surface spheres (some solid, some hollow) had a more crinkly facing; one hollow example showed a collapsed shrinkage hole. The spheres obtained from the combustion area were distorted and very rough, probably due to the corrosive effects of the gases (CO_2 and SO_2); they could almost be described as of a chunky/pebble shape.

The internal structure of the spheres was either a solid dark core magnetite Fe_3O_4 with an outer shell of red haematite Fe_2O_3, or completely hollow. The rough surface spheres were elementally analysed by EDX (see page 235) and showed the presence of Fe, Si, Mn, Cr and Cu — all constituents of the diesel engine cylinder liner; their internal structure was also highly dendritic in form (almost like rounded stalactites or stalagmites). The inside of those 'melting' spheres which were hollow, were much more rough in appearance although the metallic constituents should have been the same as they had been formed when processing a cylinder liner in an alloy cast iron block.

Thus Jin & Wang 1989 indicate that spheres are produced by melting and rapid cooling of micro portions of metal. Even the initial wear, they consider is due to melting of the high spots and asperities where high surface flash temperatures exist. The rough, more severe, wear, again seems to involve the molten 'casting' because of the internal and external form of the spheres. Although they do not totally reject fatigue as a means of formation, they do not have any evidence for it.

The suggestion that the spheres come from fatigue is a much older idea. Several papers were published in the early 1970's on the subject. It is suggested in the Wear Particle Atlas 1982/1991 that some spheres originate from bearing fatigue cracks, and can, in fact, be the earliest signs of the fatigue; these spheres are usually smaller than 3 μm and come in quantities up to several million in the course of a complete bearing fatigue failure. The spheres are bright, shiny and smooth, and are considered formed by the rolling action of the bearing balls or rollers in the vicinity of the fatigue crack. (It is conceivable also that white metal spheres are possible from plain bearings from grinding as well as from melting). The examination of these spheres by optical microscope is discussed on page 242.

After various forms of 'melting' and 'fatigue' a possible third spherical debris particle is the non-metallic one. There is the glass bead associated with peening. This is more clearly recognisable due to its glassy appearance. There are also formed semi-solid polymer spheres made up from the oil constituents; in their case they have a distinct colour, and a little external heating will cause them to collapse.

Naturally in other applications which may use doses of particles, the particles may, in fact, be spheres. Latex spheres are used for calibrating some particle counters, for instance. However, these do not usually come into contact with the machinery.

SHAPE 2 — DISTORTED SMOOTH OVOID/PEBBLE

These are not really ovoids or ellipsoids. They are smooth pebble like chunks; the sort of egg-shaped particles which, at extreme size, we see so much of on the beach on the sea coast. They are quartz or silicaceous, and semi translucent, revealing attractive colours under a polarised light. Their origin is usually the dust in the quarry atmosphere. Just as the beach pebble comes with many varieties of indentations and outlines, so this shape is somewhat random. They are normally smooth, although occasionally the particles would have been broken in two, exhibiting at least one sharp edge.

Such particles should be prevented from entering a mechanical system by suitable air and breather filtration. However, unless a system is truly sealed, there can be no guarantee that such debris will not enter and be the cause of wear and blockage.

Fig. 3.2 Pebble shape

SHAPE 3 — CHUNKS AND SLABS

The previous shape might be called a 'chunk' except the word does not really apply to pebble-like stones. A 'chunk' is a rough all round well proportioned lump, with width, breadth and length not too dissimilar in dimension. It is sometimes referred to as 'granular'. It is usually metallic, but could be rock. Fatigue spalls, pitting and break-up produce chunks of sharp rough metal, usually with at least one edge showing fatigue lines. Where there are no fatigue lines, then the action would be more of secondary break away.

Rough 'Slabs' are the thick flattish pieces of metal with rough edges. They are almost a combination of platelet and chunk. Usually they have been generated by sliding action involving very high temperatures due to overload with inadequate lubrication. Like the 'chunks', they originate from metal surfaces due to fatigue, but they have come from a hardened surface like a tooth face and will have very sharp brittle type edges and a temperature colouring of straw/brown/blue depending on the severity of the fault.

Fig. 3.3 A Chunk

Chunks of brass or bronze may be confused with tempered steel at first sight, but they can be identified because their scuff markings are more random (also brass tends to be more yellow and bronze more pink). Whilst steel, in the untempered state, shows as a bright silvery reflection, chromium is even more intense.

Machining swarf, at the rough cut stage of manufacture, may cause this type of debris. It is likely to have colouring from excessive heat generated during the operation. Again there could be a smooth and a jagged face.

The 'rock' content will depend on the environment where the machine is operated. Anyone who analyses an oil should enquire about this from the person supplying the sample; for instance chunks of amorphous coal are a familiar sight in coal mining oil samples with a rather indistinct outline, almost fuzzy black. Another black chunk is shale, but this is shiny and hard, and has curved lines on it where it has been separated from the main mass.

It should be noted that the crystalline form of coal can be mistaken for metal, however it is a more grey than silver. This is a fine distinction and needs a keen eye and plenty of practice to ensure reliability in the identification.

Corrosion byproducts may produce an agglomeration of chunky shape. Various patches of colour will be present, and, if rust is the problem, rough brown regions will be predominant. (See also Flakes, below.)

SHAPE 4 — PLATELETS AND FLAKES

Perhaps 'flakes' is the better word for the total category, because 'platelets' implies flat and smooth, whereas 'flakes' are thin but may be slightly twisted. However, 'platelets' is a common medical and metallurgical term which is often used of particles in fluids, and so the term continues to be used.

Fig. 3.4   A Platelet

Platelets in engineering oils come from, firstly, running-in and normal wear. They often exhibit rubbing lines on one surface and have a quite random outline. They are likely to be thinner at the edges of the 'plate'. Below 25 μm these flakes are very common and normally acceptable. Above that size severe break-up is occurring or beginning to occur.

Another source is ingestion of flakes of paint or rust. (Or even the rusting within the system due to excessive condensation or inadequate pretreatment.) These are often obvious because of the distinct colour they have — maybe all green (or whatever colour paint is used), or reddish brown for rust. The will tend towards the chunky shape with a bigger ratio between the thickness at the centre to that at the edges, than that shown by metallic platelets. They will be soft, and hence a 'softer', less jagged outline. Copper flakes at 80 μm size have been identified where a copper fitting grease has been used.

SHAPE 5 — CURLS, SPIRALS AND SLIVERS

These is the more typical machining debris which, at the larger size, is often seen around a lathe or miller. They are mainly produced because of temperature differences across the debris causing the particles to distort. In the case of the small particles in an oil, if these have been produced in the original manufacture of a component then heat colouring may be present due to the high stresses involved. These are normally very brittle particles, and hence sometimes only a small part of the spiral is present in a sample.

Fig. 3.5 Typical Spiral debris

The spiral debris is also produced by a 'ploughing' action in valves and bearings where a harder part of a component, or a sharp ingested particle, digs into a softer surface. This occurs where a fit is too loose and movement occurs, such as with a poor housing to a bearing race, which then begins to 'spin'. This debris is soft and will not show any heat colouring.

SHAPE 6 — ROLLS

A roll is both a spiral and a plate in the sense that it might commence as a platelet, and then be rolled over to give a curved effect, or change its shape because of temperature effects. It is difficult, therefore, to place it in either of the two previous classifications. In origin, it is likely that the discussion relating to platelets is more appropriate.

Fig. 3.6 A roll

SHAPE 7 — STRANDS AND FIBRES

These are the high aspect ratio particles like needles and strings. They are not normally metallic, but, if they are, their colour and reflectivity would indicate so. The most common source is polymeric, ie. man-made fibres absorbed from the atmospheric dust. Cottons and wood fibres also are possibly present.

Fig. 3.7   Fibrous strands

The oil from one rubber company was found to contain vast quantities of rubber strands of high aspect ratio. The problem with long, very narrow, strands is that the filters used in the oil circuit may be totally ineffective, and so they are likely to continue to abound in number if they are present at all.

Metallic needles or strands may be present in the oil because of being built in at the finishing stage of manufacture. Deburring and fettling may generate this type of shape.

SMALL SIZE

The range of size for all the shapes above is almost limitless. However, it should be pointed out that fine debris, below 5 μm, is very common in mobile machinery used in quarry conditions. The hard ingested dust causes a multitude of small scores to occur on the metal faces generating this fine metallic silt debris.

Fretting of surfaces without adequate lubrication can also produce a similar sized debris in great quantity. It is caused by excessive loads, often point contact, producing black oxide of iron Fe_3O_4 (magnetite) at the high temperatures.

METHODS OF SIZING

So far we have dimensioned particles in terms of μm (micron), ie. 10^{-6} metre. For the spheres, this was no problem as the diameter was meant.

But what about all the other particle shapes, the chunks and platelets, the spirals and fibres?

There are a variety of dimensional methods enabling us to define the size of non-uniform shapes. Some relate to 2-dimensional analysis, others to solid 3-dimensional shape. We will examine these shortly, but bear in mind that 'visual' dimensional analysis is not the only means to determine the magnitude of a particle. For instance, the 'strength' of a particle may be more important, eg. how able it is to wear away machinery. Another attribute is the ability to block; again with strong connotations with real machine problems. Yet another could be its metallic content, ie. how much of it has actually come from a component within the machine, rather than been absorbed from the atmosphere outside. These features are used by a variety of non-optical particle detection methods as described in Chapter 6.

2-Dimensional Size

We are assuming here that a single numerical value is to be given to define the size. Obviously if there is a complex shape, a multitude of dimensional data would really be needed to describe the size; but this is not very convenient to measure or to express, particularly if comparisons are to be made.

There are four well known single figure 'diameters' as they are called. (There is an unfortunate tendency to want to hark back to the simple circle or sphere in dimensional analysis, and hence the use of the word 'diameter').

LARGEST DIAMETER

This is the longest dimension, ie across the shape from one side to the other. The orientation of the shape is not taken into account.

FERET'S DIAMETER

This is normally the maximum distance between two parallel lines, set at a fixed angle, which just touch the shape in the position it takes. (Orientation does matter, but the measurement angle must be stated.)

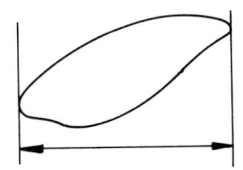

MARTIN'S DIAMETER

This is the breadth of the shape where a fixed angle line bisects the shape into two equal areas. The position the shape has taken up, is critical.

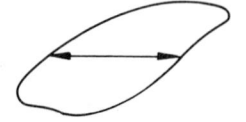

PROJECTED DIAMETER

This is the diameter of a circle whose area is the same as the cross-sectional-area of the shape. This is sometimes called the Equivalent Circle Diameter.

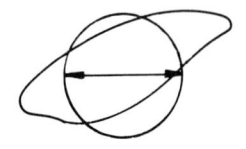

Shapespeare 1991 rightly points out a number of inconsistencies in the above ideas. For instance, although the measurements of the Largest Diameter and the Projected Diameter are the same no matter which way round the particle shape is viewed, this property of invariance is not true with the others. Even Martin's Diameter, except in the case of circles, will have an infinite number of different length lines which bisect the shape.

The word 'equivalent' is quite often used in size analysis, because, as mentioned above, we feel comfortable about a shape which is totally recognisable like a circle. Several of the definitions below use the equivalence to a sphere.

3-Dimensional Size

The 3-dimensional size of a particle can be computed from the 2-dimensional projected area or shape. This is done by simply rotating the shape shown. This may sound grossly inaccurate, and so it would be if there were only one particle, however, if a large number of particles are being examined, the statistical variation in orientation would help to get a reasonable average 'size'. This assumes, of course, that there is no forcing of the particles into any preferred plane as would be the case where they are caused to lie flat on a membrane filter, or where they have to follow stream lines in a sensor; both of these processes must cause a certain amount of error from the true statistical value.

However, there are several specific size definitions for the 3-dimensional case, again all related to diameter. Two of these methods are non-optical, looking at the interaction between the particle and the fluid —

VOLUME DIAMETER

This is the diameter of a sphere whose volume is the same as that of the particle. It gives the Equivalent Spherical Diameter. It is not dependant on particle position.

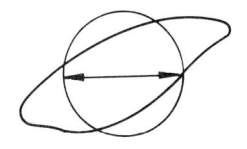

SIEVE DIAMETER

This is the width of a (usually) square grid which just allows the particle to pass through it. It may depend on particle orientation if no force is applied.

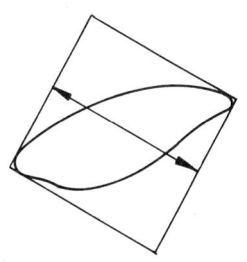

SPHERE OF IDENTICAL SURFACE AREA DIAMETER

This is the diameter of a sphere whose surface area is the same as the particle, including the roughness and porosity effects of the particle.

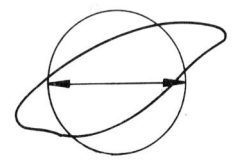

STOKES' DIAMETER

This is the diameter of a sphere of the same density as the particle, which, in a chosen fluid, has the same free-fall rate as the particle.

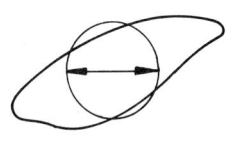

BROWNIAN MOTION DIAMETER

This is the diameter of a sphere which will move randomly at the same rate as the particle, when bombarded by molecules in the fluid. Only applicable to particles < 3 μm.

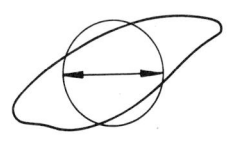

Endecotts 1977/1989 rightly points out that the 'equivalent sieve aperture diameter of a particle' must refer to the second largest dimension because particles are rarely spherical. In other words, for high aspect ratio particles, the sieving technique will give a significantly smaller 'size' value than other techniques.

Other Parameters

Using image analysis a whole new range of size measurements can be made. From observations of the image seen under a microscope, the following assessments can be made about individual particles — note that some are dependent on the current attitude of the particle at the time of measurement :

SIZE:
    Area
    Maximum Diameter (Length)
    Minimum Diameter (Width)
    True longest dimension [along shape of particle]
    Perimeter [real perimeter within a certain surface roughness]
    Convex Perimeter [no account for indentations]
    Perimeter to surface area ratio
    Centroid Chord Horizontal
    Centroid Chord Vertical
    Feret's Horizontal
    Feret's Vertical
    Maximum Feret (Longest Dimension)
    Breadth [Feret measurement at right angles to Maximum Feret]
    Axial Ratio [Breadth/Maximum Feret]
    Equivalent Circle Diameter

SHAPE:
    Aspect Ratio [length to width ratio in straight lines]
    Elongation [length to width ratio along the particle shape]
    Feret's Aspect
    Centroid Chord Aspect
    Hole area (Porosity)
    Circularity [$2(\pi a)^{1/2}/p$, a = area, p = perimeter]
    Roundness
    Macro-roughness Perimeter [including only bigger indentations]
    Micro-roughness Perimeter [including minor surface asperities]
    Lumpiness (Rugosity)
        [(Micro-roughness Perimeter)/(Macro-roughness Perimeter)]
    Shape Factor (Form Factor) [$4\pi a/p^2$, a = area, p = perimeter]
    Sphere Diameter [$2a/p$, a = area, p = perimeter]

DENSITY:
Optical density (Intensity)
Centre of mass (based on variation in optical density, possibly)

All of these features are used in one or another image analyser (see page 243 and Appendix 1 section on Image Analysers). The range of possible descriptions can thus be seen as very complex if taken in total, and the user must try and identify which characteristics are most applicable to the kind of debris he or she is likely to meet.

The range of shape features gives a good indication of the complexity of the particle in comparison to a true sphere. 'Shape Factor' alone can be defined in other ways, depending on the type of particle, which extends the range still further. An examination of sphericity and roundness was reported by Huller & Weichart 1981. They were concerned primarily with the behaviour of particles in engineering systems and the fact that the shape materially affects the movement, however, they could not associate either of the parameters with defined size.

The features Shape Factor and Sphere Diameter are both helpful in assessing the closeness of the particle shape to a sphere. Thus the Shape Factor is a measure of circularity, with a value of unity being a true circle. Sphere Diameter gives the diameter of a sphere whose bisected cross sectional area is equal to the area of the measured particle — thus the value can be used in techniques which only take spheres as the assumed particle shape.

The use of additional parametric descriptions as exampled above, is valuable, but only as a means of describing appearance rather than in suggesting size.

WHICH 'SIZE' IS RIGHT?

There is no problem with spheres. All the descriptive methods would give the correct answer; they would even give the correct answer no matter what the orientation of the sphere was. But how does one describe a non-spherical particle in a way which is understandable to anybody else? Basically one must admit that no way is actually 'right', so there can be no absolute guarantee of what 'size' really means except by using one of the stated methods and stating what method was used.

But even that is not enough. How do you calibrate an instrument which has to be used to determine the 'size'? British Standards, involving calibration of Particle Counters, usually suggest that the only accurate particle to use for calibration is the one that is to be examined later, ie. for debris analysis, actual known sizes of similar debris should be used. They may be mono-sized (all the same size), or they may have a distribution of sizes but who is going to produce this debris and how can it be measured? It is too difficult, so a compromise is reached where a regular dust distribution (such as AC Fine Test Dust comprising mainly silicaceous material), or monosized latex spheres, are used. But this is a compromise.

One may feel that an optical examination of a particle would result in a precise decision regarding its size. But consider, if we assume size means volume, do we mean SOLID volume with full deduction for any dimples or hollowness, and could we see these even if we wanted to? We may not even be able to rotate the particle so that we can observe its exterior appearance all round! Fluid displacement measured in some way (such as the electrical sensing zone principle) could provide a reasonable assessment of solid volume.

It is important that this double warning is appreciated for debris particles:

1. There is NO ABSOLUTE method for defining SIZE
2. DIFFERENT SIZING METHODS are NOT COMPARABLE

However, take heart! We do not need to have the accuracy in the sizing that might be thought necessary. Contamination control standards which cover particle counts from the ones and twos right up to the hundreds of millions, look only at orders of change. In other words a doubling or quadrupling of count has to occur before any concern need be shown. Now if the 'size' related to that count is 20% in error for some reason, then the decision making is hardly affected at all!

The important point is that of being consistent. Because the great majority of ingested particles are chunky, it is fairly commonplace to use an Equivalent Spherical Diameter, or largest dimension. However, this may not be the case with wear debris which tends to be flatter. Already we must sense some confusion developing; and this complexity develops even more when we are involved with particles which deviate into the more exotic shapes!

This need for a deeper understanding of what is being measured will be discussed in a little more detail when examining some of the instruments available (in Chapter 6 for instance). The 'response' type of magnitude measurement may be more acceptable, particularly if it relates to what the particle is doing in the system.

CLASSIFICATION BY ORIGIN

There is another way to classify particles. This is by describing their origin. This is undoubtedly the best way, but one doubts whether it is always right. In a sense, it looks at all the comments made earlier in this chapter and reverses them.

The problem is that there are few cases where one can be sure of the cause of the particular debris. So often one needs to rely solely on the 'appearance' and suggest from that appearance, a likely source. However, for completeness in this book, a typical five types of origin would be as that which follows based partly on the Ferrographic work of Bowen & Westcott 1976:

1. Rubbing Wear, Normal Wear

    This are the generally accepted regular wear particles which are formed from between lubricated sliding surfaces. They would take the shape of 'platelets' up to 15 μm wide and 1 μm thick, down to 0.5 μm by 0.15 μm.

2. Cutting Wear

    These particles are either formed by the metal parts digging into each other, or by encased hard contaminant particulate gouging out grooves in the metal surfaces. The form of these particles can vary just like full size machining swarf — chunks, spirals or even wire like strands as thin as 0.25 μm.

3. Rolling Fatigue

    Spherical particles appear quite strongly under this heading, but see the earlier comments made in this chapter relating to 'spheres'. It would have to be assumed that the spheres come from fatigue cracks in bearings. Chunks of metal can appear from fatigue, as large as you like! (Hopefully in condition monitoring, nothing larger than 100 μm is observed, otherwise there may be serious trouble.) Another form of particle is the platelet, perhaps up to 50 μm dimension.

4. Severe Sliding Wear

These are also large particles, depending on the magnitude of the sliding action load and speed. Usually in the form of platelets with surface stress striations on the wearing surfaces. The higher the stress level, the larger the ratio of large particles to small particles.

5. Combined Rolling and Sliding Wear

All these particles would be too large for SOAP methods, ie probably greater than 10 μm. They are caused by scuffing between gear teeth producing wear and fatigue. Thick platelets or thin chunks are generated with smooth surfaces, ie. they are really 'slabs'.

Beerbower 1975 includes a fascinating list of over 50 ways of describing a failure within a helicopter gearbox, with the origin of the fault in mind. (He mentions that this list also originated with Bowen.) The headings of his brief descriptions were:

Gear tooth or spline — 1. Wear:

Destructive Wear, Abrasive Wear, Galling, Scoring, Frosting, Corrosive Wear, Interference Wear, Burning, Discoloration, Misalignment, Surface Treatment Worn Through, Oil Absent, Corrosion — Other.

Gear tooth or spline — 2. Surface Contact Fatigue:

Destructive Pitting, Spalling — Fan Shape, Arrested Pitting, Pitch Line Pitting, Addendum Origin, Dedendum Origin, Case Crushing.

Gear tooth or spline — 3. Breakage:

Fatigue, Wear, Overload, Misalignment, Quench Cracks, Grind Cracks, Impact.

Gear tooth or spline — 4. Debris in Mesh:

Moderate Damage, Heavy Damage.

Bearings:

Spalling, Surface Distress, Corrosion, Dented, External Debris, Internal Debris, Broken, Grind Cracks, Rubbing Cracks, Defective Material Cracks, Smearing, Glazing, Wear, Grooved, Brinelled, Fretting, Creeping, Spinning, Incorrect Installation, Disassembly Damage, Discoloured due to Temperature, Scratching.

DISCUSSION

This chapter has introduced many ideas, which will have confused the reader at first reading. This is on purpose, because the subject of classification is complex, but it is so often only given a cursory glance. The old adage of reducing all particles to spheres is just not good enough for engineers. It may be all right for chemists and polymer scientists, who deal mainly in spherical entities, but the debris analyst needs to be more selective.

As far as the engineer is concerned, he wants to know what is happening to his machine, or what will happen to his machine. As the debris in the oil is examined, there needs to be a means of describing it. The description may be helpful for management to appreciate, but the basic reason is so that appropriate corrective action can be taken.

The debris analyst must make a choice.

> Do I describe debris by its likely source?
> Do I describe debris by its likely effect?
> Do I describe debris by what it looks like?

It is the old argument between the pure mathematician and the applied mathematician. Both have their place, but in the wrong place they are useless. It is fascinating to give debris all sorts of names and sizes and shapes, and I must add to be fair, these do give the lead into the real world, but ultimately we are concerned with reliability in engineering.

How then does all this merge together? Fortunately the later chapters on methods and instrumentation, indicate techniques which do arrive at a destination. It may not be the whole way exactly, but the monitoring methods are already achieving much success as exampled in Chapter 14. By looking at the possibilities of classification, we have prepared a background to help in the understanding of those techniques, and why some are likely to be better than others — in particular applications.

CHAPTER 4

Contamination Control

Although so far there has been an emphasis on 'debris' as being generated from machinery, there has also been a mention of the concept of particles entering a system from outside. This debris is considered a 'contaminant'.

'Contamination' is a nasty word. It conveys something which is unwanted. But the word carries a much stronger connotation than even that — disaster is imminent!

A recent book published by HMSO has the title 'An evaluation of contamination monitors'; but it is only when the contents are examined that one finds out that the 'contamination' is restricted entirely to radiation contamination. For those interested in 22 different hand-held radiation monitors it is probably very valuable. 'Contamination' in this chapter, however, relates to the solid foreign bodies which enter fluid power and lubrication systems, and cause destruction of the mechanical plant. It is not surprising, considering the seriousness of the effects which result, that a special discipline has grown up with this subject in mind called 'Contamination Control'.

Contamination control is a science all on its own. Many papers have been written on it, even conferences held, and far from being a luxury of an elite few in fluid power, it is an essential ingredient of every modern hydraulic system which claims efficiency and reliability.

Maybe it is not clear what the contamination or contamination control precisely is, or why it is included in this book at all.

> CONTAMINATION CONTROL is the process of keeping the quantity of particulate contaminant in the working fluid, to a level where it ceases to have any damaging effect on the machinery.

The complete subject deals with the entry and detection of particulate contamination, how to reduce its effect on the system, and, finally, with the means of reducing it to an acceptable level. The last part, involving filtration, is only briefly mentioned here as it is covered so extensively in other manuals. However, the other features are all very appropriate to the contents of this book.

The subject of particulate contamination monitoring is tied up very closely with Wear Debris Analysis. Both involve the assessment of particles in the liquid medium. Both know that particles mean trouble. The difference is that it has already occurred with debris monitoring, but with contamination monitoring it can be expected to occur in the future.

CONTAMINATION may be the cause of disaster
WEAR DEBRIS is the disaster.

Contamination control is well proven in the fluid power industry and, historically it has been its exclusive right. Day & Tumbrink 1991 compare the findings of two fluid power surveys; one in 1971 and the other in 1983, where it was concluded that 55% of the problems reported were attributed to the presence of dirt in the hydraulic fluid. No doubt, due to the increased publicity given to the subject, significant improvements in reliability had taken place in the period between the two surveys. The reported findings were as follows:

Breakdown of problems reported in

1. the 1971 BHRA survey and

2. the 1983 Department of Trade and Industry survey

(numbers in brackets relate to systems 2 years old or under) —

	1971	1983
Total no. of systems	15 (9)	114 (33)
No. of problems reported	1 (1)	25 (18)
Real no. of systems with problems	14 (8)	89 (15)
Non-hydraulic faults	–	6 (1)
Material failures	–	12 (6)
Minor leaks	1 (1)	13 (4)
Serious dirt failures	5 (5)	9
Short term problems	8 (2)	20 (4)
Long-term problems	–	29

The last three lines refer to dirt related problems. They are the number of breakdowns being experienced, ie. 'serious' would imply around 3 or 4 per year, 'short term' perhaps 1 or 2. The 'long-term' problems were where the operator expected failure after a few years as the result of 'normal ageing' (but probably due to insufficient contamination control!). The actual failures described were such as accelerated wear to components like pumps and actuators, and jamming of electro-hydraulic valves.

55

In some senses the very disasters produced in hydraulic systems by contaminant, can be just as likely to happen to lubricating systems and fuel systems. Indeed the tribological field has realised, from its own tests that such contaminant has a serious effect on the system.

This chapter will deal firstly with the effects of contamination, secondly with detection of the particulate and, finally, with the reduction.

THE EFFECTS OF CONTAMINATION

The three main areas of distress caused by this contaminant are wear, jamming and blockage.

Wear

Mention has already been made in Chapters 1 and 3 concerning the way sharp hard particles can gouge out grooves in metal surfaces. This normally only happens when two metal faces are brought close together under pressure, and there is relative shearing movement between them, and there are particles trapped in between the surfaces.

In lubricating systems this effect is seen in such common applications as bearings. Several examples are given in the literature of tests which have shown a dramatic reduction in wear when small particles are filtered out of the lubricating oil. For instance, the Glacier Metal Company, in their 1985 brochure, give the following two examples of the difference in wear rate achieved when different sized particles were present. Rather than the larger particles (29 μm) being the most serious, it was those in the region of 3 μm–13 μm; and when they were removed the improvement was indeed dramatic (Fig. 4.1).

SKF, the bearing company, have similar examples using actual bearings which have run in a contaminated oil compared with those in a thoroughly filtered oil. E. Ionnides and B. Jacobson 1989 in the SKF Ball Bearing Journal Special Issue of 1989, review many cases of bearing wear from particles. B. Fitzsimmons and H.D. Clevenger 1975 undertook wear tests on tapered roller bearings, they "found that:

1. The wear was proportional to the amount of contaminant

2. Taper roller bearings will continue to wear as long as the particle size of the contaminant is larger than the lubricant film thickness between the bearing surfaces.

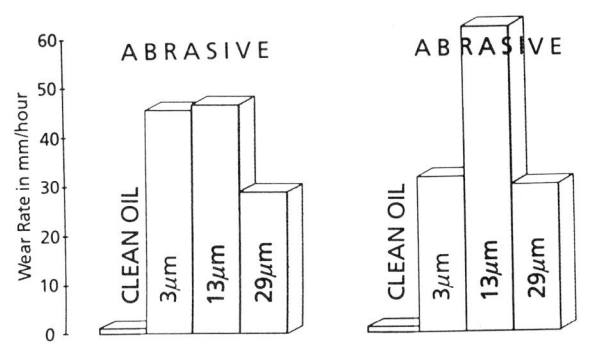

Fig. 4.1 Wear rates in the presence of three abrasive particle sizes with an Al/Si pin on two iron discs (Glacier Metal Co Ltd 1985)

3. For bearings to wear significantly, the contaminant particle hardness has to be greater than or equal to the hardness of the bearing material."

Another discussional point made in the SKF paper followed from the work published by R.S. Sayles and P.B. Macpherson 1982. In their case real debris from a helicopter gear box was passed into a bearing test rig with different levels of filtration. The filters were the conventional fabric cartridge types with various 'absolute' ratings from 1 μm to 40 μm. (Filter ratings are discussed later in this chapter.)

The figure below shows the life effect against different levels of filtration.

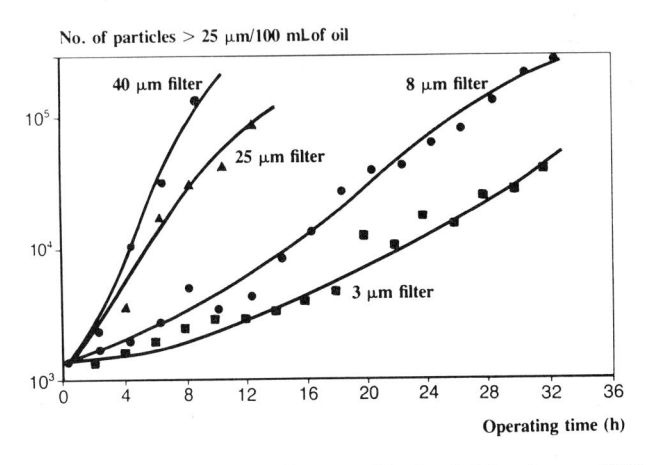

Fig. 4.2 Bearing life with filtration (Sayles & Macpherson 1982)

They came to three very important conclusions:

1. Oil film thickness has very little influence on bearing life when high levels of contamination are present.

2. Finer filtration gives much longer life, "typically 6 times longer L_{10} life with a 3 μm filter than with a 40 μm filter".

3. Early failure of the bearing will occur even if only a short period (half an hour, say) of high level contaminant is experienced.

The SKF review continues "This indicates that the mechanism of failure is associated with a change in surface contact properties rather than with the continued presence of a third body".

A final illustration from the SKF review is worth repeating. Although it relates to one specific test regime, it does highlight the influence of hardness as well as size on the damaging effects on bearings. The material used for the different hardness of the debris was cold drawn brass, mild steel, annealed copper, pure aluminium and various plastics. Not surprisingly, the effect of the very soft plastics was almost negligible, but the size was critical with the other debris. The complete plot is reproduced in Fig. 4.3 below.

Wear in fluid power systems, due to contamination, is a little different to lube systems. It is seen, primarily, in the enlarging of clearances producing inefficiencies. Fluid power relies on very close tolerances, where the fit between a cylinder and a piston or spool is only a few microns.

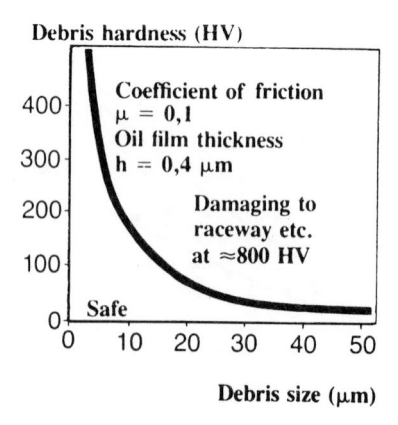

Fig. 4.3   Effect of Hardness/Size of particle (Ionnides & Jacobson 1989)

Fig. 4.4 New and worn pump control spools

This close fit is essential in order to operate with the high pressures necessary, eg. 210 bar is very common, and sometimes considerably higher pressures are used. An example of the drop in efficiency due to a pump control spool wearing because of a grossly contaminated hydraulic fluid, is seen in Fig. 4.4. The photograph shows a new spool (the lower one) compared with the worn spool (approximately 50 μm radial wear of most of the lands with further erosion of the body); the drop in efficiency from new to worn was 11% — and that is only the control valve effect, there is the rest of the system to consider as well! (The pump wear at the same time produced another 22% drop in efficiency from the main output pistons.)

The level of contamination was high in the above tests, about 300 mg/L, but the pump had only run for about 4 hours before that wear was experienced. With a reduced level of debris in the hydraulic fluid, the same result could still have been achieved although it would have taken somewhat longer.

Contamination in fuel also causes wear. The Lantos team (see Asesoramiento Técnico en Lubricación Industrial in Appendix 2) have much experience in the examination of diesel fuel systems. They are aware that the damage caused to engines by contaminated fuel costs more than the fuel does! Fig. 4.5 shows the example of an injection nozzle when new and the same nozzle after 2000 hours running. The erosion was due to catalytic fines and other debris in the fuel; the injection pumps were also damaged.

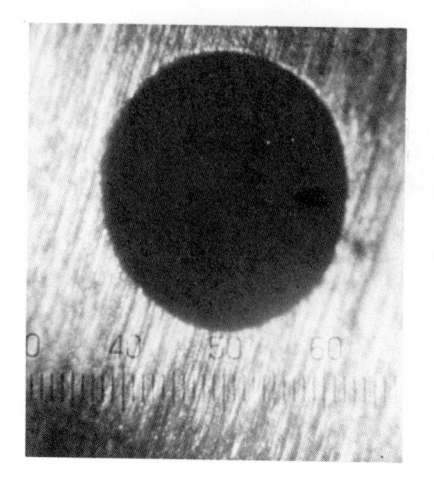

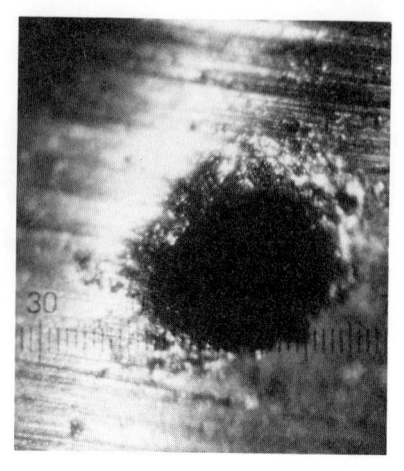

Fig. 4.5 Damage to an injection nozzle — new and after 2000 hours (Lantos)

One problem with diesel fuels, as Lantos reports, is that the filtration of the fuel is not quite as easy as with mineral oils and hydraulic fluids. Centrifugal filters may be used, or even blockage filters, but the solid contaminants in fuel oil are colloidally dispersed, adhering to the nuclei of asphaltene micelles. Centrifugation only separates the biggest micelles, and the remnant micelles, including absorbed solid contaminants, are of a glutinous consistency which tends to clog finer filters. Finer particles, thus remain in the fuel and cause the damage which is so costly.

Jamming
The word used in the lubrication application is usually 'seizing'. We must be careful, though, not to identify all seizures with debris effects. Most would be due to lack of lubrication and local intense heat. But debris can cause seizure of rotating lubricated components when it is of sufficient size. Again one must be wary of suggesting that the debris is necessarily the prime cause; it most often will be the secondary effect, something else having caused the initial break-up and then the debris produced causing the jamming and seizure of other components.

In fluid power experience, the debris can be the first salvo of disaster — even quite small debris. Jamming is a great worry to spool valve users. Spool valves are sliding valves which have to have radial clearances of only a few micron in order to prevent leakage and maintain pressure. One particle of size close to the clearance, or a little greater, is quite sufficient to stop the spool moving across. It must be understood that although there are very high pressures present, these are not the means of forcing axial movement of the valve and hence cannot push aside the debris; there will probably only be a solenoid, or a relatively low pressure pilot line, to provide the movement.

Most solenoid manufacturers insist on very fine filtration up-stream of their valve. In some cases a mesh filter will be integral with the valve. If either of these is missing the guarantee is void. But particles still get through, and the odd one will jam up the valve from time to time.

It is for this reason that 'contaminant insensitive' components are beginning to hit the market. Instead of sliding valves, poppet or flap valves are being recommended in situations where high debris levels are likely; their mere lifting and squashing action, with a measure of flushing, has no problem with the average particle.

Blocking

Blocking, like jamming, is only a problem with the lubrication engineer when the debris is really large, or has accumulated over many running hours. Usually the flow of oil keeps the oilways open and prevents any gradual build up of the debris.

Blocking is not the same as jamming. By 'blocking' is meant the partial or complete reduction of a small orifice or port. This can be a problem again in Fluid Power, in certain valves, but this time in the sense of reducing the flow or pressure rather than completely stopping the action.

This effect may not be just due to a single particle, although that has been found occasionally; more common is the accumulation of many fine particles in the sense of 'silt'. Silting is a gradual process, almost innocuous at first, but then insidiously building up to cause a serious change in circumstances. This is particularly evident where there is a slow or zero flow on the inside surface of pipes, and in dead-ends. It is generally felt that around and below 5 μm is the silting particle size. (This is emphasised by the International Standards Organisation (ISO) whose 4406 Standard uses 5 μm and 15 μm as the two key sizes to evaluate contamination in an hydraulic oil; the 5 μm size being the one which causes silting and the 15 μm size being more related to wear effects. See page 277.)

Blocking also occurs because of monitoring! Magnetic chip collectors, if not frequently examined, cause an agglomeration of fine debris to be formed. If the flow forces acting against this debris become too large, a mass of particulate is suddenly released, rapidly filling up and blocking an orifice! (Or, indeed, causing jamming.)

Disasters
Examples of disasters, because of the presence of debris in the fluid, are difficult to publicise. Most companies are not keen for their names to be mentioned! Another reason is that the evidence is so small, it is easily lost during the strip of a component. Certainly examples of this are known to the author regarding spool valves; the particle has been removed at the time of investigation, only to be gently brushed aside into oblivion by an unseeing hand or arm!

One West Country example concerns flushing. Flushing is the exercise of passing precleaned liquid through the system, at the time of build or after a particular experience, to get rid of the particulate ingrained in the system. (This is discussed further on in this chapter.) The debris is removed by means of filters as the flushing fluid passes round the system. The problem, in this case, was that the maintenance engineer did not realise how dirty the system was at the start. When, therefore, he found the filter element was being indicated very frequently as fully loaded (by a pressure drop sensor built into the filter), he assumed the sensor was in error, and certified the system 'cleaned'. Within one month two pumps in the system had been destroyed. An analysis of the oil indicated the presence of pump debris, but also significant particulate which could be sourced to the original manufacturing process!

Whether the stoppage of a machine through contamination is a disaster or just a passing nuisance, is affected by two highly important aspects:

the fineness of the engineering process, and

the seriousness of failure.

To some extent, the second (the seriousness) has been exampled in the previous paragraph. But there are other engineering reasons relating to serious effects other than loss of hardware. Consider underground mining. The fitting of a ram pump in a coal mine, apart from its actual purchase price, could cost many thousands of pounds because of the man-power involved. This is tripled if it has to be removed, renovated and replaced due to a fault developing. It is not just the fact that it might have to be dismantled to be moved; the inconvenience of a long low height mine shaft; the inability to use certain tools due to safety factors; the manhandling necessary where a machine loader cannot be used. This is costly, and a failure is serious.

The seriousness of failure is also well known to those industries involving 'continuous processes' like mills and production lines. Pilkington Glass, if they are not to incur heavy financial penalties caused by having to remove and reset the glass on the extensive production length, must keep their 'Float Glass' process line running continuously. Each part of the process needs monitoring, and spares need to be available for instant replacement should a fault develop or be anticipated. The hydraulic fluid is monitored on line with a Fluid Contamination Monitor to ensure that the fluid cleanliness is kept to an acceptable level.

Another highly important aspect of failure is where it concerns human life. Aircraft, both fixed winged and helicopters, are critical, particularly where they are in a civil application and no parachutes are available. Sadly, there are many examples of such debris related failures, sometimes due to ingested contaminant (often from airfields), sometimes from initial break-up of a secondary component. Take, for instance, the remarks of Stuart Bell QC in the fatal accident enquiry on the November 1986 Chinook helicopter disaster off the Shetland Islands:

> "I am advised that condition monitoring is not a new principle. Its introduction would be expensive. Considerations of expense should be far outweighed by the considerations of safety for persons travelling in helicopters." [The Independent 2.7.87]

The 'fineness of the engineering process' really means the opposite to how rough the engineering is. A slow moving, rather slack, piston, in whatever application, is doing a job, but it is unlikely to be bothered by debris in the oil. However, a high pressure, high efficiency piston pump is certainly a candidate for disaster if there is debris around. That 'fineness' gets even finer with precision measurement, control and manufacture.

A visit to a small machine shop which has been operating for a good few years, often reveals vast quantities of dust and general machine debris on the floor and on the machines. The cutting stations are cleaned with rags, and that is about it, apart from the once a week sweep. To suggest contamination control to them will not only be quite foreign but it is also likely to be unnecessary. The output has been acceptable to the customer for many years, and although there has been the odd tool breakage, the machines are "as good as when we bought them".

On the other hand a visit to a servo valve machine shop in Japan, for instance, will reveal a totally different situation. Here is an atmosphere closer to a high class office, except instead of carpets there is an immaculate glossy floor with the machines reflected as on glass! They know, as many other valve manufacturers do, that contamination destroys their machinery, their product and their livelihood, unless it is removed at source.

THE DETECTION OF CONTAMINATION

This is, of course, the prime subject of this book. An analysis of an engine lubricating oil or a hydraulic fluid will reveal both wear debris and contamination. The purpose of the analysis will be to identify in some way, which is which. That does not necessarily just mean the elemental identification of particles; nor indeed that elemental analysis is an essential. Other features, such as size and shape and quantity may well be sufficient.

The investigation of contaminant is with the ultimate purpose of ensuring long life for the system by removing its unwanted presence. It does not matter whether the particles are what might be called 'wear debris', or are ingested from the atmosphere, or are a combination of the two, such as original machining and welding debris. The effect on the system will be similar, as shown in the subsections above — wear, jamming and blocking will occur due to the presence of all sorts of solid particle.

Contamination detection, therefore, tends to be concerned with quantity and size. It has to be based on statistical confidence or likelihood of occurrence. We all know that one single particle, of suitable size, in an otherwise totally clean fluid, can completely destroy a system! But the likelihood is remote, to say the least. There are fine scientific instruments, particularly where the failure of such would bring financial disgrace (such as in space) where every conceivable action is taken to completely remove all gremlins and bugs and any foreign body. Restricted entry clean rooms and the wearing of overclothes made of non-shedding material become a routine way of life for that sort of contamination control.

The statistical aspect of the detection is taken care of by the use of Standard Classes of Contamination. In other words, with a reasonable degree of confidence it can be stated that provided the level of contamination (in terms of quantity or count) is below a certain level appropriate for that system, then an acceptable working life will be achieved. Actual standards are discussed in detail in Chapter 12.

THE CONTROL OF CONTAMINATION

Although standards, as described in Chapter 12, are available for a whole range of different systems, it should be borne in mind that the levels are not just quoted as checks. They are there to be met. By the means described below it is possible to reduce the contamination quantity until it reaches or is better than the quoted level.

Mention has been made that contamination is concerned with quantity. If we look at the numbers of particles acceptable to one industry as against another we get a chart something like the following:

MECHANICAL HANDLING	80,000
MARINE INSTALLATIONS	40,000
MOBILE MACHINERY	20,000
AIRCRAFT TEST STANDS	10,000
MACHINE TOOLS	5,000

Number of particles >15 μm per 1 litre of hydraulic oil acceptable in different industries (approx. only)

However, oil from an ordinary drum as supplied from a manufacturer may enter a machine at levels far in excess of the required cleanliness. Not that the manufacturer has produced the debris with the oil, but when the drum of oil arrives at a firm, the cap could have been left off for a while, dirty pipes or a dirty container could have been used to transfer the oil to the system. For the cleaner end of the scale, even the normal drums used by the supplier may be too dirty.

So even before we generate any particles by using our machine, the acceptable level may have already been exceeded. It may also have been exceeded from what was built into the system at the initial build. Key aspects such as this need to be addressed seriously, therefore, the rest of this chapter deals with five actions for cleanliness, which, if taken, will enable the system to achieve its acceptable level. The build and operation will be discussed first, followed by the design of systems, to make them more able to cope with contamination problems. A final section deals with filtration.

Five Actions to control the contaminant level
Action 1 concerns the manufacture and supply of the individual components used in the system. We will describe the action necessary for the most critical contamination control, less stringent action can be taken with other systems –

ACTION 1

On completion of the manufacture, the components should be washed several times in an ultrasonic bath using precleaned fluid until the contamination level of the washing fluid, after the wash, is below the maximum permissible level.

The working surfaces and fluid cavities and bores should then be sealed by suitable non-particle-shedding capping and film.

Additional information is available from BFPA/P48:1988 'Guidelines to the Cleanliness of Hydraulic Fluid Power Components' which will help in ensuring that the components are of sufficient cleanliness when supplied and before build commences.

The components are then ready to be transferred to the build environment. Here is Action 2, which in some ways may affect Action 1. What level of contaminant exists in the atmosphere? It is totally useless to expect that the final build will be clean if the atmosphere in which it is being undertaken is loaded with debris.

The discussion on spheres in Chapter 3 highlighted the ease with which welding debris enters a system. It is not a 'light haze' in the air which results from welding — it is a storm of debris destroying machine and man! Beware. Grinding is another similar atmospheric debris-generator. It must be clear that it is essential for such processes to be isolated from the area where the system build is taking place.

As with Action 1 we will describe the next action with regards to the most refined of operations, eg. the building of spool valves.

ACTION 2

Ensure the atmospheric contamination level is reduced to a level which will not adversely influence the final build.

The use of a 'clean' room incorporating a 'clean cabinet' is recommended.

The clean cabinet is a laminar flow cabinet which draws air in from the outside through a special HEPA air filter. The flow pattern, such as produced in the cabinet in Fig. 4.6 from the Bassaire range, ensures that the great majority of the cabinet volume is swept clean by the cleaned air. The cabinet also acts as a means of providing a positive pressure in the adjacent room ensuring that atmospheric debris is pushed out rather than sucked in. Additional features such as clothing to cover hands and feet, sticky mats and dual dooring may be necessary in applications involving ultra-clean assembly or testing.

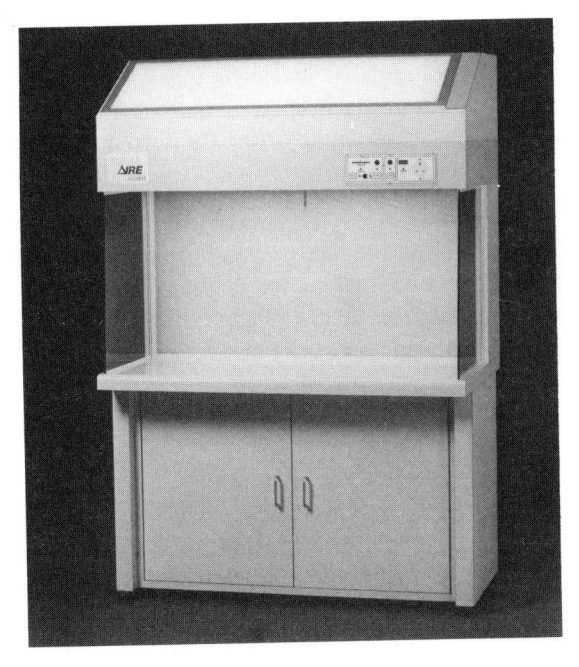

Fig. 4.6 A laminar flow clean air cabinet (Bassaire)

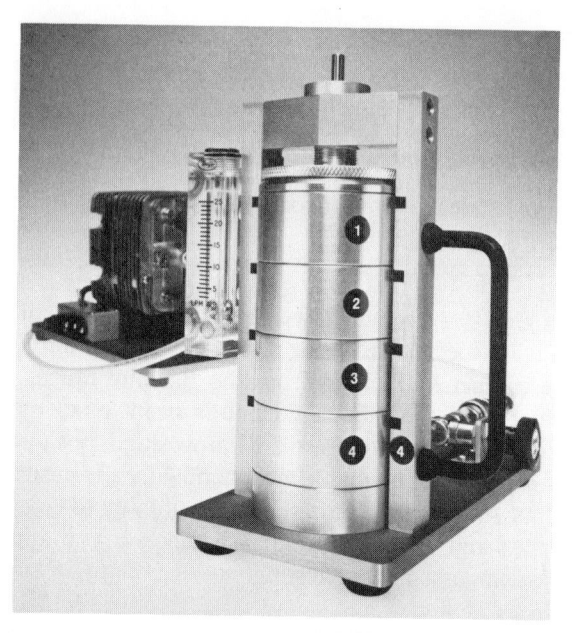

Fig. 4.7 Dry particle detector (California Measurements)

The atmospheric pollution (particle contamination) can be measured by dry particle analysers. One device from California Measurements will not only trap particles but automatically set them for SEM analysis (see Figs. 4.7 and 4.8).

Action 3 is the flushing out of the system. No matter how well the build has been undertaken, invariably some debris will have entered or been generated by the connecting up of the components.

ACTION 3

Before fitting super-critical components, blank off their connections and fill the system with a suitable oil for flushing.

Force the oil through the system as much as possible, using pulse flows and flows higher than are likely to be experienced. Continue until the fluid leaving the system has reached the objective level of cleanliness.

Fit all components and reflush with precleaned oil. Operate the system normally and as violently as possible to dislodge all debris.

Continue until the fluid reaches the objective level of cleanliness. Replace the filter element immediately each time the filter pressure drop indicator shows that a change is necessary.

If a special flushing oil has been used, drain off as much as possible and fill with the final fluid or a rust inhibitor. Seal all open connections.

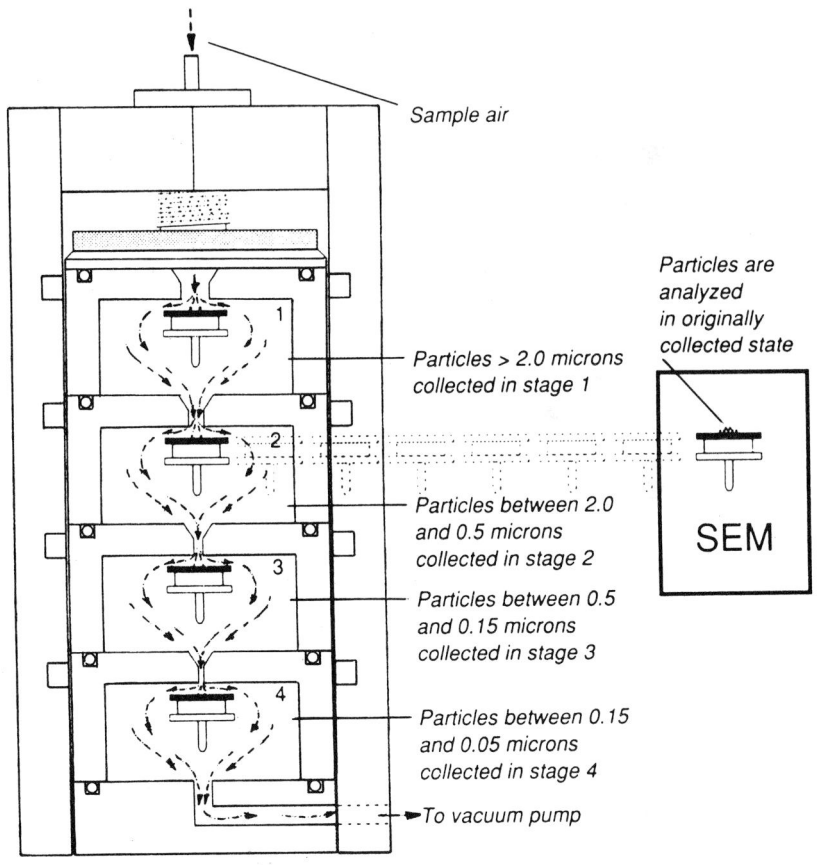

Fig. 4.8 Working operation of the dry particle detector (California Measurements)

The type of fluid or oil used will depend on the final application. The flushing fluid must be compatible with the eventual fluid. If anything, it is helpful to use a lower viscosity fluid for the flushing because the more turbulent the flow the greater the effectiveness of the flushing. (The final fluid may be used if no other fluid is available.) Another point to note is that the flushing fluid must be precleaned to well below the level which is the objective, possibly two Contamination Classes better.

A fine filter must be used where the fluid is being recirculated. Perhaps a 3 μm or even a 1 μm. The filter must have a 'clogging' indicator (pressure drop), and action to replace the filter must be taken religiously — it may be necessary to replace it several times before the system is finally acceptable, and this could take many hours or days. Obviously a high dirt-holding capacity filter would be helpful. Further detail is given in BFPA/P9 1992 'Guidelines for the Flushing of Hydraulic Systems', which also outlines seven levels of application to which flushing needs to be applied to a varying degree — right up to the clean room condition with a separate flushing circuit attached.

Action 4 is the filling of the system with the operating fluid. A little earlier we mentioned the possibility of a large quantity of debris entering the system. However, before accusing the oil manufacturer of negligence, just think why that initial oil has entered at such a high contamination level. True, it could be the manufacturer, and indeed he will not necessarily have prefiltered the oil to such a fine specification unless he is asked to do so (and paid accordingly). However, there are other dirty links in the chain. How about the answers to these questions -

What was the drum like before it was filled?

Has the drum been left outside in a vertical position, in a moist and dirty environment?

How was the oil withdrawn from the drum?

Was any filtration used in the oil transfer?

Basically what is being said is that oil transfer from source to machine is another critical path, where contamination can easily be ingested into the machine. So

ACTION 4

Oil drums should be kept horizontal so that no atmospheric debris or moisture settles close to the entry/exit cap.

Use an oil transfer means which incorporates a suitable filter.

Care should be taken not to let any debris on the outside of the filler tube enter the system.

Always recap the drum immediately after use.

There are many oil drum pumps for removing the oil from a container. They make much of the ability of being able to be fitted easily to the drum, but most of them make very little reference to where the oil is to go! The arrangement should be an 'oil transfer' system with particular note on the filter included in the system. This filter is critical. It needs to not only remove the debris from the oil, as it passes from the drum, but it also needs to be changed as soon as it becomes loaded. This means the system should include a means of signalling when the filter needs to be changed, by using a pressure drop indicator. One example of an oil transfer trolley is that supplied by CJC Napier Limited.

Oil filling had become such a problem in some mining applications that in an increasing number of cases the filler cap is positioned on the side of the tank instead of the top, so that it can only be filled by special above atmospheric-pressure fillers with integral filters. (It also prevents items being hung in the reservoir in order to warm them up for lunch!)

Indeed the filling of oil reservoirs and sumps is a major hazard. Tests undertaken on motor vehicles revealed that the debris in the sump significantly increased each time the sump was 'topped up'. What was added was not just the 'clean' oil but also the contaminant in the immediate vicinity including the rag used to wipe the dip-stick and any filler can or funnel which happened to be lying around. (It must be repeated, the as-supplied oil, even in a sealed can, is not designed to be super clean, unless a specific quality has been requested.)

Although a system may be satisfactorily cleaned at the start, there is no automatic conclusion that it will remain so. Certain actions need to taken during the working life and these are outlined in the next action which, in a way, is more concerned with management.

ACTION 5

Decide on a maintenance programme and ensure it is kept. Include not only the machine checking (what, how, trends, etc.), but who is responsible to put it right.

Have regular maintenance report sessions with the Management.

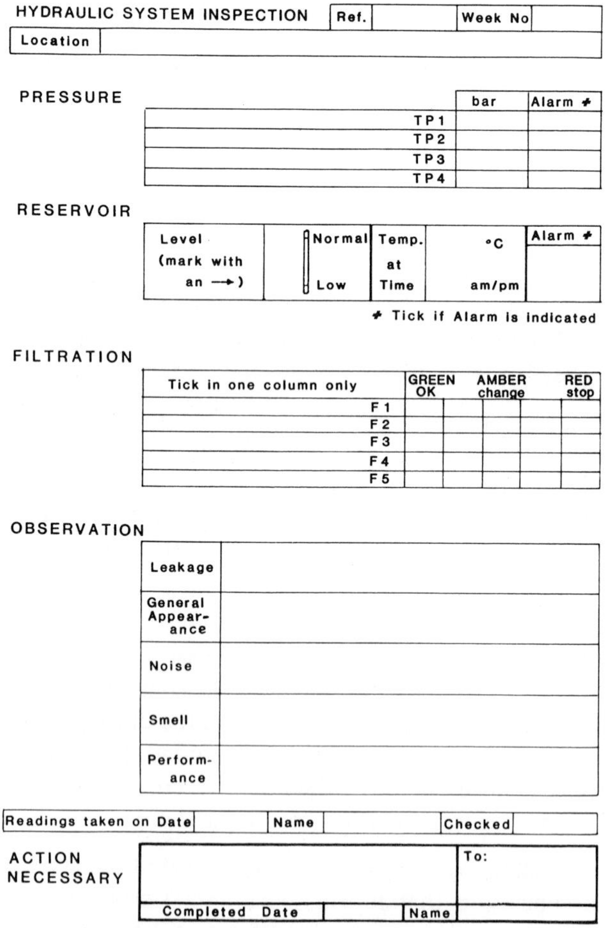

HYDRAULIC SYSTEM INSPECTION	Ref.		Week No	
Location				

PRESSURE

	bar	Alarm ✱
TP1		
TP2		
TP3		
TP4		

RESERVOIR

Level (mark with an →)	Normal / Low	Temp. at Time	°C am/pm	Alarm ✱

✱ Tick if Alarm is indicated

FILTRATION

Tick in one column only	GREEN OK	AMBER change	RED stop
F 1			
F 2			
F 3			
F 4			
F 5			

OBSERVATION

Leakage	
General Appearance	
Noise	
Smell	
Performance	

Readings taken on Date		Name		Checked	

ACTION NECESSARY			To:	
	Completed Date		Name	

Fig. 4.9 A typical maintenance chart

The checking should include the use of condition monitors, where appropriate, the checking of oil levels and contamination classes and the general condition of the equipment. It is quite common to find filler caps off, or the filter inside removed, seals not replaced when a leak begins, filter elements (air and oil) not replaced when due.

A maintenance chart such as shown in Fig. 4.9 can be drawn up for a system and used regularly with very little effort but considerable reward.

The final section in this Control of Contamination section relates to very heart of the system — its design. What is included, how it works, how it reacts, can all influence the final ability to last effectively in the real world. In this case there are a number of features which can be used, the more the better. Filtration, breathers and the design of components are discussed.

Filtration

Standards of contamination or cleanliness class have been mentioned in Actions above. We will find in Chapter 12 that standards vary in the size ranges they check. The wider range, going up to 100 μm, was necessary, and still is in many cases, because of inefficient filtration. A 100% efficient filter will trap all particles greater in size than that quoted. However, such an ability is not yet achievable, and the proportion of over size particulate which passes through the filter can be considerable. There is even a coding to define that proportion, the Beta-ratio, ie

$$\beta_N = \frac{\text{No of particles} > N\ \mu\text{m upstream of the filter}}{\text{No of particles} > N\ \mu\text{m downstream of the filter}}$$

where N is the size in micron (μm)

Thus a Beta of 2 means 50% of the particulate is trapped at each pass
a Beta of 10 means 90%, etc.

It is quite common to take a Beta of 75 as the best for a filter, ie. 98·7%, and call it the 'Absolute Rating', although at least one maker does produce a Beta of 200. ie. 99·5% efficiency.

With these more able filters there is a reasonable cut off of size at say 15 μm or even 5 μm, and so there is no point in checking particles up at 100 μm to compare the count with a standard class. The international standard, the ISO 4406, currently looks just at 5 μm and 15 μm, which as said before, are the two sufficiently dissimilar sizes of importance,

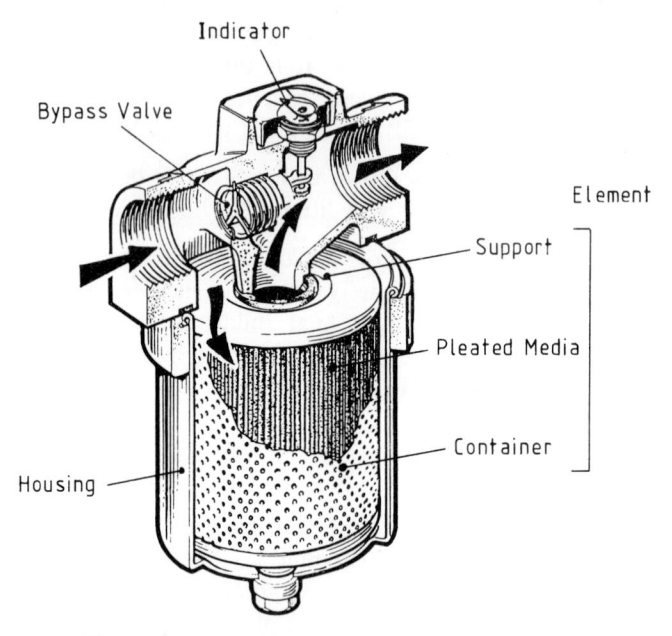

Fig. 4.10 *Typical high pressure blocking type filter*

relating to silting and to wear and blockage. (This coding may be extended to 2 μm as well, when it would thus be described using three numbers, eg. 16/12/9, meaning at 2 μm/5 μm/15 μm.)

Filters are quite complex. They come in a variety of forms and can be fitted in different ways. The blocking filters contain some means to trap the debris as the oil passes through. The high pressure designs usually have removable, replaceable insides (Fig. 4.10) — the 'elements'. The low pressure units may just be spin-on cans. The actual working fabric of the filter may be such as cotton strands, glass fibres, metal mesh, sintered balls, etc. A filter manufacturer/supplier would advise on the best for any particular application.

Other filter types include centrifugal, magnetic, electrostatic and ultrasonic. Whilst the magnetic filters are obviously designed with ferrous debris in mind, they often include a secondary blocking filter to remove non-magnetic debris also; in other words, the basic blocking filter is present with a separate magnetic core to trap the larger ferrous particles

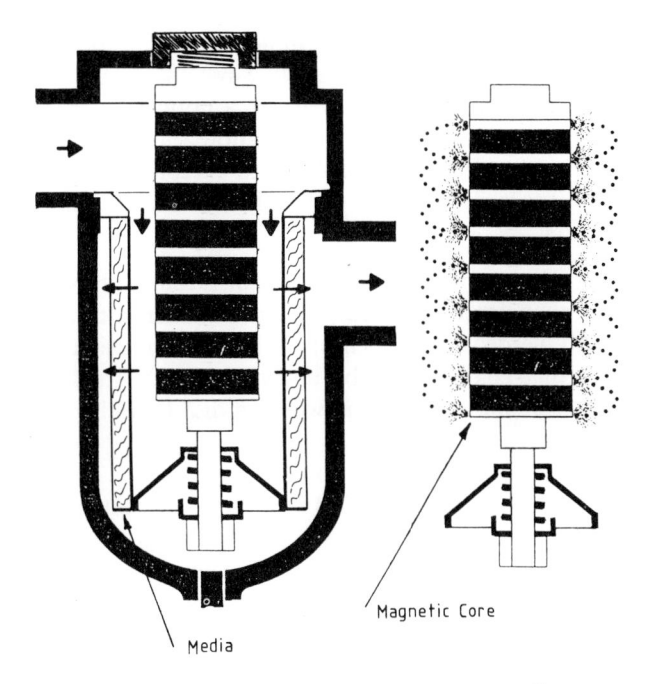

Fig. 4.11 Combined magnetic and blocking filter

first. (In this case the flow goes from within the filter element to the outside, see Fig. 4.11. Alongside the cross-sectional view is shown the type of ferrous particle build-up on the magnetic sections — these build-ups add, to a certain extent, to the filtration of the non-ferrous particles.) The other types mentioned are of considerable value as regards a large quantity of debris removal because the removed debris does not block the flow of the oil as in the case of the blocking filter.

Blocking filters, naturally, have a limited dirt holding capacity. Beyond that capacity they will begin shedding debris downstream or, even worse, break up under the increased pressure and cause the filter media as well as collected debris to hit the system. Except for critical systems (which should be monitored thoroughly) most filters would also include a bypass to allow fluid to pass through unhindered should the maximum dirt holding capacity be reached. Visible indicators on the filter housing will inform the maintenance team when a filter is nearing this blocking position, and hence should be replaced. (Some indicators can be wired into the main control system.)

The electrostatic liquid cleaner has an important part in off-line contamination control. As with all types of off-line systems it cannot be used in the sense of a protective filter upstream of a critical component. It removes a small proportion of the contamination at each pass and gradually brings the contamination level down to exceptionally low levels. It is described in more detail on page 90, along with the ultrasonic and centrifugal filters.

Filters are not the only way that contaminant is removed from a system. The action of changing the oil can provide that change to a better level which is rapidly needed. It is much like a blood transfusion. But, as has been stressed, that change must be undertaken with great care, just a using a dirty needle for a transfusion will be a disaster rather than a help.

Breathers

'Breathing' has been mentioned. This is a necessary feature of any system which is not open to the atmosphere. (It is also true of open systems as well, but that is more obvious!) The machinery contracts and expands, and with it the level of liquid in the tank will rise and fall. The action of 'falling' has, by definition, to suck extra air into the tank to allow the oil to enter the system. There are three ways out of this:

Separators
Enlarge the reservoir
Fit an adequate breather filter.

A separator seals the atmospheric air from the reservoir and, hence, from the oil at that position. (Air can, of course, enter at other places such as seals.) The separator can be

a flexible tank or component in the fluid circuit, enabling the reservoir to be sealed,

a compressible bag on the surface of the tank which in itself can breath but it is sealed to the edge of the tank to prevent the air touching the liquid. (This is called a 'Breather bag' or 'Separator' by one supplier and can be used either inside or outside the reservoir, eg. Fawcett Christie Ltd.) Or by

a pressure filler breather cap. This is designed specifically for a tank under pressure. When the pressure decreases and the liquid level drops, air enters the tank through a vacuum breaker via a small filter. When the fluid returns to the tank, the air inside is gradually

          type SFN   type SFP

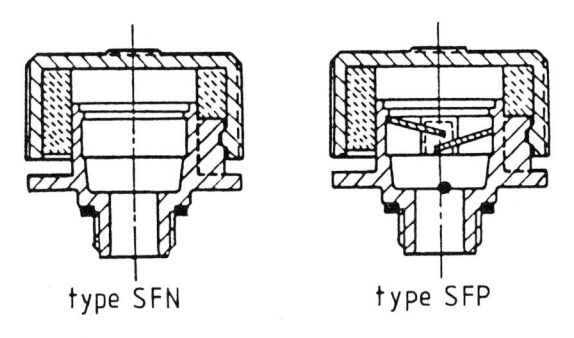

Fig. 4.12 Two typical filler breather caps (Elesa)

pressurised until it reaches a set relief pressure (say, 0·5 bar). At this setting it is released to the atmosphere. Small changes in the liquid levels are contained by the automatic change in the size of the air bubbles in the liquid rather than by extensive unnecessary breathing. This type of cap is particularly important for extremely dusty conditions, eg. Elesa shown in Fig. 4.12. The type SFN is especially suitable for use in reservoirs and power packs where rapid changes of volume are likely to be experienced — requiring high air flow rates through the breather. Type SFP has an integral patented splash guard device for use where violent agitation is expected and without the device there would likely to be a loss of oil through the cap.

To enlarge the reservoir, is a way of spreading the problem in order to make it insignificant, hopefully. However, this is not an economical suggestion because of the size of the system necessary and the quantity of fluid which will have to be used.

A breather filter is, however, the most common way out. The filter must cope with both the quantity and rate of air experienced due to the changes of oil level in the reservoir. The filter must be of sufficient capacity and efficiency to cope with the atmospheric level of pollution in the specific environment.

Design of system components
An hydraulic system can be designed with contaminant insensitivity built in. The objective being to recognise that contaminant will get into the fluid to a certain degree, but to use components which are not too worried by the presence of that contaminant. Some ideas would be where

Fig. 4.13   The Holmbury flat faced coupling.

sliding components can be replaced by flapping features (like poppet valves), where small orifices are shaped so that contaminant is easily washed off rather than jams in, where surfaces can either crush the contaminant or totally absorb it. All these will enable a higher level of contamination class to be tolerated, and thus be more easily achievable.

The placing of the filters upstream of important parts is another invaluable design asset for a reliable system.

A less obvious idea, but nevertheless just as valuable, is the use of hose couplings which are designed to prevent the ingress of atmospheric contaminant. The flat faced couplings from Holmbury Limited for both bulkhead and hose-to-hose use are shown in Fig. 4.13. In their particular case the main advantages are

> The flat faces are easily cleaned before coupling occurs, preventing the ingress of contaminant (which otherwise might enter via a conventional quick release coupling that has a wide open female end).
>
> No air intrusion on connection, no oil loss on disconnection.
>
> Streamlined internal flow path minimising pressure loss and heat generation.

CHAPTER 5

Methods of Collecting Debris

Probably the most unreliable part in the debris analysis process is the way the debris is extracted from a system.

It may be removed on its own or in the oil. Filters or magnetic chip collectors will trap debris for direct removal. Alternatively, an oil sample may be taken and the debris measured in the sample.

Although this is the primary hazard with off-line analysis, it is also just as important in in-line and on-line sampling. A sample of liquid has to get from the system and into the chosen monitor for analysis, and in that transferral the debris must neither be lost or be altered.

The monitoring device fitted to the mechanical system must 'look' at a truly representative quantity of oil if the decision taken by the monitor is to have any meaning. The sample becomes unrepresentative if the liquid withdrawn does not contain the actual debris evidence. It is also unrepresentative if it contains debris from sources extraneous to the liquid, such as from the sampler, the bottle or the atmosphere.

Debris does not become part of the liquid; it is in the liquid but it does not conform to the liquid. It is not homogeneous, and does not normally mix naturally. Therefore, some care must be exercised in ensuring that sufficient time has elapsed and mechanical/fluid mixing is occurring before a sample is taken. And that is not always an easy decision. For instance, as has just been said, one would usually wait for a system to stabilise (perhaps 30 minutes) before taking a sample, but it may well be in the early stages of running that excessive wear or contamination is a problem; there may be a sudden influx of debris which has no chance of becoming thoroughly mixed — and these are the critical moments as far as the machine is concerned.

This chapter examines the important features to be taken into consideration, in order to be sure that the debris withdrawn, or the sample taken, is as truly representative of the debris in the liquid as possible. It also looks at the hazards which nullify the results if care is lacking in the preparation and procedure necessary in the sampling.

In-line and on-line instruments will often have their own fittings for connection to the liquid circuit. These will be discussed when the individual instruments are explained (Chapters 7, 8 and 11). For off-line sampling, the sampling points, for extraction of a liquid sample, could be the same as used for the on-line sampling, but with suitable adapters. This sampling is also described in this chapter.

We look first at the characteristics of liquid flow and, then, how the particles move in the liquid, before examining the practical means of collecting debris or liquid for analysis.

LIQUID FLOW

The turbidity of the liquid is a major factor which affects particle movement. By turbid, in this context, is meant the churning-up and thorough mixing of the fluid. ('Turbid' can mean dense and muddy, which is the visual impression caused by the mixing.) The opposite is laminar flow, where the fluid tends to move in uniform parallel lines with no disturbance between the layers of flow. The change-over between laminar and turbulent flow can be calculated using what is called the Reynold's Number (R_e), reported by Professor Osborne Reynold in 1883. Reynold's Number for flow in a pipe is calculated as follows:

$$R_e = \frac{v\,d}{\nu}$$

where v = liquid velocity (m/s)
        d = internal diameter of pipe (m)
        ν = kinematic viscosity (m²/s)

The change over point from laminar to turbulent is often taken as being at 2300. Thus

if R_e is < 2300 the liquid is laminar
if R_e is > 2300 the liquid is turbulent.

However, this may not always be true in practice due to variations in pipe geometry and surface. To be absolutely sure that turbulence has been achieved, a safe value is taken at $R_e = 4000$.

[EXAMPLE: v = 1 m/s,
            d = 20 mm ≡ 0·02 m,
            ν = 10 cSt ≡ 0·00001 m²/s

then

$R_e = 2000$, ie laminar.]

In most circumstances the oil circuit will have been designed to be laminar. (Flow resistance and pressure losses are greater with turbulent flow, and hence the system would be less efficient if the fluid were allowed to remain in a turbulent form.)

Turbulence, however, does occur naturally in the vicinity of discontinuities. Edges and bends and sharp changes in diameter alter the uniform pattern. A rough surface can also be a disturbance. It is not surprising, therefore, that special 'disturbers' or 'flow mixers' can be fitted to enable oil sampling to be more effective. (These should not be confused with the more common 'flow smoothers' which are necessary upstream of special sensors such as flow and temperature transducers.)

Laminar flow usually sees a variation across the cross-section of a pipe. Wall friction tends to drag the flow close to the pipe walls (the solid/liquid boundary), and thus cause a distribution with a greater velocity in the centre, and any attempt at sampling close to the pipe walls would provide grossly misleading results. Turbulent flow has less of a variation and because the debris is churned around, it makes little difference where the sampling probe is placed. (It used to be considered that a pitôt-type tube, pointing into the flow, would provide the most representative sample; however, this was not often allowed because of the difficulty of fitting and the restriction to flow which was caused.) A general view of the variation in a typical pipe bend emanating from a pump is shown in Fig. 5.1.

Another important variation in flow is due to dead-ends. A 'dead-end' is a pipe which normally does not have a flow in it. It may be a pressure line to a gauge or a control, where instead of flow, it is pressure which is being transmitted; it may be a line which is only used rarely, even a sample line! (If it is a 'monitoring' sample line then it is essential practice to allow a sufficient quantity of liquid to be ejected to waste before the liquid is bottled.)

Thus for the fitting of any sample point, care must be taken to avoid long lengths of line which are used solely to provide a pressure pulse and only use those which experience a regular flow. And, at the same time, a turbulent flow should be encouraged, at least in the vicinity of the sample point.

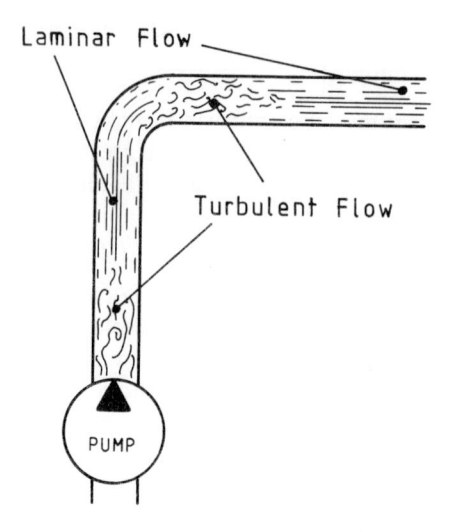

Fig. 5.1 Liquid flow distribution in a pipe

PARTICLE FLOW

Because the particles of debris or ingested dust and rock are of different physical property to the liquid they form a non homogeneous mixture with the liquid. This is not solely due to the density differences, or even the non-liquid state; features such as porosity and surface tension, electrostatic potential and dimensional ratio can vastly change the motion of the particles in the liquid, particularly in the flowing situation, but also in the static state.

The sub-micron particles have a motion of their own. This is called Brownian Motion, or Brownian Movement after Robert Brown who discovered the concept in 1858. It was a discovery really concerned with the detection of molecules, rather than particles, because, by observing the particles it was apparent that some other forces were present in the liquid. The particles were being jostled around by some unseen bombardment (the liquid molecules moving). However, the scientist is able to use the idea in reverse, to help in the detection and sizing of very small particles.

Molecules in a liquid or gas move randomly. This phenomenon is highly important because it allows, among other things, the equilibrium of pressure of a fluid in a sealed container. Should, however, a small particle get in the way of the molecules, then it will be pushed along, one way or the other. In a very short space of time the small particles become thoroughly mixed by the molecules. By 'small' we mean particles with a spherical diameter of less than about 0·8 μm, and the smaller the particle the greater the rate of random movement. (However, it is possible to consider the same action up to possibly 3 μm, or even 5 μm, with suitable corrections.)

Any particles larger than 0.8 μm will tend to resist the molecular bombardment to a certain extent, and move in accord with other forces, such as gravity, bodily liquid flow, molecular attraction or electrostatic attraction. (Van der Waal determined an equation of state which corrected the so called 'perfect' laws by including the mutual attraction between molecules and the volume of the molecules themselves. Such attractive forces exist between atoms and molecules of all substances. The forces occur because the electrons in adjacent atoms or molecules move in sympathy with each other.) As to which of these is the prominent factor in the flow rate and direction, will depend on other features such as viscosity of the liquid and the density ratio of particle/liquid.

Another complicated variable is the shape of the particle. Chapter 3 showed a variety of possibilities, and, for example, where there is a strong plate-like shape then motion will be devious (eg. side to side).

The movement of these larger (greater than 0·8 μm) particles under gravity has been studied by Sir G.G. Stokes. His complex mathematical analysis concerned the resistance to flow of a sphere when falling in an infinite deep liquid in an infinitely large container. The terminal velocity (v) was calculated as:

$$v = \frac{2 g a^2 (\rho_P - \rho_L)}{9 \mu}$$

where
- g = gravitational acceleration (980 cm/s^2)
- a = radius of particle (cm)
- ρ_P = density of sphere (g/cm^3)
- ρ_L = density of liquid (g/cm^3)
- μ = dynamic viscosity of liquid (Poise)

[EXAMPLE:

$a = 0.01$ cm
$\rho_P = 7.0$ (g/cm^3)
$\rho_L = 0.9$ (g/cm^3)
$\mu = $ {for 100 cSt and a ρ of 0.9 g/cm^3} = 0.9
{Note: Poise = Stoke × Density
 = (cSt/100) × g/cm^3}

then $v = 0.1476$ cm/s]

Now this refers to particles which are spherical. Very few particles of debris or ingestion are genuine spheres, and many of them are not as smooth as a 'sphere' should be. This means that Stokes' Law is not exactly true for the real situation, with genuine debris. However, there are some general ground rules on the accuracy:

Provided the particle has an aspect ratio (length/width ratio) of not more than four then a reasonable closeness will be apparent — for particles up to, perhaps, 200 μm maximum dimension in lubricating oil. It can be appreciated that when any body passes through a field there is generated a turbulent region behind the body if the velocity is too great, and the body will be 'held back'; hence the suggested 200 μm figure above. If the fluid is more viscous (thicker), then the particle will fall slower and the likelihood of turbulent drag is reduced. (Fitch 1979, in his Encyclopedia, suggests the maximum particle size is only 150 μm if a liquid of viscosity close to water is used (ie. 1cSt).)

Stokes' Law indicates the movement is dependent, for one, on the size of the particle, the larger the particle the faster the fall. Particle movement in a real confined situation (as against an infinitely large container) means that there are other influences. Wall boundary effects, when there is laminar flow in a pipe, cause the larger particles to dawdle along close to the tube wall, moving slower than the smaller particles which are in the more central position.

Lantos (see Appendix 2) makes a tentative suggestion as regards a correction for Stokes' Law where "hindered" settling is involved. By this he means the settling of particles in diesel fuel oils where a certain amount of adherence of the particles to the nuclei of asphaltene micelles occurs. His correction reads –

v(corrected) = v(calculated) x $(1-w)^n$

where, w = mass fraction of disperse phase (ie. mass of micelles divided by the mass of the fuel),
and n = an empirical exponent.

In practice it had been found that n = 4·65 (approximately) for extremes such a slurry or mud.

A word about agglomerations. Agglomerations are the opposite to separations. They are the binding together of particles in what may appear to be a single large particle or mass. They occur because of the attractive forces between particles mentioned above, and because of the surface tension of the liquid. They are a nuisance to wear debris analysis and particle detection, and they need to be broken up in order to achieve correct identification of the particles. This is another reason for the importance of thoroughly mixing the particle/liquid suspension before sampling occurs.

Before leaving the subject of particle flow it is essential to be aware of the problems associated with liquid containers (be they called sumps, tanks or reservoirs!). Tilley 1984, of the Fluid Power Centre in the University of Bath, undertook valuable research on the movement and settling of particles in a static reservoir. (A mobile reservoir has different characteristics because of the jostling motion of the support.) He discovered that if particles were allowed to settle out, in stagnant regions, far from resulting in an improved system situation, they actually made the system worse. This was because they could be suddenly jerked free by a fluid pulse and a serious agglomeration could be reingested into the system. He, therefore, proposed a preferred design which encouraged the constant mixing of the particles which remained at a reasonable concentration; if necessary to be filtered out elsewhere in the fluid circuit. The advantage of this type of reservoir to the analysis of wear debris is immediately obvious; the debris evidence, instead of dropping to the base of the reservoir, remains in suspension and can be detected at a circuit sampling point. The key parts of his design are shown in Fig. 5.2.

ACHIEVING TURBULENT FLOW

It has been mentioned above that turbulence is helpful, but only in the region of the sampling point. A means is, therefore, necessary to cause a local disturbance without excessively reducing the main flow in the system.

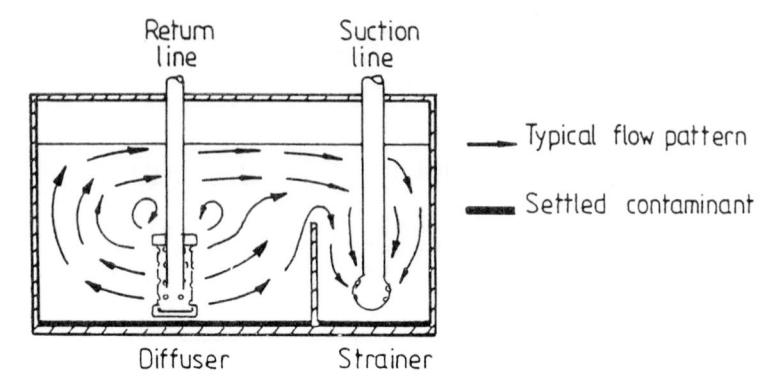

Fig. 5.2 *Preferred features in a reservoir for maximum mixing (Tilley 1984)*

The positioning of sample points in a circuit is discussed in Chapter 14. These points are defined with maximum turbulence in mind. However, there will be other positions in the circuit from which samples need to be taken; or, maybe, there is no choice of position except as regards accessibility. In these cases an artificial turbulence generating device should be installed.

Basically the objective of the device is to cause maximum disturbance to the fluid (to mix the debris content) whilst at the same time permitting the least restriction to the liquid flow (to maintain maximum efficiency). It needs, therefore, to be of minimal length but to influence all the liquid which passes.

It is interesting to note that quite often the only special fitment that is recommended is that of a U-shaped bend in the pipe run, up-stream of the sampling point. However, the bend actually introduces a centrifugal action on the mixture which, therefore, can separate rather than mix; and depending on which side of the U-shape is chosen, either the larger or smaller particulate will be sampled.

Complex reverse angled baffles fitted right across the pipe bore, so that the fluid has to actually come back on itself, will certainly produce excellent mixing, but this will have to be balanced with the extra power necessary to drive the liquid through the region.

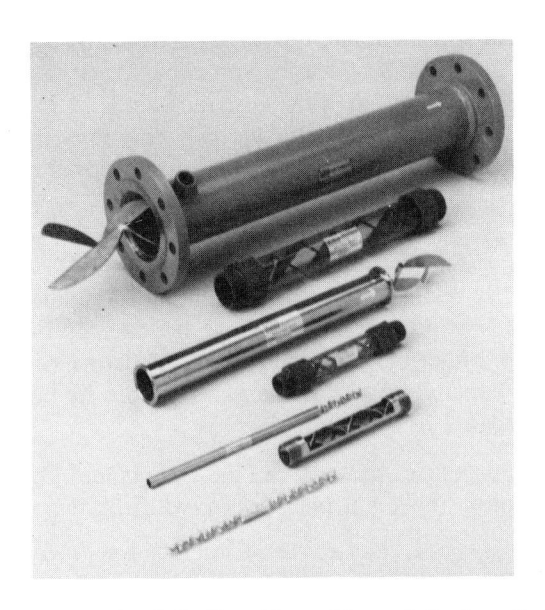

Fig. 5.3   The Koflo Flow Mixer (Pump & Package Ltd)

One commercial flow mixer is that produced by the Koflo Corporation which makes use of reverse screw baffles, ie. the flow is caused to rotate clockwise and then anticlockwise as it passes down the pipe. The sudden jolt apparently works well but the smooth design has the added bonus of providing the minimum of friction loss. Another important feature is that there are no moving parts, and spares are unnecessary. The company describes the action of its multi-element mixer as

> "The mixing process divides the stream of product, forcing the stream to the outer wall which then transfers the product in a single direction back to the centre line of the second mixer element; both lines of product then come together causing a vortex and a shearing action to take place. The step re-occurs but in the opposite direction of rotation taken during the passing of the product through the first element. This re-occurs as many times as the number of elements fitted to the mixer. The in-line static mixer with its compactness, high efficiency and simplicity allows for a quicker continuous operation than is possible for example from a mixer tank and stirrer."

Examples of the design are shown in Fig. 5.3.

Another series of similar in-line mixers is produced by Sulzer, primarily for the processing industry and using somewhat larger diameters. Whatever type of mixer is used, as with those mentioned, it is always best to only use devices which are non-moving. Not only does this ensure long life, but it is, after all, the liquid which needs to be moved not the mixer.

ACHIEVING PARTICLE MIGRATION

Whilst not exactly relating to the classical liquid sampling method, particle migration is important in sampling. It concerns the trapping of particles, not by a filter blocking the path of the debris, or by taking out a liquid which hopefully contains the particles, but by moving the particles into a prepared reception area.

Vickers Tedeco produce what they call the Lubriclone® Cyclonic separator/Debris Monitor. This is a twin device, with the purpose of forcing

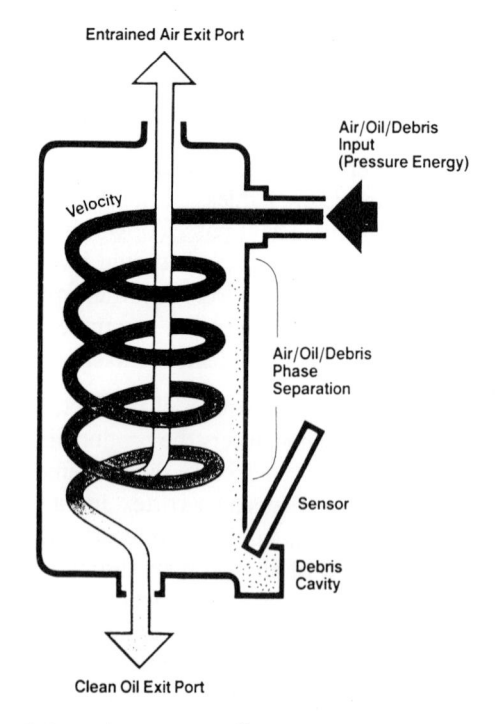

Fig. 5.4   *The Lubriclone® separator (Vickers Tedeco)*

the fluid into a cyclonic path which tends to fling the debris particles out to the perimeter where they are trapped by a magnetic monitor. They have found it is excellent for particles above 200 μm, and even at 50 μm they have achieved a 40% efficiency of collection. (A bonus is the concurrent removal of air from the oil.) No moving parts are involved, the design causes the fluid to rotate automatically due to the tangential entry. Centrifugal forces 700 to 1000 times greater than gravity separates oil/air/debris, because of density differences. The debris travels to the outer wall and drops to a sensor chamber at the base of the device. The basic idea is shown in Fig. 5.4.

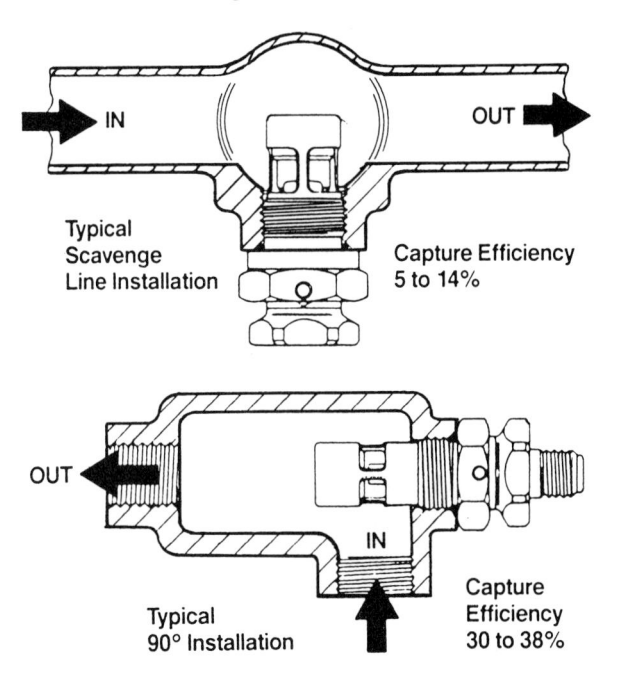

DEBRIS MONITORING LUBRICLONE

Capture Efficiency		Maximum Allowable Pressure Drops
1000μ	99%	20 PSID
500	95%	
250-300	70%	

Fig. 5.5 *Capture efficiencies of magnetic plugs for ferromagnetic particles depending on type of installation. (Vickers Tedeco)*

Another, simpler, device from Vickers merely produces a right-angle change in flow direction causing the larger particles to be flung down on to a magnetic collector. They use it also as a protection device with an inbuilt mesh filter. (Note that this is where 'separation' rather than 'mixing' is the essential feature.) The different efficiencies of capture are illustrated in Fig. 5.5.

Another possibility on similar lines, where a large amount of particulate is removed, is the centrifugal filter. This type of filter causes the fluid to violently rotate (up to 15000 rpm) with the result that the particles, even submicron particles to a certain degree, are thrown out to the perimeter for later collection (and analysis if required). One example is shown in Fig. 5.6.

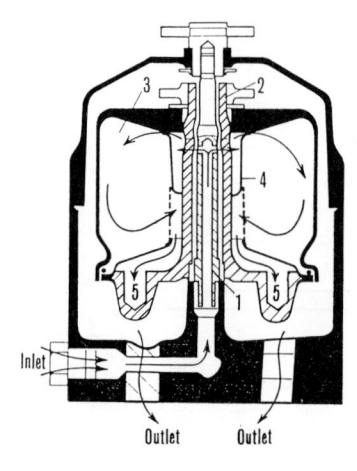

1 - Hollow Spindle
2 - Central Pillar
3 - Rotor
4 - Stand Tube
    (dividing cleaning chamber from driving chamber)
5 - Nozzles (diametrically opposed)

Fig. 5.6 Cross-section of a centrifugal filter

Particle migration can also be achieved by other means, such as electrostatics and ultrasonics. Both are used for filtration of systems, but because they bring all the contaminant together for collection they are of

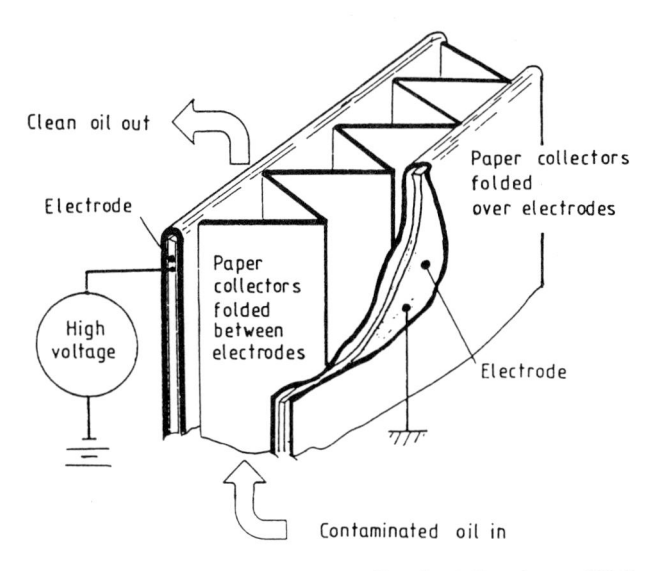

Fig. 5.7 Electrostatic Filtration (Sasaki & Dunthorne 1984)

considerable benefit in the discipline of sampling as well as filtration. They are briefly described below:

The Electrostatic Cleaner manufactured by Kleentek in Japan and supplied by United Air Specialists (UK) Limited, uses the principle of electrostatically charging particles so that they have an affinity to an opposite electrical pole. As the liquid passes up alongside a vertical filter element, the 10 kV applied voltage charges the particles so that they move either to the positive or negative pole. The collector element blocks this migration and collects the particles. When filled, the collectors are removed and the contamination analysed, if required. A visual idea of the electrostatic filter is shown in Fig. 5.7.

The Ultrasonic Filter is not the same as an ultrasonic cleaner. Ultrasonic cleaning of components causes the debris to be separated using an ultrasonic frequency around 20 kHz–40 kHz. The ultrasonic filter uses the much higher frequency around 1 MHz–2 MHz and uses it to draw particles together which then agglomerate and, as much larger masses, fall under gravity to be collected. This idea has been developed by Sonofloc in Vienna, Austria.

THE MAGNETIC CHIP COLLECTOR

Before looking at the removal of oil samples, we will examine the much simpler debris removal by magnetic chip plug. This is effective and has been in use for many years particularly in transport vehicles including aircraft.

Strictly speaking the 'collector' is different to the 'detector'. The true idea of the 'detector' is that of being able to signal the presence of debris on a magnetic chip plug — for instance, by means of electrical circuitry. However, the expression MCD (Magnetic Chip Detector) is sometimes used of the simpler collector which just gathers debris for removal and examination.

Fig. 5.8 is an example of the chip collector. It is constructed in two parts, one part fixed into the oil system, and the other part being the

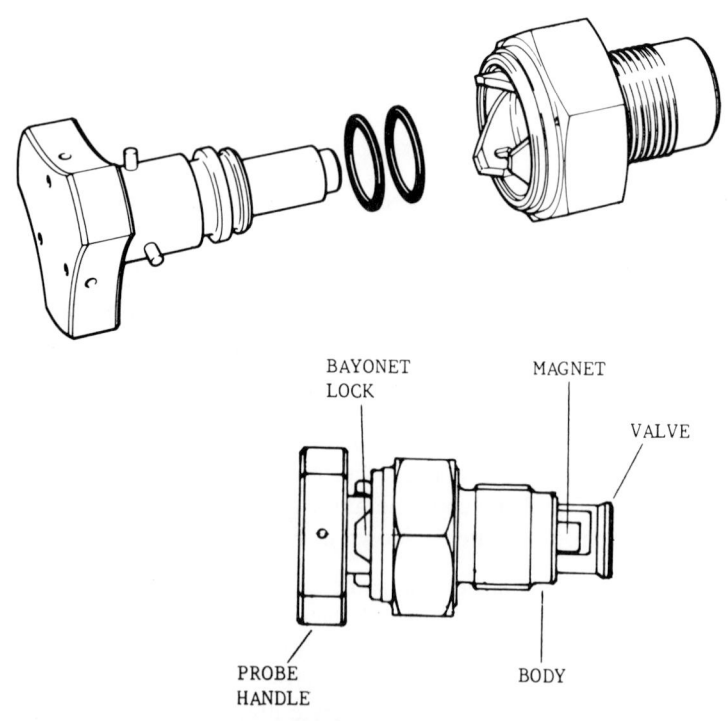

Fig. 5.8 Two views of the Magnetic Chip Collector (Muirhead Vactric Components)

removable plug containing the magnet. The chip collector can be used as a stand-alone device, where the probe is removed, assessed on the spot and replaced. Or, for more detailed work, the probe is taken away for the debris analysis; in this case an exchange probe replaces the removed one.

Magnetic chip collectors used by Muirhead Vactric Components are widely used in civil and military aircraft, including helicopters, but they are also suited for many other types of power transmission systems.

It is common practice when debris is to be removed for analysis, for the plug to first be 'dried' by means of a solvent to remove the oil. The debris is then eased off the chip probe by means of transparent sticky tape. The evidence on the tape can be stored or compared with previous 'removals' by means of an instrument like the Debris Tester or Particle Quantifier

SAMPLING POINTS AND VALVES

Where the liquid to be sampled is under pressure, then a valve can be fitted. The magnitude of the pressure will decide the type of valve, not only in the sense of its strength but also whether it can provide an acceptable flow to issue into a sample bottle container.

A very satisfactory design comes from Hydraulic System Products in Wakefield. It comes in two parts. First, there is a pipe coupling which includes an integral self sealing test point. This is permanently installed in the system. The second item is a screw-on probe connected by flexible hose to an adjustable flow control valve. Caps are provided for all ends to prevent extraneous contaminant entering the sampler when not in use. The type shown in Fig. 5.9 is capable of 420 bar upstream pressure, and the probe connection can be made with the circuit at full system pressure. The figure shows the arrangement of the self sealing coupling and the form of the test probe.

There is no need for the probe to project into the flow, indeed this would be a disadvantage for two reasons. Firstly it would interfere with the flow and, secondly, it would be difficult to fit reliably. With turbulent flow, it can be assumed that the distribution of particulate size is reasonably uniform, and hence it is only necessary to have an opening at the edge of the tube or pipe.

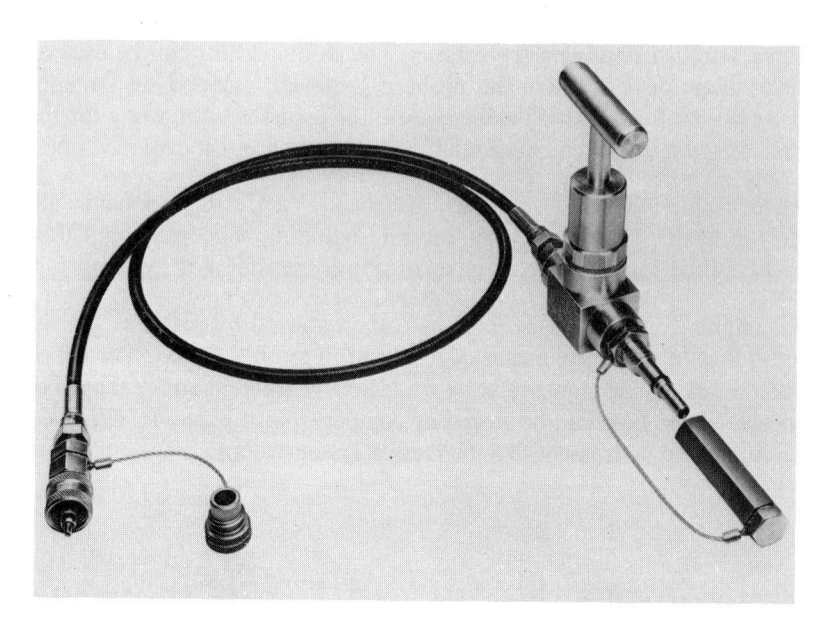

Fig. 5.9 HSP Sampling Probe and coupling

A similar sampling point is provided by Hydrotechnik with its Minimess system of test points, probes, flexible hoses and couplings.

Should the sampling point be attached at the side, top or bottom of a horizontal pipe? Does it matter? Is a vertical pipe preferable? There are two ways of looking at the position problem, depending on the situation. If it is expected that chunks of debris might be released, and that it is important to detect each one as much as possible, then a sample point should be on the underside of a pipe. In this way the 'dead end' acts as a catchment region, allowing debris to settle until the sample valve is opened; each particle is virtually individually analysed.

If, however, the debris level which is currently occurring is required, The situation is different. In order to be able to detect any increase in wear rate, then the fluid must not be influenced by debris which would have been passing earlier. In this case the valve should be fitted to the side or top of the pipe, preventing any settling and retention of debris. Alternatively the sampling can be undertaken on a vertical upwards pipe where the angle has no meaning.

THE BOTTLE SAMPLE

'Bottle' is a fairly arbitrary term. Any container, but not 'any old container'. The difference is the cleanliness of the container and the ability to keep the container free from solid particulate contamination before and after the sample has been taken. This will become clear as we discuss the use of an actual bottle. (Note we are not considering biological contaminant which may be destroyed by heating. Solids need actual removal, and biological 'automatic cleaners' are not normally able to achieve this.)

For contamination control in hydraulic systems, the bottles have to be precleaned to a very high standard. Typically the bottle should not contain more than 500 particles greater than 5 μm for contamination control applications in hydraulic fluids. This can only be achieved by a careful sequence of washing and drying. For the most consistent results the operations should be conducted under Class 1 clean air conditions using a suitable laminar flow cabinet (see BS 5295). There are many slightly different routines; the following is one suggestion:

> Using a 250 mL wide neck, flat bottom, glass bottle with screw fitting (often called a Powder Round)
> Remove the cap
> Wash the bottle with warm tap water and a liquid detergent (perhaps with a small brush)
> Rinse with luke warm running tap water
> Rinse with distilled (or deionised) water
> Rinse with prefiltered (0.45 μm) propan-2-ol (Isopropyl alcohol)
> Rinse with a prefiltered (0.45 μm) solvent which is compatible with the liquid to be sampled (eg. Petroleum Ether for mineral oil)
> Allow to drain for a few seconds
> (An alternative is to leave a few millilitres of the solvent within the bottle to create a slight positive pressure when the bottle is capped.)
> Fit a precleaned cap in a clean cabinet or similar environment
> (Note that if a non-shedding polypropylene cap is used, it needs to follow a similar procedure of cleaning before it is placed on the bottle. An alternative is to use a suitable non-shedding sheet seal, eg. PTFE, under an ordinary cap.)

In official Standard applications the bottles would then have to be checked for cleanliness levels, to ensure that they comply with the Standard (eg. ISO 3722, see page 296). This is done by 80% filling one in five bottles with precleaned solvent, capping, shaking thoroughly by hand, then counting with an appropriate instrument to determine the number of particles per 100 mL of liquid. To be completely sure, the 5 μm count should not exceed 200 if the cleanliness level for the 250 mL bottle is not to exceed 500.

Obviously, wear debris applications do not need this level of cleanliness for the bottle, but they do still need to be cleaned to a level which is well below the levels being monitored in the system. If Standards are used, such as ISO 4406 (see page 277), then the cleanliness should be at least two classes better than the significant level in the system. (In actual particle count this amounts to a four times difference.)

If precleaned bottles are not available, and say, more general gear box sampling is being undertaken, flush the bottle with a solvent filtering dispenser, making sure that all surfaces within the bottle, the threads and the cap are covered (Fig. 5.10).

Allow sufficient fluid to pass to waste from the sampling point, to clean away any debris residing in the flow exit (Fig. 5.11).

Fig. 5.10   Flushing sample bottle (Millipore)

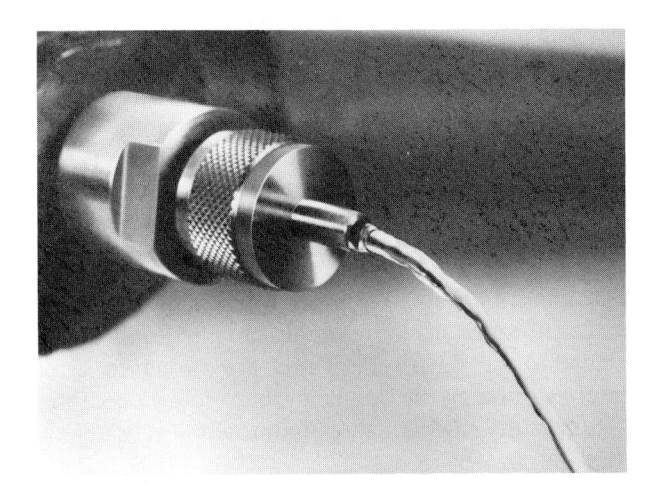

Fig. 5.11 Initial flow to waste

Not more than 80% of the bottle should then be filled. More than this quantity will prevent full mixing of the debris in the oil when the time comes for the sample to be shaken and used.

The bottle needs to be identified clearly so that it is not opened until required. The following was designed for use by the Fluid Power Centre at the University of Bath (Fig. 5.12):

```
CONTENTS  Date.........Ref.........
Fluid _____
Source _____      MAX.
Remarks _____     LEVEL
_____

          CLEAN BOTTLE
     (< 500 Particles > 5µm)
    OPEN ONLY TO FILL/SAMPLE
              ● FluidPower
                 at Bath
```

Fig. 5.12 Labelling of a Sample Bottle (Fluid Power Centre)

DYNAMIC SAMPLING/RESERVOIR SAMPLING

All the above sampling has been with a pressurised and flowing system. If possible it would have been undertaken when the system reached equilibrium, ie. the temperature is stable and the debris would have become well mixed. This is called Dynamic Sampling. This is the best and most representative type of sampling; it can be repeated exactly on later occasions so that precise comparisons of debris level can be determined.

However, not all dynamic sampling can be undertaken in that closed pressurised manner. Two other types of dynamic sampling, which provide invaluable information, ie. reservoir sampling and drip feed sampling, are now to be described. These are prone to error due to inaccuracies of positioning the sample probe, and also because often the very act of taking the sample causes more debris to enter the system. (See ISO 4021 on page 297.)

As regards position, the best place in a reservoir is at half oil level in the stream of oil returning from the system, be it a steady flow or drip feed. Here the debris is mixed and should be truly representative of the system. Air may be ingested at the same time, but that can be removed later.

As regards the sampling action causing debris to enter the system, this is entirely due to poor 'housekeeping'. The very method is somewhat lackadaisical, and so the operator may well see no point in taking extra precautions. Wiping the surround of the filler cap hole, or dip-stick hole (used as an orifice for inserting the sampling probe) with an old rag, may well push in debris from the outside; even a clean rag may not improve the situation if used badly.

This debris entering feature is even more likely when new oil is added to the system (at an oil change). In earlier trials (Hunt 1985) I found a significant increase in particle count with just a top-up of oil in a transmission gear box. (Pumping the oil in through a filter would greatly improve this situation, but whilst this is always a preferred technique for fluid power systems, it is rarely done, if ever, on automotive applications.)

STATIC LIQUID SAMPLING

As the title implies this is where the liquid is non-moving. It should nevertheless be at a point in time when the debris is well mixed, eg.

immediately after a full run of the system. (In contamination control situations it would also apply to the need to assess the particulate levels of the oil in as-supplied drums; in this case shaking is virtually impossible, but some mixing could be attempted by turning over the drums or by circulating the fluid by pump. It should be remembered, also, that the fluid will be filtered as it is pumped into a system reservoir, so it could be sampled at that time.)

This sampling in a system would normally be undertaken at the reservoir (see section immediately above). When the fluid is static even greater care must be taken. Larger particles settle very quickly and temperature stream lines can confuse the position of other particles. The results of such tests do not normally carry much credence because of the large extremes which can be obtained, which can be most misleading.

Samplers able to remove liquid from a reservoir and place it in a bottle are available. They simply consist of a small vacuum (maybe a hand pump or press bulb) and an automatic one way valve. The liquid is then sucked up by the vacuum and transferred to a connected bottle. A sampling kit used by SGS Redwood Oilscan is shown in Fig. 5.13.

Sample packets are also available, such as supplied by A.I.M. München Vertriebs from Industriebedarf Rehm. As well as precleaned bottles and

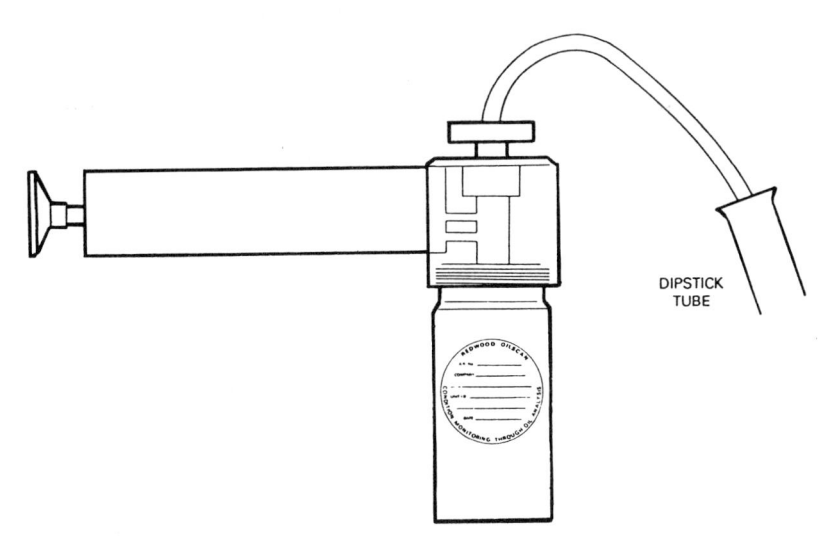

Fig. 5.13 *A hand sampling device (SGS Redwood Oilscan)*

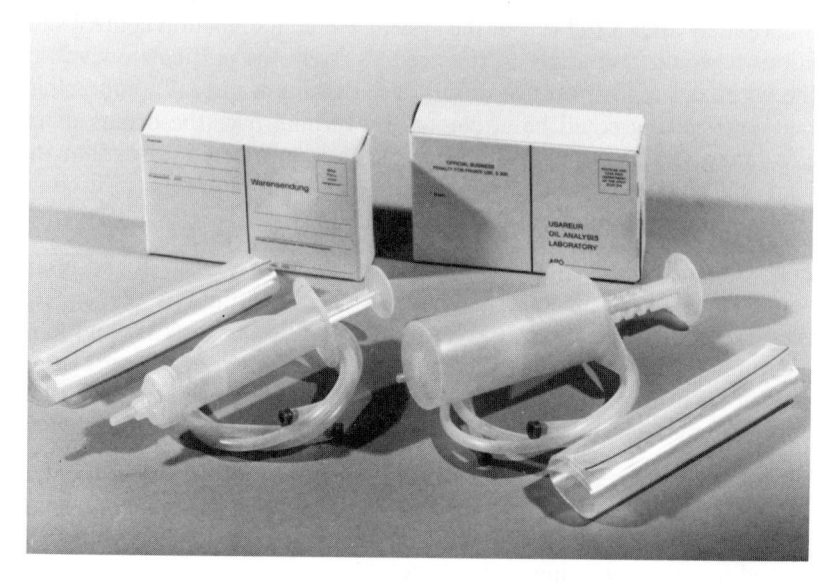

Fig. 5.14 *Oil sample extractor/containers (Industriebedarf Rehm)*

appropriate packing, A.I.M. München also supply two sizes of precleaned syringe and suction hose able to take either 45mL or 100mL of sample oil, see Fig. 5.14. Packing is also supplied.

Price of sample bottles in the UK tends to about £3 for each for small quantities, and maybe only £1.50 for syringe kits. Each laboratory or supplier will have his own scale of prices, and care must be taken not to pay more than is necessary. Highly guaranteed cleaned bottles for ultra pure applications are not needed for general gear box testing.

MAKING A MEMBRANE FILTER ANALYSIS PATCH

It is possible to attach a membrane filter (in a support) to the sample probe. In this way a sample of, say, 100 mL of liquid can be passed through the filter and the debris retained on the membrane. Millipore supply prepared analysis monitors containing 0·8 μm membranes on their own or in a special Fluid Sampling Kit with the necessary fittings to attach to the fluid line. In their case an upper pressure of 7 bar is allowed.

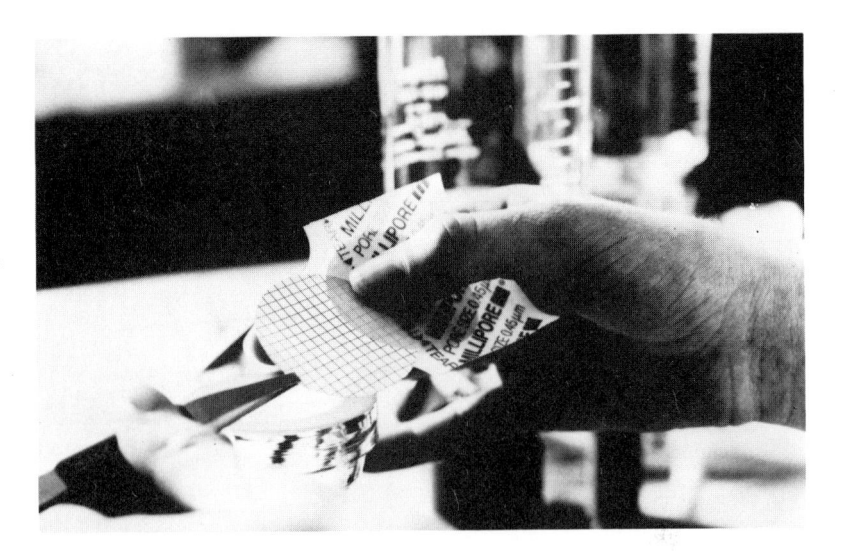

Fig. 5.15 Removing a membrane filter from a packet (Millipore)

The process of making the patch from a bottle sample is as follows (all the photographs were kindly provided by Millipore UK Limited):

All the glassware in the process needs to be thoroughly cleaned using prefiltered solvent.

A membrane filter should then be carefully removed from a pack (Fig. 5.15), preferably in a clean cabinet.

The membrane is then placed on the centre of a Millipore, or similar, membrane glass filter holder (Fig. 5.16).

The glass filter funnel is then clamped onto the filter holder, clamping the membrane tightly onto it. 100mL of the oil is then poured into the glass funnel (Fig. 5.17).

It will then probably be necessary for a vacuum to be applied to pull the oil through the membrane. particularly if it begins to block earlier on due to a high level of debris (Fig. 5.18).

Finally prefiltered solvent is gently dispensed around the glass funnel to ensure all debris has reached the membrane. It also causes the membrane to be cleaned of its oil and partially dried.

The membrane filter can then be removed after unclamping.

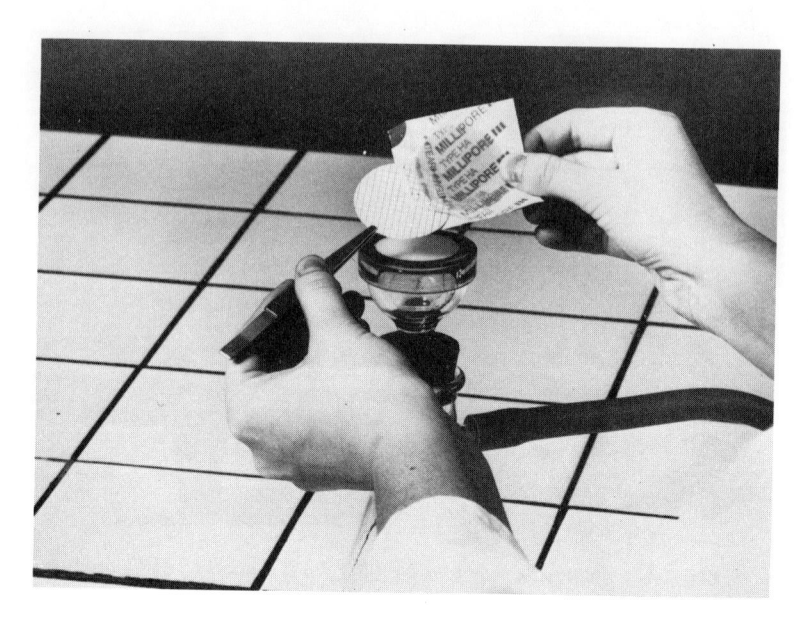

Fig. 5.16 Placing the membrane on glass filter holder (Millipore)

Fig. 5.17 Pouring the oil into the glass funnel (Millipore

Fig. 5.18   A vacuum arrangement (Millipore)

Using similar analysis monitors, and a similar kit, Thermal Control Products of Howden Wade provide the 'Conpar' monitoring kit. This includes, in addition, a small comparative microscope (reflected and transmitted including an optional polarising and magnetic stage described by Goldsmith 1984) and prepared fixed comparison slides. Rather than counting particles on the membrane, the sample membrane is compared in optical density with Master slides. The maximum pressure is stated as 12 bar, but also the minimum should be at least 1 bar to force the liquid through the membrane. As with the Millipore kit, a small hand vacuum pump is provided to dry out the membrane ready for analysis.

The membrane filters are normally 0·8 μm in cellulose nitrate which are acceptable for most mineral oils, but for polyglycols and phosphate esters an alternative is advisable. However, it is always important to check the compatibility with the supplier of the membrane filters (eg. Millipore, Whatman LabSales). (In some cases a 5 μm membrane filter can be used particularly if there is only need to examine the larger particles.)

GREASE STATIC SAMPLING

The sampling of grease is relatively new. For a long time it was felt that it would be totally unrepresentative of what was really happening, and probably there would be more influence from the outside atmosphere rather than the debris generated within the lubricated bearing. However, that has since found to be not the case.

Ferrographic workers seem to have been the first to have some success in grease lubricated systems, reported by Bowen, Bowen & Anderson 1978. However, although this primarily only examined the ferromagnetic particulate, it did introduce the technique and show that viable results were possible.

Jones 1990 outlines a more comprehensive system of grease sampling and analysis. The process is basically threefold — removal of grease by means of a sampling gun, preparation of the sample by dilution and finally analysis by emission spectroscopy with the aid of a rotrode pre-sparking to give an even coating. The actual taking of the sample required the use of a nylon tube attached to a sampling gun inserted into the loaded region of the bearing; about 60 mm of grease was sucked into the tube with several strokes of the gun. The tube was then cut and sealed ready to be diluted and analysed in the laboratory.

He describes the preparation of the analysis sample as follows:

    a. Force a weighed amount of grease from the tube into a beaker.

    b. Add to the grease three-times its weight of base oil (Conostan 75 clean base oil standard has been found most suitable for dilution). Thorough mixing results in a semi-liquid sample suitable for analysis.

    c. Pour the mixed sample into a specimen cup for emission spectrographic analysis. The pre-sparking on the rotrode ensures an even coating of the graphite disc, prior to analysis.

Jones then goes on to give an example of its use on a strip mill taper roller thrust bearings. The slow speed (2 rev/min) of these grease lubricated bearings (almost 500 mm outside diameter) meant that vibration analysis was impossible. Grease was extracted by tube through the drain plug hole every two weeks. The analysis eventually showed up a definite increase in nickel and chromium which was identified as the high alloy steel of the bearing. Over a nine month period five bearings were identified (and later confirmed) as faulty, resulting in the redesign and use of a higher load capacity bearing.

DEBRIS IN FILTERS

A more laborious, but nevertheless invaluable, technique for debris acquisition is achieved by removing in-line oil system filters elements. Cumming 1990 mentions this as a back-up technique for civil aircraft engines in the United Kingdom. Maier, Jantzen & Schröeder 1991 explains how this is done in M.T.U. and the German Aerospace Research Establishment. It seems unlikely that such a technique would be extensively used, but it is worth bearing in mind if, in emergency, it is required to check back on what the system has been generating.

CHAPTER 6

Methods of Detecting Particles

A particle can be described from many different view points. Chapter 3 indicates the enormous variations possible, not only in external appearance (the visual domain) but also in the vast array of non-visual effects. These 'effects' require the operator to use different domains, or sensory planes, to define the presence and value of debris.

Human beings have five basic physical sensory features — touch, sight, smell, taste and hearing. We also have the unseen spiritual awareness. It is not surprising, therefore, that debris condition monitors use different sensory techniques to detect the presence and magnitude of a particle; some are physically visible, others less obvious. There is no ultimate qualification or agreement that one sensory method is the best, or even, with much confidence, that one is better than another. Another obvious conclusion must also be that the 'magnitude' displayed by instruments will vary depending on the sensor used.

Maybe one is tempted to say that the visual sense is best. Many mechanical engineers have great difficulty in accepting electronic functions — because they cannot be seen! We may feel more comfortable, for the same reason, in suggesting that a visual impression is more genuine than one which requires complex instrumentation. We may well find that we are wrong! Ultimately, the only real confirmation is whether a technique has been able to successfully detect a change in the particulate content, relating to the condition of the system. If it has been able to consistently detect even quite a small change, then so much to the better; it will be able to give more advance warning of an impending disaster.

So we can say, that providing a reasonably acceptable method of detection is used, and the results are able to be successfully compared with what is actually happening in the system, then it does not really matter what technique is employed. Note, however, that critical word 'successfully'. The various techniques described in this chapter are not all relevant to all systems; for instance, there is no point in using magnetic detectors when the metallic components likely to fail are non-magnetic. Conversely, when a specific type of debris is to be detected, maybe it has been irradiated, then the detector must only sense that debris, eg. by using a scintillation counter. Success in one system does not necessarily imply success in any other system; each system has its own idiosyncrasies, and it is as important to consider the system operation as it is to assess the potential ability of a method or technique for debris detection. Although the chapter deals only with the detection of particles, always have in mind the application.

IN-LINE, ON-LINE AND OFF-LINE

Fig. 1.6 highlighted the three basic positions for monitoring a system — in-line, on-line and off-line. It is worth defining these in more detail because of the importance of the difference in deciding the value of a device for a particular system:

IN-LINE — The monitor examines all the system oil that passes through the machine circuit at that point.

ON-LINE — The monitor draws off a proportion of the flow, a sample, to analyse and then returns it to the system. It need not return the flow direct into the same line; instead it could be returned to the tank or reservoir, or even to waste.

OFF-LINE — Using, maybe, a sampling point or perhaps a specially attached withdrawal tube, a sample of fluid is taken away for analysis remote from the machine. It involves liquid transfer and removal in a bottle or some other container such as a syringe.

In-line

'In-line' appears to have the edge over the other methods in that a total examination is being conducted and no external influences are involved. However, in practice, it could be iatrogenic, causing a greater problem to the machine system, by severely restricting the flow in the circuit. More knowledgeable manufacturers of this type of monitor also recognise that not all the fluid is actually tested, and they give a likelihood of detection factor. The reason for this will vary with the different techniques, but basically it is a case of flow lines in the vicinity of the detector causing the debris to be positioned differently depending on size and shape of the particles. (This is further discussed in Chapter 5.)

On-line

'On-line' is a compromise, but a reasonably acceptable compromise. There is no need to possibly spoil the results by involving bottles, but also the system flow need not be disturbed significantly. The attachment of such an instrument can be made after the system is running, and hence the same unit can be used in a number of different positions, where a maintenance engineer carries a portable analyser from one point to another, and records the values determined. In order to be reasonably realistic, a fair proportion of the flow should be sampled, otherwise the genuine evidence can be quite easily missed, and the statistical significance will be severely reduced.

Some devices use a very small proportion taking only a few millilitres from hundreds of litres of oil (occasionally only taking a sample from a sample). This could be quite acceptable for a thoroughly mixed fluid where the commencement of appearance of debris is indicated by a very large quantity of well distributed particulate. It has been successfully used on power station turbines.

Another type of on-line testing is really a mixture of on-line and off-line analysis. Here instead of the 'static' off-line sample being tested elsewhere, the instrument takes a small flow from either a supply container (into another container) or from the actual system reservoir. Because the liquid has not had to be removed via a bottle, the process can be considered on-line. It is always conceivable that such a method can be attached to a system, although that may not be its primary use.

Off-line

'Off-line' is the cheap-to-start alternative. It certainly gives the result that is needed if great care is taken in the obtaining of the sample (see Chapter 5). It is more long-winded, and if samples have to be sent away, can cost a great deal in wasted man hours not only awaiting the results of the analysis but also in continuing to run machinery when its operation should be curtailed. It does have some advantages, though, the particles, if detectable, will have been trapped for a later full analysis, such as visual and spectrometric as described in Chapters 9, 10 and 11. In certain applications the collection of the evidence is essential, so that confirmation of the earlier findings can be verified (eg. Mucklow 1984 who writes on behalf of the aircraft fleet operators).

SENSORY DOMAINS

Before detailing the various domains chosen to describe the different techniques, it is important that one understands the relationship of the variety of physical concepts involved in optical techniques. The optical approach has already been mentioned as, perhaps, the most fundamental. It is also the most complex.

Optical illumination techniques

Optical instruments are a far cry from "shine a light in the fluid and see what size the particles are". In most cases the particles cannot be seen by human eye — they are too small. But the other major deviation is that particles can respond to light in numerous ways. And they do not just respond in one way or another; there will be a combination of responses. Consider the following six important responses:

Reflection
Absorption
Refraction
Diffraction
Scatter
Transmission

(A simple definition of each may help at this stage:

Reflection — light comes away from a surface at an angle, where the angle of reflection is the same as the angle of incidence.

Absorption — light apparently (or does) get absorbed by the surface, because it is no longer detected in the same direction.

Refraction — light entering the surface comes out ahead at a different angle because of the particle/air refractive index.

Diffraction — light just skimming the edge of a particle is pulled out of a straight direction by surface forces.

Scatter — light apparently randomly bursting away from a surface because of the interaction of the light with a moving surface.

Transmission — light passes straight through a particle because of its transparency or porosity.)

Apart from absorption, each of these will differ with angle to a certain degree, eg. 180° reflection (ie. straight back) is different to 90° reflection.

Again scatter could be at any angle including forwards, and it could be associated with any of the other responses. It is this interaction which raises a number of questions; for instance, is forward angled light always called 'diffraction', or can you have forward 'reflection'? (In my categories below I have chosen these two ideas to be different.) These difficulties in understanding optics are seen in the various descriptions given by manufacturers to describe their techniques. Quite often the technique used is couched in manufacturer's jargon, because the method is not really just one simple idea. The words 'light interruption' are sometimes used to describe a technique, but this is not specific enough — all the optical techniques interrupt the light!

The instrument maker has, if he wants to use an optical technique, to choose one (or more) means of detecting response which he feels will provide the necessary accuracy and precision required for measurement. It should also be mentioned that there are not just two features involved in optical techniques — the particle and the response — there are three. The light source is another variable. The source could be monochromatic light (basically a single frequency, such as a laser) or white light (tungsten or halogen), or even ultra-violet or infra-red, to increase the complication; in addition, the light could be moving instead of the particle.

In order to help the reader make some sort of valued comparison between the different optical instruments, the following headings have been chosen. They are categories which are sufficiently different from each other, so that the instruments within them bear some degree of similarity:-

a. FORWARD REFLECTANCE (providing a size count)
   The light is reflected forward in a narrow angle band (ie. 6.34° to 18.95°). The intensity within such a near-forward angle depends on the surface area, which depends on the size of the particle.

b. FRAUNHOFER DIFFRACTION (providing a size distribution)
   Also called Forward Scatter, it causes a pattern of concentric rings (dark/light) which can be detected.

c. LIGHT OBSCURATION (providing a size count)
   These are variously termed Light Blockage, Light Absorption or Light Extinction (although, usually, the light is not absorbed nor is the light extinguished!) They cast a shadow which is detected in the negative sense, straight ahead of the light beam. Perhaps LIGHT SHADOW would be a more easily understood description.

d. LIGHT SCATTER (not Forward or Brownian) (providing a distribution)
Uses three separate theories of light scatter — below 0·1 μm, Mie theory and over 3 μm. Usually detected at 90°.

e. NEPHELOMETRY or TURBIDITY (providing a general level)
The idea is that of looking at the overall cloud appearance within the liquid because of the particles. The means of detecting this varies considerably, from straight through, 90° scatter, 180° reflection, multi-angle scatter, etc.

f. PHASE/DOPPLER SCATTER (providing a size count)
Uses the Doppler shift provided by the particle on the move, as seen by two or more detectors.

g. PHOTOMETRIC DISPERSION (providing a general level)
This looks at the fluctuation of the light levels around the mean level.

h. PHOTON CORRELATION SPECTROMETRY (providing a size distribution)
Relates to Brownian Motion and is sometimes termed Dynamic Light Scatter. Here it is the motion of the particle due to molecular bombardment, which is being detected by being highlighted by a light beam.

i. TIME OF TRANSITION (generally a distribution is obtained)
Either blockage or back scattering is used from the light of a rotating laser beam. The time of scanning across a particle is the key feature of the detection method.

It should be mentioned that there are other optical detection methods which relate to particles, but which are not covered in this book, because they refer more to the property of the particle rather than to its presence. For instance, its velocity is one such feature. Another is its zeta-potential, where the difference in electrical potential in an area between the surface of a colloidal particle and its surrounding liquid, is measured. The optics, in this case, involve the use of laser light directed onto the particles when an electric field is applied; electrophoretic mobility (the particle velocity divided by the applied electric force) being dependent on the zeta-potential.

The list of domains
The following are the different domains used for the techniques which are examined (in alphabetical order). It should be noted that some techniques involve more than one method, or they overlap to a certain degree, or are a slight variation on one another type. It should also be understood that the detection limits vary considerably, from submicron particles to millimetres, and hence the suitability for wear debris applications will differ; however, for completeness, the survey covers most particle detection methods relating to the presence of the particle. The purpose of this book is to explain and discuss the measurement of particles in liquids, however some of the optical techniques described, and others, can also be used for the measurement of particles in air streams or sprays; some, in fact are designed for that use (such as the Insitec Inc. forward scattering particle analysers) but can also be used for monitoring particles in liquids. (Common alternative names, or variants, are given in brackets):

1. Capillary Hydrodynamic Fractionation
    (Hydrodynamic Chromatography,
    Sedimentation Field Flow Fractionation,
    Flow Field Flow Fractionation)
2. Dielectric Constant
3. Electrical Conductance
4. Electrical Sensing Zone
    (Coulter Principle,
    Electrical Resistance,
    Electrical Resistive Pore Sensing
    Electrozone)
5. Filter Blockage
    (Pore Blockage,
    Mesh Obscuration,
    Pressure Potential Difference)
6. Gravimetric
    (Weight)
7. Image Analysis
    (Machine Vision
    Microscope Examination)
9. Magnetic Attraction
    (Ferrography,
    Magnetometry)

10. Optical — Forward Reflectance
 (Near-angle Light Scattering)
11. Optical — Fraunhofer Diffraction
 (Forward Scatter)
12. Optical — Light Obscuration
 (Light Absorption,
 Light Blockage,
 Light Extinction)
13. Optical — Light Scatter (not Fraunhofer or Brownian)
 (90° Reflection)
14. Optical — Nephelometry / Turbidity
 (Absorption/Transmission Ratio,
 Multi-angle Scattering,
 Optical Reflectance,
 Ratioed Forward Scatter,
 90° Scatter,
 180° Reflection)
15. Optical — Phase/Doppler Scatter
16. Optical — Photometric Dispersion
17. Optical — Photon Correlation Spectroscopy
 (Brownian Motion,
 Dynamic Light Scattering
 Quasi-Elastic Light Scattering)
18. Optical — Time of Transition
 (Scanning Laser Beam)
19. Radioactivation
 (Surface Layer Activation,
 Thin Layer Activation)
20. Sedimentation
 (Disc Centrifuge Photosedimentation)
21. Sieving
22. Silting
23. Ultrasound
 (Ultrasonics)
24. Visual Appearance
 (Microscope Analysis,
 Patch Test)
25. Wear
 (Thin Film Sensor)

Where relevant, additional commercial information on the most popular types of detection devices is given in Chapters 7, 8, 10 and 11 with the special features associated with each instrument; company profiles are also briefly given. A detailed listing of the commercial instruments available is also given in Appendix 1, compiled from information kindly supplied by the companies. The relative time of analysis and cost of the instruments are not discussed in this chapter, due to the large variations with each technique; some of that detail is given in the Appendix where it is applied to specific instruments.

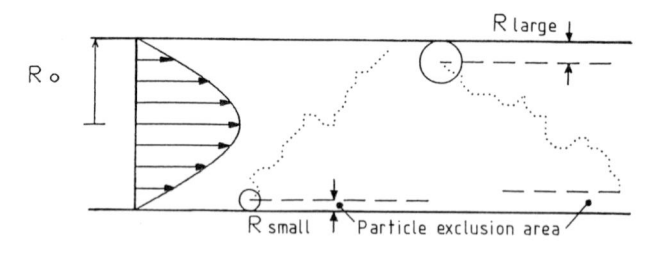

1. Capillary Hydrodynamic Fractionation (CHDF)
(Hydrodynamic Chromatography, Sedimentation Field Flow Fractionation, Flow Field Flow Fractionation)

(eg. Matec Applied Sciences 'CHDF 1100')

Hydrodynamic chromatography (HDC), in basic terms, is the separation (fractionation) of substances (different sized particles) when forced through small diameter tubes (capillaries) due to the characteristics of liquid flow under pressure (hydrodynamics). Unlike liquid chromatography, where small molecules issue first, followed by larger molecules, HDC causes the particles to appear in ever decreasing size.

Particles close to 1 μm and below have a particular type of motion when the liquid in which they are contained is passed through capillary tubes. A combination of Brownian motion and liquid boundary effects causes the larger particles to travel faster than the smaller and a velocity profile is formed. Ordinary HDC has limitations such as particle retention by the column packing material but the CHDF uses much smaller capillaries allowing a more consistent flow. The set of capillaries is normally chosen to cover the submicron particles only. The method does not require the density of the particle to be known.

Fig. 6.1 Capillary Hydrodynamic Fractionation

Sedimentation Field Flow Fractionation (Sedimentation FFF) can achieve separation down to about 0·01 μm, whilst Flow Field Flow Fractionation (Flow FFF), which involves a forced pressure drop across the flow greater than along the flow, resolution down to 0·001 μm is possible.

APPLICATIONS: Normally only used off-line. Primarily suitable for the measurement of submicron particles in a super-clean fluid. Distribution is calculated rather than measured.

2. Dielectric Constant
(eg. UCC International 'Oilcheck')

A Dielectric is an insulator. The Dielectric Constant (or specific inductive capacitance, or permittivity) is the ratio of the capacity of a condenser using the substance (oil) as the dielectric to the capacity of the same condenser with a vacuum as dielectric. (Air has sufficiently close enough properties to a vacuum for this application.)

The 'dielectric constant' of a fluid is changed as its physical and chemical content is altered by use and by contamination; it may increase or decrease. Such a technique is thus able to detect whether a change has occurred in an oil, but it will not be able to identify the reason. It is conceivable that two different effects may cancel and produce a null reading. Common effects which alter the oil characteristics are oil oxidation, water, acids, mixed fluids and gross wear debris. In fact, such a technique is used to measure the water content in an oil in one commercially available instrument. It is, therefore, important for the user to have some secondary awareness of what is likely to be in his oil.

APPLICATIONS: Used in the off-line mode only. Needs to be calibrated each time with 'clean' oil. Normally unable to differentiate between liquid debris and solid debris, but gives a measure of oil acceptability.

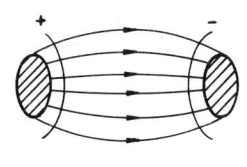

Individual Poles   Concentric Poles

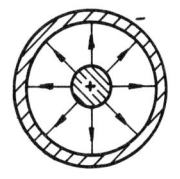

Fig. 6.2 Dielectric Constant

3. Electrical Conductance
(eg. Vickers Systems Tedeco 'Electric Chip Plug')

Electrical Conductance is simply the ability of a substance (particle) to allow an electric current to flow through it.

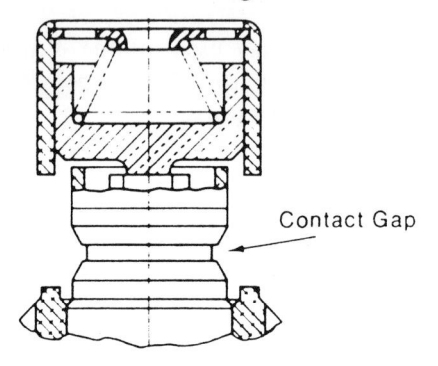

Fig. 6.3   Electrical Conductance

The landing and retention of ferrous particles between two electrical terminals will eventually cause a short circuit, thus indicating their presence. It could be one large particle or many small particles. Normally the system operates once only (ie. particles cause the short circuit across the terminals or poles, the magnet is removed, inspected and cleaned before resetting). However, it is possible to remove some of the finer debris by 'burning' it off, and in this way to differentiate between the presence of major wear debris and the presence of fine silt wear. With suitable design, constant on-line monitoring is possible.

APPLICATIONS: The technique is used either in-line or on-line. The particles must be electrically conducting and the 'oil' must be non-conducting. Where a magnetic attraction technique is used (as in the magnetic chip-plug), the particles must also be ferro-magnetic. The process requires action immediately a 'capture' is signalled; the sensor is cleaned and reset, to prevent particles building up on it and being released into the system as a conglomerate.

4. Electrical Sensing Zone
(Coulter Principle, Electrical Resistance, Electrical Resistive Pore Sensing, Electrozone)
(eg. COULTER® 'Multisizer' series, Particle Data ELZONE® '282PC')

This method relies on the change in electrical impedance in a conducting liquid, when one or more particles of different conductivity to the liquid lie in the current path.

By causing a certain quantity of liquid to be forced through a small orifice between two containers, the momentary change in electrical impedance can be monitored, as each particle passes through. The changes are presented as a series of voltage pulses, the height of each being basically proportional to the volume size of the particle. Particles vary considerably in electrical resistance, but the great majority can be measured accurately. The technique is further described on page 192.

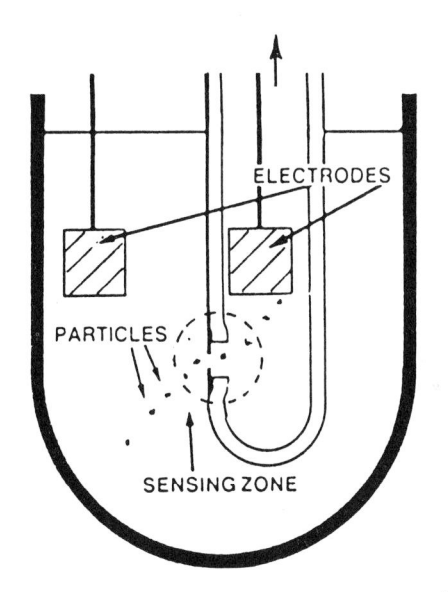

Fig. 6.4 Electrical Sensing Zone (COULTER®)

APPLICATIONS: Only used off-line at present. Only electrically conducting liquids may be used, therefore it is not suitable for mineral oils directly. The mineral oil requires diluting into a solvent-electrolyte solution (organic solvating electrolyte) so that the particles can be counted and sized, or the solid debris particles can be filtered out of the oil and resuspended in a suitable electrolyte solution for analysis.

5. Filter Blockage
(Mesh Obscuration, Pressure Potential Difference)

(eg COULTER® 'LCM II', Diagnetics digital 'Contam-Alert'®, Lindley Flowtech 'FCM')

This technique relies on the change in pressure/flow characteristics when particles block one or more orifices through which a liquid is passing.

In practice many thousands of tiny orifices are normally used, all of near identical dimension (in a mesh, screen or similar filter material). Any particle in the liquid, exceeding the dimension of the orifice it approaches, will cause some blockage to the flow. To assist in the analysis it is usual to either have a constant flow (and accurately monitor the change in pressure), or to use a fixed upstream pressure (and monitor the flow or quantity of fluid passed through). The technique is further explained on page 196 with some extra detail on page 249.

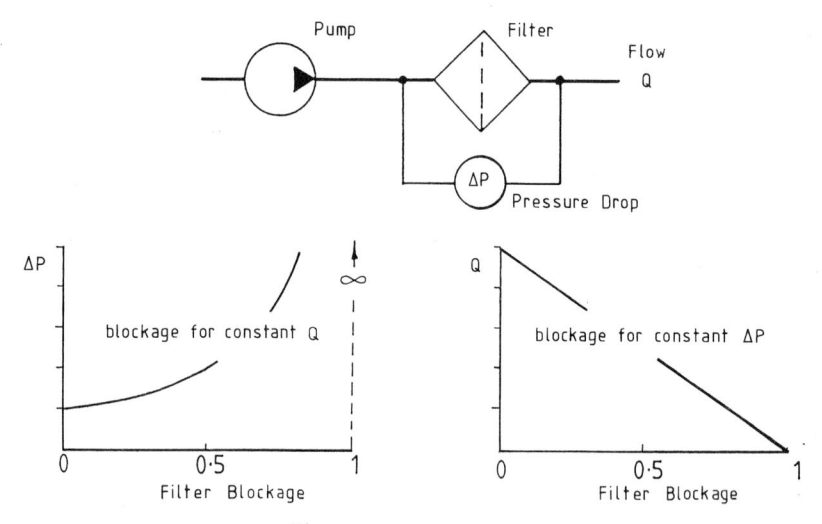

Fig. 6.5 Filter Blockage

APPLICATIONS: The technique works independently of the fluid, ie. mixed or coloured fluids or air have virtually no effect. It measures all solid debris. It is useable on-line or off-line. Some devices are one-pass only (others are continuous). Most do not normally store debris for inspection. The 'count' instruments can only strictly speaking determine a count at the mesh sizes fitted; extrapolation or estimation of slope (with a single size) can be used where the distribution of particle sizes is known.

6. Gravimetric
   (Weight)

(eg. Millipore Corp.)

Gravimetric analysis concerns the measurement of the weight of all debris, above a certain size, in a given volume of liquid.

The method gives a good overall awareness of the quantity of debris in a sample above the agreed size. It does this by causing a certain quantity of fluid to pass through a fixed size membrane filter. The rise in weight of the filter, with suitable compensation for the fluid effects, provides the gravimetric level. The process and the means of compensation are described on page 143. Some liquids pass through the membrane by gravity, but most would require a vacuum pump, particularly those which are heavily contaminated or are very viscous fluids, like heavy gear oils.

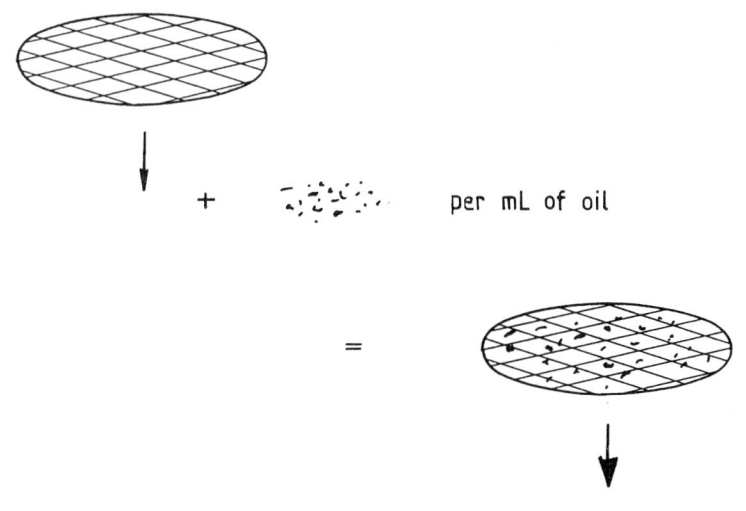

Fig. 6.6   Gravimetric

APPLICATIONS: This method is only suitable off-line. The method only gives a weight-per-mL and hence does not indicate what distribution is present or even what size of particle is the main constituent, except that the particles are greater in size than the pore size of the membrane filter.

7. Image Analysis
(Machine Vision, Microscope Examination)

(A separate list of Image Analysers is provided in Appendix 1)

It is necessary first to have an image to analyse. This may be in the form of a membrane filter on which particles have settled and have been dried, or it may be that the particles have been collected as a dry 'powder'; individual particles can be examined qualitatively.

The second requirement is a video camera which produces an electronic digital map of the particle outlines on a displayed 'image'.

The third part of the process is the computer analysis of the particle 'map'. This varies considerably from instrument to instrument, but most would have the basic facility of counting to a predetermined 'size' and 'greyness' (or colour). Additional features could relate to the particle contour, porosity, length-to-width ratio, sphericity, etc. Image analysers are explained in more detail on page 243.

APPLICATIONS: Image analysis is normally only available off-line; however, one system does include image analysis of 'frozen' pictures of debris as it is being counted on-line by another technique. Although

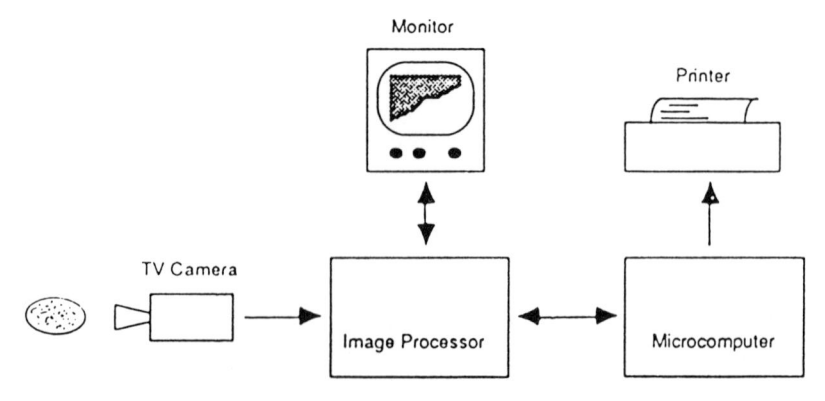

Fig. 6.7   The Image Analysis arrangement

excellent for powder applications, the processing of debris in liquids can be relatively long winded although very rewarding.

8. Inductance

(eg. Staveley NDT Technologies 'Debris Tester', see also the next category)

Inductance, or magnetic induction, is the modification to the magnetic flux caused by the presence of a substance (particle) in a magnetic field. (There is an interaction between the magnetic flux and the atoms of the substance.)

If an inductive coil is placed around a tube or container, through which the liquid is passing, then individual large particles can be detected in-line. Suitable electronic correction circuitry can be arranged to discard signals from air bubbles and non-metallic particles. By suitable electronics it is possible to differentiate between ferrous and non-ferrous metal (see page 153).

An alternative use is to actually remove particles from a collector (such as a magnetic chip plug) and place them in the vicinity of an inductive circuit. This will give an indication of the overall value of ferrous particulate content amongst the total debris in an off-line sample. Further detail is given on pages 154 and 155.

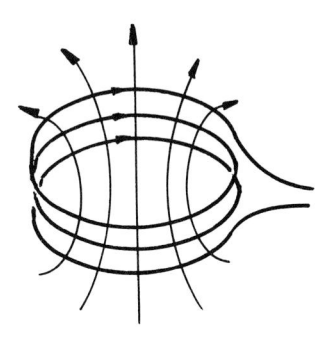

Fig. 6.8 Inductance

APPLICATIONS: On-line use is limited at present to particulate greater than 100 μm due to extraneous electrical signals. When used off-line the instrumentation gives an overall index value. Normally only ferrous wear is detected.

9. Magnetic Attraction
(Ferrography, Magnetometry)

(eg. Ranco Controls 'CDM', GasTOPS 'FerroSCAN® 210', Muirhead Vactric Components Magnetic Chip Collectors, Swansea Tribology Centre 'PQ' & 'RPD', Vickers Tedeco 'QDM', Tribometrics Inc 'Wear Particle Analyser', VD Amos Ferrograph REO 1)

Although some of these devices use inductance to sense the presence of the particles (see previous category), they are special in the sense of using a magnetic field to attract the particles. In other words they all make use of the response of the particle to a magnetic field. They use many ingenious techniques to detect and isolate the particles, some being as follows:

When a sample of liquid slowly flows over a surface (by gravity or centrifugal force), a magnet beneath the surface pulls out ferro-magnetic particles and holds them on the surface for analysis. Bigger particles are separated from the small particles because the magnetic force is arranged to vary over the liquid flow path. This is the Ferrographic principle.

Another technique determines the total ferrous content by means of a fixed magnetic field which traps the particles; this field is then modified by the presence of the particles. The amount of flux change is constantly monitored and at a set time, or when a set quantity of oil has passed, a reading of the level is displayed. This level is related to the ferrous debris level.

There are different devices available to work in the in-line, on-line or off-line modes. Most of the instruments are described in Chapters 7, 10 and 11.

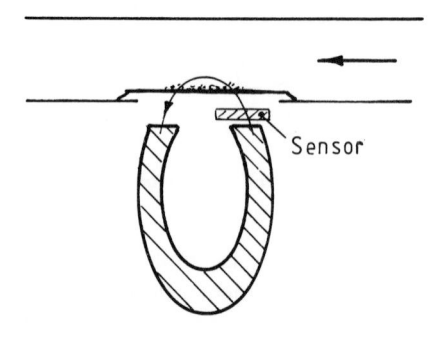

Fig. 6.9 Magnetic attraction

APPLICATIONS: Only ferro-magnetic debris can reliably be detected, although a small proportion of other particulate will be trapped. Limitations vary with the different types of device.

10. Optical — Forward Reflectance
(Near-angle light scattering, Near forward angle light scattering)

(eg. Pacific Scientific HIAC/ROYCO Submicron Liquid Sensor, Met One Liquid Particle Counter, Spectrex SPC-510 Laser Particle Counter System)

This is a very shallow angle of incidence, and can only just be called reflectance of light. Reflected light occurs at an intensity of light proportional to the surface area which is struck by the light. However, if that region is shallow to the light beam and the light impinges at, say, between 81·98° and 87·17° to the normal (this is the example given in the Spectrex Counter), it is found that the intensity in the angle band ahead (ie. 16·04° to 5·66°) is proportional to the size of the particle.

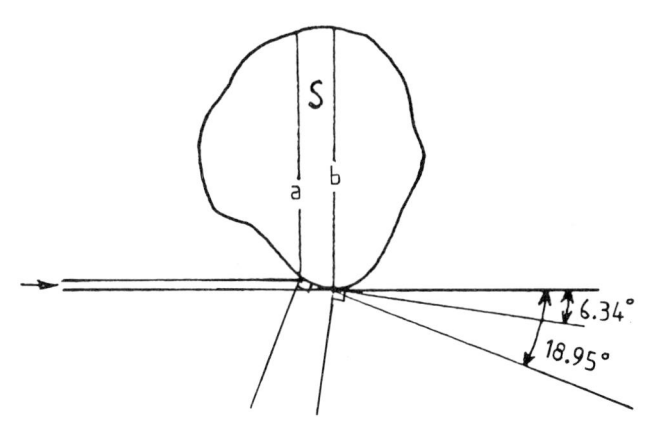

Fig. 6.10   Forward Reflectance

APPLICATIONS: The size range which can be detected by this means is quite large, perhaps 0·5 μm to 100 μm or even 600 μm, or restricted to 0·2 μm–2 μm, but the counting is limited by a low concentration level. This can be overcome either by dilution, or by the fact that the unit is used primarily on quite clean systems with considerable success.

11. Optical–Fraunhofer Diffraction
(Forward Scatter)

(eg. COULTER® 'LS Series', Fritsch 'Analysette 22', Horiba 'LA-900', Insitec 'EPCS', Leeds & Northrup 'Microtrac 'SRA' & 'FRA', Malvern 'Mastersizer', Monitek 'Spatial Filtering System', Sympatec GmbH 'Helos')

Light diffraction is that attribute of light which forms dark and light bands at the edge of a cast shadow. Where the subject (particle) is very small in comparison to the distances of the light source and the receiving panel, the type of diffraction is called Fraunhofer.

A number of concentric rings is produced. Larger particles give less scatter and the small particles the large angle scatter. A multi-array of detecting diodes is thus able to record the presence of a variety of particles sizes at the same time. A distribution size of particulate is achieved rather than an individual particle count.

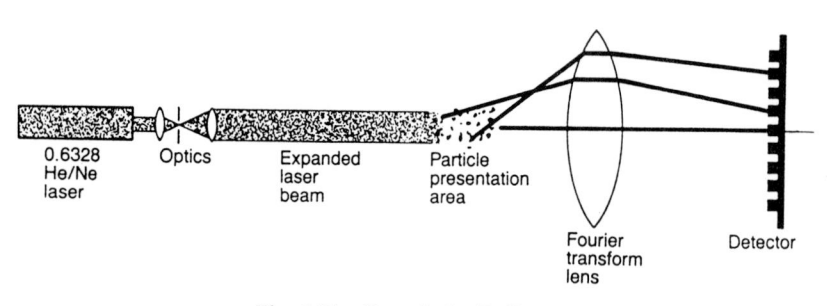

Fig. 6.11 Fraunhofer Diffraction

APPLICATIONS: The range of sizes covered with the single setting can be as high as 0·1 μm to 2000 μm although this will necessitate the use of some special means of dealing with the particles smaller than the wavelength of the light (eg. by use of polarisation). An off-line sample size of only 1mg can be analysed with an appropriate accessory. Normally only used off-line. Only clear or semi-clear liquids are possible. A distribution is given. See page 168.

12. Optical–Light Obscuration
(Optical Absorption, Light Extinction)

(eg. Climet CI-1000, Pacific Scientific HIAC/ROYCO '8000', Hydac 'Purity Controller', Malvern 'ALPS 150H', PMS 'LBS' with 'IMOLV' sensor, PMT '3120' & '2120', Rion 'KL-01', UCC International 'CM 20')

The description of the technique varies as to whether the method is described from the view point of the light source or from the receiver. The method is simply that of detecting the change in light intensity in a straight line, when a particle obscures the beam.

The static devices, which look at a membrane filter on which particulate has been deposited, merely look at the total drop in intensity and relate that to a comparative slide.

Where a flow of liquid is involved (as with the majority of instruments which examine the flow in or from a bottle sample) the intention is to have just one particle in the light beam at any one time. This can never be guaranteed, but by arranging the flows and concentrations a statistical compromise with, say, 95% confidence is achieved.

The output from a photodiode, which detects the change in intensity, is calibrated to indicate a specific size of particle. The calibration is only true for a certain flow rate and type of particle; recalibration (or a change in upstream pressure) is necessary when the flow is modified because of a viscosity change.

This type of instrument has carried Standard approval for particle sizing for many years. See page 181.

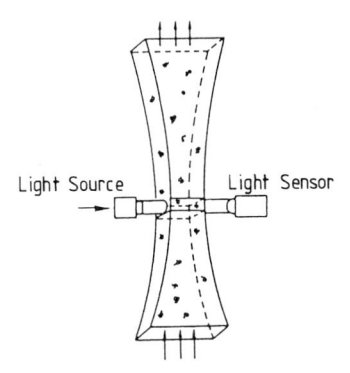

Fig. 6.12   Light Obscuration

APPLICATIONS: Due to aperture limitations most instruments only work off-line (one large particle could get jammed into the light sensor). Some are possible on-line in selected cases where permanent blockage of the sensor is unlikely. Size ranges covering 0·5 μm–8 mm are claimed, with a common dynamic range per sensor of 50 (eg. 2 μm to 100 μm). Clear or semi-clear uniform density liquids only (ie. air bubbles and additional non-compatible liquid, such as water in oil, will be counted as particles). The unblocking of the sensor will require a strip, clean and rebuild (possibly only available at the manufacturers).

13. Optical–Light Scatter (not Fraunhofer or Brownian) (90° Reflection)

(eg. A^3 BS Series, Climet Instruments Co 'CI-2000', Horiba 'PLCA-520', PMS Volumetric and In-situ, Polytec, Rion 'KL-20' and 'KL-22')

Light scattered at right angles to a white light beam, has an intensity which is proportional to the square of the diameter of the particle for most scattering angles for particles above 4 μm. Below 4 μm diameter down to about 0·1 μm, the Mie theory of scatter applies. (Below 0·1 μm, the light intensity is proportional to the 6th power of the particle diameter, ie. the Rayleigh region.)

The white light gives a much smoother relationship than monochromatic light, and the intensity of the scattered light is almost linear against size of particle. See page 186. Laser light is used for the smaller sizes and the scattered light pulse is detected by a photomultiplier tube. Forward as well as right angled scattered light is involved.

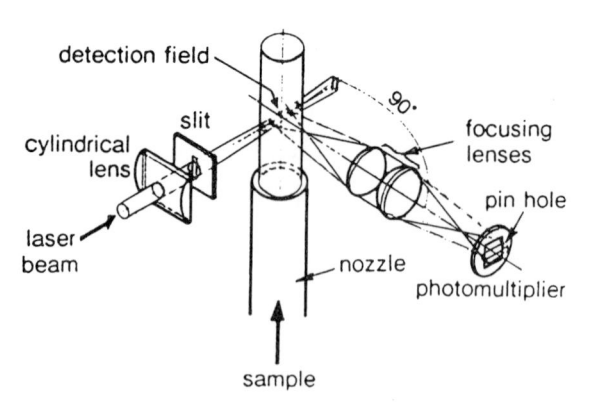

Fig. 6.13 Light Scatter

APPLICATIONS: A typical size range is 1 μm to 300 μm with white light, or 0·2 μm to 10 μm with laser light. The method can be used on-line with velocities up to 20 m/s. It can also be used in the off-line mode. Clear or semi-clear liquids are necessary. A change in refractive index could be a problem.

14. Optical–Nephelometry/Turbidity
(Absorption/Transmission Ratio, Multi-angle Scattering, Optical Reflectance, Ratioed Forward Scatter, 90° Scatter, 180° Reflection)

(eg. Climet Instruments Co 'CI-1500', Hach Turbidimeter, Monitek Technologies Inc 'CLAM' and 'Turbidimeter', McVan Instruments 'Analite', Partech DRT Turbidimeter, Wyatt Technology Corp. 'Dawn®')

Opacity is the change in clearness of a liquid due to the presence of darker particles, usually *en masse*.

'Turbidity' appears the same as an opacity change, but comes from the thought that particles or debris have been disturbed and have clouded an otherwise clear liquid. In fact, turbidity is defined even more precisely than that; Spair 1990 repeats an earlier Standards definition giving "turbidity ... an optical property that causes light to be scattered and absorbed rather than transmitted in straight lines through a sample".

'Nephelometry' is the measurement of clouds, ie. in a similar way to turbidity relating to clouds of particles. Both words are used in the standard level called an NTU — Nephelometric Turbidity Unit.

Although turbidity can be measured direct from the 90° scatter, and in that it bears a considerable resemblance to other optical techniques, this special series of devices is more often concerned with ratios, ie. the ratio of transmitted white light to reflected or scattered light. They relate only to total concentration of debris, usually in still samples (ie. non-flowing). At least two optical sensors are used in the same instrument to give the required ratio (eg. 90° scatter and transmitted). Forward Scatter is also used at an angle of between 11° and 35°. See page 155. Multi-angle sensors can cover from 5° to 175°.

In order to be able to measure 180° reflection, the transmitted light beam and reflected light beam are concentric or close together and parallel. If optical fibres are used, the accuracy is greatly improved. The result is comparative or ratioed, as with all nephelometry.

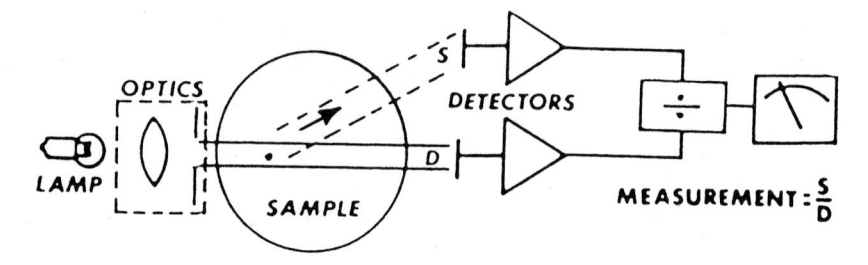

Fig. 6.14 Nephelometry

The device used would be zeroed first by taking a reading in an uncontaminated liquid of the same type, eg. water/distilled water.

APPLICATIONS: The instruments give either a total ppm (parts per million) perhaps over the range 0·001–100,000, or the standard NTU. Turbidity in terms of NTU can be detected down to 0.1 mNTU (including particles down to 0·01 μm in size); one range may cover as much as 20–20,000 NTU. Concentration Units (CU) is another type of output.

One method, using a single probe inserted in the liquid, can be used in either the on-line or off-line mode.

A variety of instruments are available covering the in-line, on-line and off-line use.

15. Optical–Phase/Doppler Scatter
   (eg. Aerometrics PDPA, Dantec Elektronik PDA)

The normal Doppler principle is based on the fact that light scattered from moving particles will be at a different frequency to the light illuminating the particle. The frequency shift being proportional to the velocity of the particle.

This principle is used in size measurement by incorporating two or more photodetectors set for different scatter angles. The phase difference between the photodetector signals is, under certain conditions, a direct measure of the particle size.

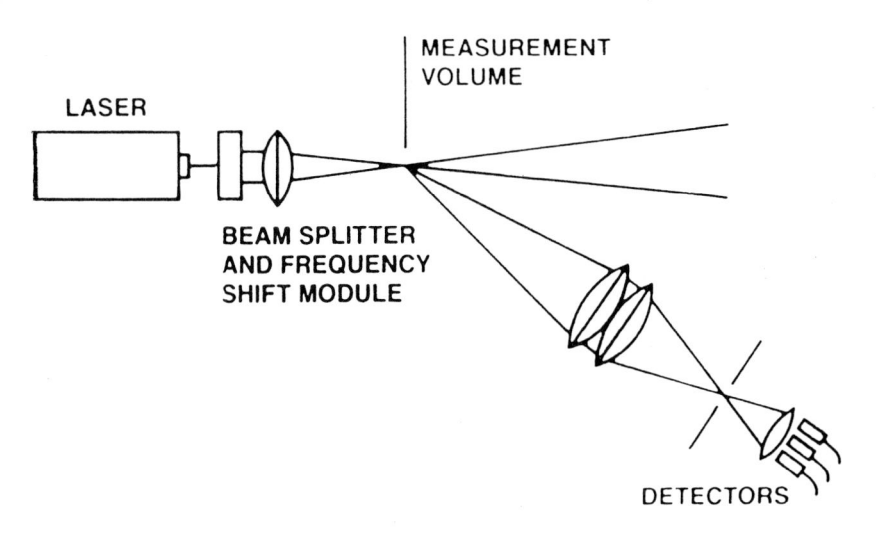

Fig. 6.15 Phase/Doppler technique

APPLICATIONS: The primary use of such a technique is with airborne particles or sprays in, such as, combustion systems. However, its exceptional feature, worth noting, is the extremely large size range from 0·5 μm to 10mm with a good remote sensing distance.

16. Optical — Photometric Dispersion
(eg. Rank Brothers 'PDA 2000')

A narrow beam of light, eg. from a masked high intensity LED, is passed through a liquid moving at a low flow rate, and monitored with a sensitive photodiode. Unlike a turbidity meter the technique is not concerned with the basic light level but with the fluctuations around the mean. Thus, both the average transmitted light (the dc component) and the oscillations caused by the particles interacting with the light (the ac component) are monitored. The intricacies of this instrument are explained in more detail on page 166.

Because the rms value of the fluctuating ac signal is related to the average number concentration and the size of the suspended particles, a definite assessment can be made of the total particle content.

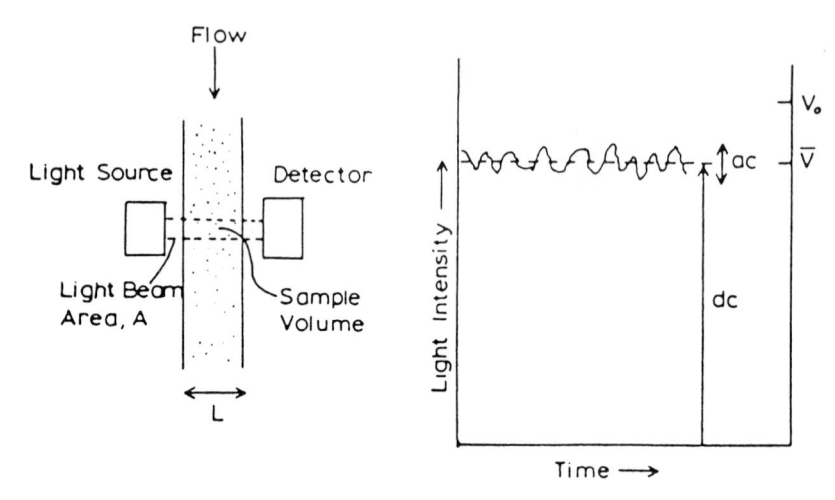

Fig. 6.16 Photometric Dispersion

APPLICATIONS: Used in the on-line or off-line mode only. It is only possible to determine a general level of debris, which is given a Particle Index. There is no indication of particle size and hence it is only really suitable for flocculation and dispersion processes.

17. Optical–Photon Correlation Spectroscopy
  (Brownian Motion, Dynamic Light Scattering, Quasi-elastic light scattering, 90° Reflection)

(eg. Brookhaven 'BI-90S', Brookhaven 'BI-90', COULTER® 'N4' series, Leeds & Northrup Microtrac UPA, Malvern AutoSizer IIc & System 4700c, Otsuka LPA-3000/3100)

The method concerns the time behaviour of scattered light emanating from small particles illuminated in a liquid. The particles produce scattered light waves which, because of interference, give a specific varying intensity at a detector. Although this intensity will depend on such features as refractivity, size and concentration of the particles, it is the movement of the particle (or phase) which is the critical feature. The movement, at micron and sub-micron sizes, is a vibratory, random fluctuation caused by Brownian Motion, inversely proportional to size. Hence the intensity can be analysed by autocorrelation to give particle size alone.

Autocorrelation is used to determine the diameters of the particles assuming sphericity. With such an assumption (which may be wrong in some cases) then the size is independent of density or refractive index of the particle. The non-sphericity of particles can be partially compensated for by observing at an angle less than 90°. See page 200.

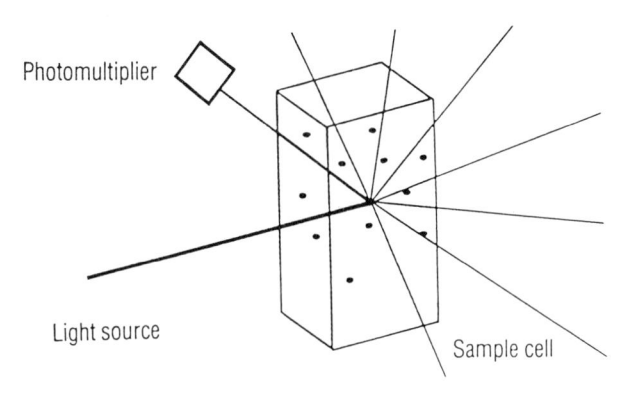

Fig. 6.17 Photon Correlation Spectroscopy

APPLICATIONS: The size range covered is between 0·002 μm to 3 μm, and the output is given in the form of a particle size distribution as characterised by the theory. Alternatively a mean size can be chosen (eg. 0·2 μm) and displayed after a set time. It is only possible to use the technique off-line. Clear and semi-clear liquids only. Because of the limitation on size, the technique is not used for wear debris analysis.

18. Optical–Time of Transition
    (Scanning Laser Beam)

(eg. Galai 'CIS-1' and 'CIS-1000', LASENTEC® 'PAR-TEC® 200' and LAB-TEC®)

The technique uses the principle of, the duration that a particle interacts with a rotating scanning laser beam is dependent directly on the size of the particle. The result is totally independent of the liquid used and hence there is no need for user calibration.

The technique, unlike some of the other optical techniques, is immune to vibration effects, thermal convection and Brownian motion. Problems associated with coincidence and orifice clogging in mechanical view windows are avoided by using interaction pulse analysis to provide an optically defined measuring zone. See pages 176 and 265.

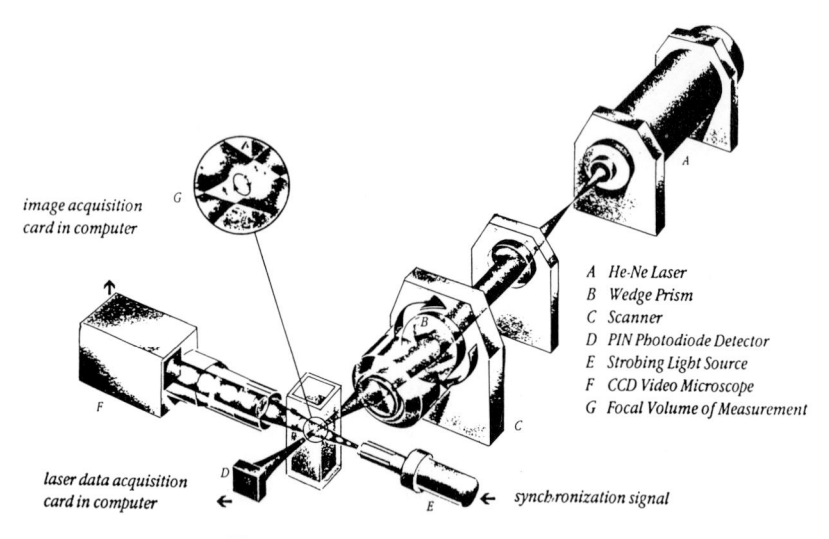

image acquisition card in computer

laser data acquisition card in computer

A He-Ne Laser
B Wedge Prism
C Scanner
D PIN Photodiode Detector
E Strobing Light Source
F CCD Video Microscope
G Focal Volume of Measurement

← *synchronization signal*

Fig. 6.18 *Time of Transition (Galai)*

APPLICATIONS: The size range on-line can go from 2 μm up to almost 4mm with suitable sensors. In the off-line mode the minimum size can drop to 0.5 μm. Concentrations of particulates can be as high as 70% (LASENTEC®) with a percentage distribution obtained in a few seconds (with this method slurries can be checked). The Galai technique requires a clear or semi-clear fluid. The result is in terms of distribution only.

19. Radioactivation
   (Surface Layer Activation, Thin Layer Activation)

(eg. Cormon 'Actiprobe', Spire Corp. SPI-WEAR®))

The principle involved is that of using a gamma-ray detection system to detect particles which have worn away from pretreated components.

A thin layer of atoms in the surface of the component is made radioactive by bombardment with a beam of charged particles, such as high energy deuterons. If material debris is later released from this surface into the liquid it may be observed with a gamma-ray detector located near the flow, or the drop in total activity can be detected at the component. The process is described in more detail on page 252.

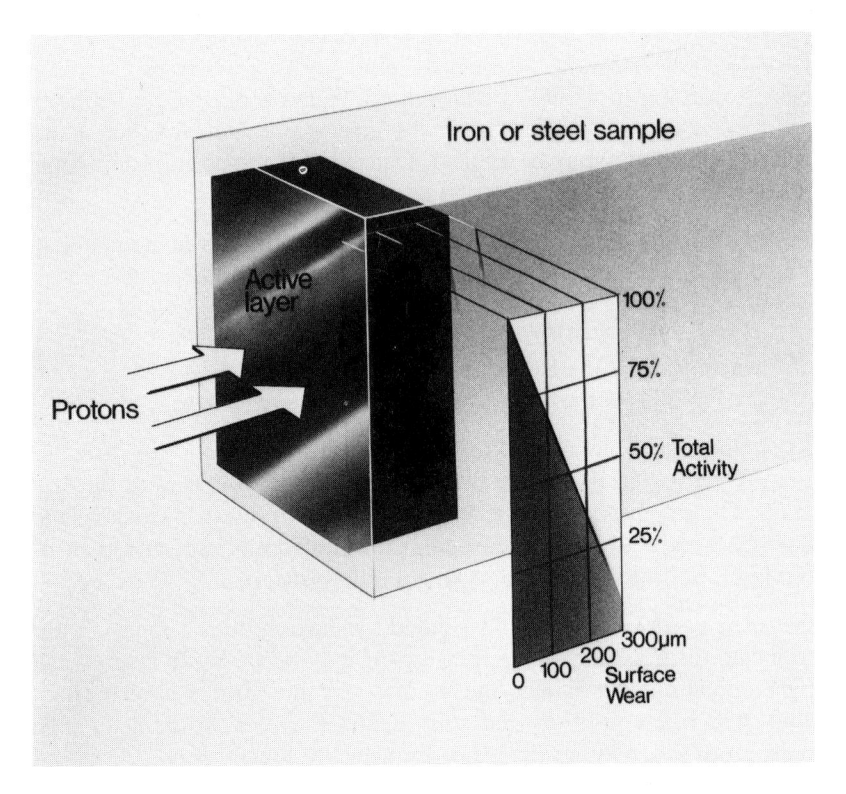

Fig. 6.19   *Radioactivation (Actiprobe)*

APPLICATIONS: The depth of the irradiated layer usually can vary between 1 μm and 1 mm. This depth determines the range of material loss monitoring and accuracy which is about 1% of the depth. The idea is particularly useful in the in-line mode. Prior preparation of the component is needed, requiring treatment at an accelerator facility. Minimal safety precautions are needed.

20. Sedimentation
(Disc Centrifuge Photosedimentation)

(eg. Applied Imaging Disc Centrifuge, Bromley Instruments 'PSA', Brookhaven 'BI-DCP', Brookhaven 'BI-XDC', Fritsch 'Analysette 20', Horiba CAPA-700, Micromeritics 'SediGraph', Otsuka LPA-3000/3100, Quantachrome 'Microscan')

Particles flow naturally through a liquid under gravity, or centrifugal field, according to a well known pattern called Stokes' Law. Although the Law is true only for spheres, the idea is suitable for other shapes, where a correction may be applied. Detection of the position of the particles against time will then reveal the size.

A suitable correction factor for Stokes' Law is the Aerodynamic Shape Factor, K (Shapespeare 1991). It is defined as follows:

$$K = 24/(R_e C_D)$$
where R_e is the Reynold's Number, and
C_D is the drag coefficient for the irregular shaped particle.

K is thus the ratio of the terminal settling velocity of the irregular shaped particle to the terminal settling velocity of a sphere of the same mass. C_D can be determined experimentally by using particles which have been carefully examined before insertion in a fluid, and then observing their free fall using a sedimentation instrument.

Once the particles are fully distributed by shaking, then a sensor is used to locate them as they move. The sensor may be an X-ray beam which, by attenuation, determines the % mass of the various sized particles while the beam position and time define the particle diameter. (The beam must move rather than the sample cell if gravity is used.)

The operation of the disc centrifuge method enables a stationary light source to be used beneath the centrifuge plate. As all the particles emanate from one central point (as the sample fluid is injected into the disc fluid) so the larger particles pass the light sensor before the smaller ones. Speeds up to 15000 rev/min are used.

For more details on these devices see page 171.

APPLICATIONS: Typical size range is 0·1 μm–300 μm for the moving X-ray beam method. Typical size range for the centrifuge method is 0·01 μm–60 μm. Only 15 mL of fully mixed sample fluid is needed. This

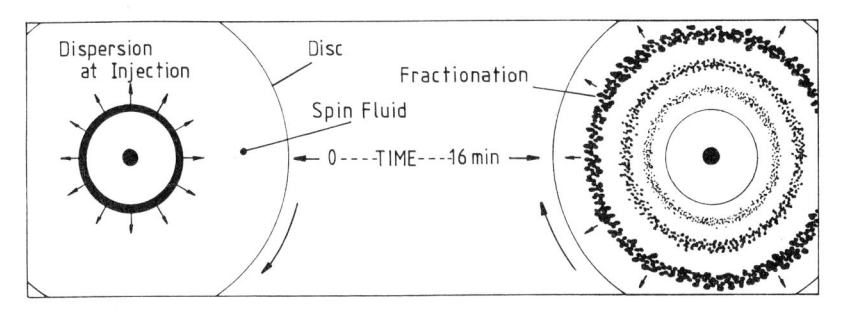

Fig. 6.20 Sedimentation

is an off-line technique only and it is only possible to analyse with clear and semi-clear liquids. A check must be made for the presence of air bubbles.

21. Sieving

(eg. Bahco Micro-particle Classifier, Endecotts, Fritsch 'Analysette 3/18')

The finest sieving is undertaken with cascade impactors which can successfully separate solids between 0·3 μm and 30 μm. Ultrasonic sieving is possible down to a size of 5 μm and up to 100 μm. Air centrifugal elutriation is possible for dry samples. Sieve Standards cover the range 20 μm–200 mm.

Sieves are made from perforated (or etched) sheet or from woven strands (usually cloth or wire). Some of the styles of weave and perforation are shown below in Fig. 6.21. In the case of woven sieves, the word 'warp' refers to wires running lengthwise of the cloth as woven, and 'weft' to wires running across the cloth as woven. In addition to the two common parameters of size of hole and number of holes, it may be important to know the 'open area' percentage, depth, strength and sharpness of edges where etching or perforation has been used. The filtration efficiency of sieves and filters is discussed in Chapter 4.

The evaluation of the particles in the sieve is a secondary operation. This may involve Visual Appearance, Optical Image Analysis or Gravimetric analysis.

The aperture width (a), wire diameter (w), mesh number (M) and open area (A) are related as follows (Endecotts 1977/1989):

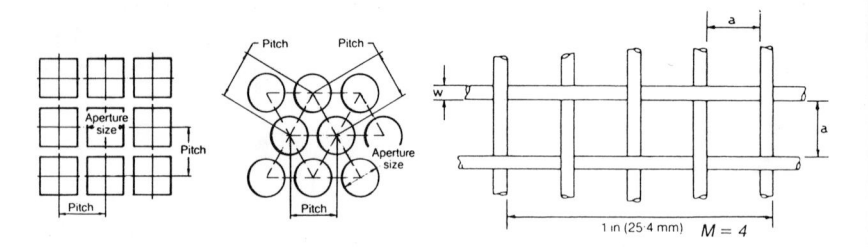

Fig. 6.21 *Sieving features (Dunn 1980 and Endecotts 1977/1989)*

$M = 1/(a+w)$
$a = 1/M - w$
$A = a^2/(a+w)^2 = (Ma)^2$.

In the case of perforated sheets with a hole pitch (p), the percentage open area (A) is obtained from

$A = (a^2 \times 100)/p^2$ for square holes
$A = (a^2 \times 90 \cdot 7)/p^2$ for round holes.

APPLICATIONS: Sieving is typically only used where there is a large quantity of particulate likely to be present. The technique can only be used in the off-line mode. It is a basic process and only suitable for gravimetric assessments or for particle separation.

22. Silting
(eg. B.H.R.Group 'On-line Contaminant Monitor')

Silting occurs when fine particles block an orifice or filter. The effect of the blockage is to restrict flow. In some ways it is similar to Filter Blockage, but it is considered separately in this book.

Various techniques have been proposed to enable quantitative measurement of particles by this means. All concern the time to block a pre-arranged 'orifice' shape or membrane filter.

The earliest idea was invented in 1966 by Dwyer & Connelly, and called the 'Silting Index'. It is described in Chapter 7 out of interest; it is no longer available.

Another automated silting monitor uses concentric cylinders a few micron apart. As particles in the fluid silt up the concentric gap, so a pressure builds up until it causes a solenoid to move the centre cylinder and release the particles. The time taken for the process cycle, relates to the concentration of particles in the liquid which are greater in size to the silt gap. This also is described in more detail on page 158; although not yet available commercially it is available for marketing.

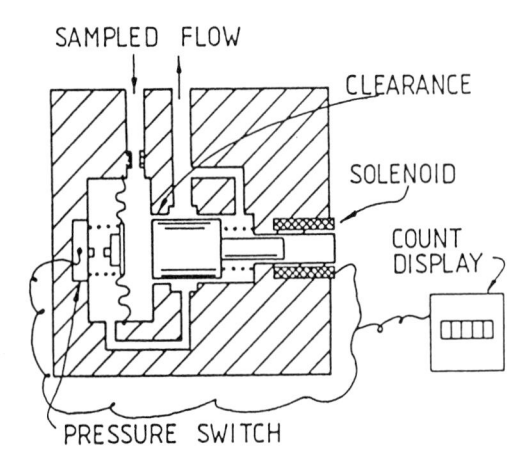

Fig. 6.22   The idea of silting

APPLICATIONS: The design available is an on-line arrangement. It is able to give a measure of the debris above one size, ie. the radial clearance size between the concentric cylinders used. Only a level code is given.

23. **Ultrasound**
   (eg. Monitek Technologies Micro Pure 'MPS-D' and 'MPS-3000')

An ultrasonic beam from an acoustic transducer is focussed into the liquid and the reflected signal monitored. The transducer acts as both the transmitter and the receiver during the monitoring process. Short acoustic pulses are emitted and the response recorded against time. If a small pulse returns early, then a particle has been intercepted, otherwise there will be just the stronger pulse returning later from the far wall. The monitor is set to just examine the reflected pulse at the time appropriate to a particle. See page 164.

APPLICATIONS: Detection can be down to 0·8 μm. The acoustic beam is not affected by sample colour or opacity. The technique is ideal for in-line monitoring, but can also be used on-line. The indication of

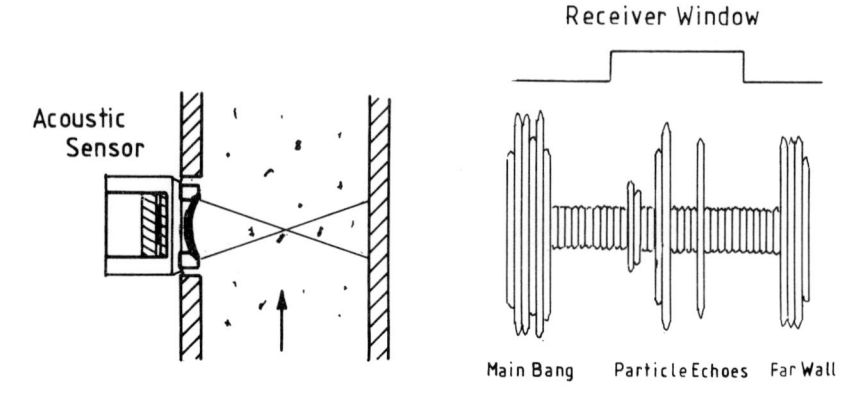

Fig. 6.23   Ultrasound

particulate given is in total concentration such as ppm, although calibration can be undertaken against known levels of contamination.

24. Visual Appearance
   (Microscope Analysis, Patch Test)

(eg. Fairey Arlon BV 'FAS-CC100', Howden Wade 'Conpar', Millipore and Parker Hannifin 'Patch Test', Oilab Lubrication, Hydrotechnik UK 'CCK 4' and UAS 'Checker Kit')

The idea of this technique is to simply look at the particles or debris, and make some kind of personal decision as to their number, size and type.

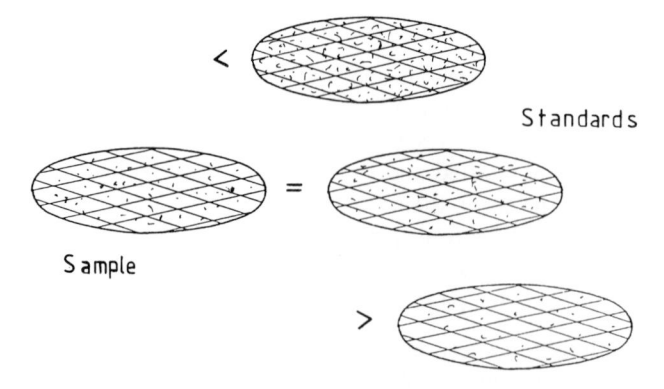

Fig. 6.24   Visual Appearance

The basic off-line kit usually consists of glassware and a vacuum pump to cause some 100 mL of fluid to be sucked through a membrane filter. The membrane (preferably dried with a solvent) is then examined under a small microscope and compared with standard slides provided. In some cases, instead of slides, a colour chart is given.

Where the liquid is in a system line with sufficient pressure (typically at least 1bar) the membrane sample can be made automatically with a special connection and membrane sample holder. For the higher pressures, an interface can be fitted (eg. Howden Wade sampling control valve). (See Chapters 5 and 7.)

APPLICATIONS: This technique is for off-line use only, although it is possible to obtain the membrane sample on-line as described above.

25. Wear
(Thin Film Sensor)

(eg. Fulmer 'Wear Debris Monitor')

A thin-film of electrically conducting material is continuously monitored for its electrical resistance. When particles, contained in the liquid which is directed on to the thin-film, begin to wear the film away, then the resistance increases. By suitable calibration a direct relationship is achieved between particle concentration and rate of change of resistance.

In order to compensate for temperature change of the liquid a second thin-film is located downstream from the wearing film, as a fixed reference resistance. See pages 163 and 272.

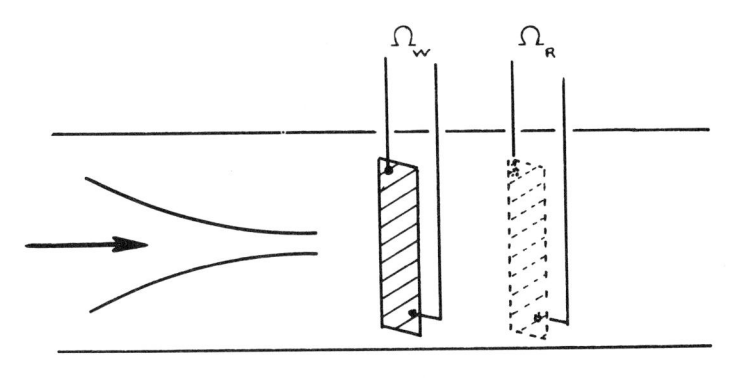

Fig. 6.25 Wear

APPLICATIONS: Very small quantities of debris can be detected, typically less than 1 ppm for the more abrasive type of particle. Needs to be calibrated for the type of particulate to be expected in the liquid, and for the viscosity of the liquid.

APPLICABILITY TO SYSTEMS

In order to be able to identify the most relevant techniques for a particular system, the chart given below is a quick reference aid. The prime point in the chart is the relationship of the instrument technique to particle size. Wear debris, for instance, may only require detection of particles greater than 5 μm, or even much larger. Silting detection will need to go down to 1 μm. Chemical processing and fluids relating to the cleaning of printed circuit boards, will require the submicron analysers.

Another feature given prominence in the chart, is the type of monitoring undertaken. It is apparent that only a very small percentage of the monitors can actually cope with all three — in-line, on-line and off-line. It will also be evident to most engineers, that if a device will work on-line or in-line, then it will also work off-line by using some form of reservoir/pump system. However, it must be borne in mind that once an extra set of equipment is installed in the process (eg. reservoir/pump with the associated transfer of liquid from one container to another), then the accuracy of the assessment will possibly be severely reduced — particularly for the ultra-clean systems. 'Off-line' is, therefore, only indicated where the technique is designed for that type of monitoring.

Because a major use of this book is for wear debris analysis, then it is right that some clear indication is given regarding the current practice. What instruments or techniques are actually used for wear debris analysis? This is the reason for indicating its appropriateness in the Wear Debris column. Obviously, all instruments which will measure particles above 5 μm could be used to a certain degree. But they are not, for three very good reasons

    a. The instruments are too expensive
    b. The instruments require highly trained personnel
    c. The instruments are likely to be damaged or blocked by the type of particle produced by wear.

These are all relative features, and the maintenance team will have to take them seriously whatever instrument is anticipated. This will be discussed more fully in Chapter 14.

The table attached below is a summary of the various types of device discussed in this chapter. It also indicates the relative range of particle size which each is able to detect. The references refer to Appendix 1 where full details of each instrument are given.

APPROPRIATENESS OF TECHNIQUE

Ref. No	Technique	Size range Sub ... 1μm ... 100μm ... 1mm +	Monitoring IN ON OFF LINE			Wear Debris	Appendix 1 Refs
1	CHF	0.015 - 1			X		47
2	Dielect. C	... not defined ...			X	Yes	57
3	Elect. Cond	-------?--------	X	X			see text
4	ESZ	0.4 -----------------1.2mm			X	?	20,25
5	Filt. Block.	5 ---150		X	X	Yes	17,30,43
6	Gravimetric	0.45 ----5+			X	Yes	53
7	Image Anal.	0.1 --------------			X	Yes	Section 2
8	Inductance	(0.02mg to 200mg)			X	Yes	24
9	Magnet. Att.	1 ----------------2mm			X	Yes	18,28,29,44, 60,67,70,75
10	Optical: - Forwd Refl.	0.1 --------------600		X	X	Yes	38,48,72
11	- Fraunhofer	0.04 ---------------------2mm		?	X		19,33,39,46, 51,74
12	- Light Obsc.	0.5 -------------------8mm		?	X	Yes	12,15,37,45, 62,63,64,66,68
13	- Light Scat.	0.2 ------------300		X	X		14,41,63,65, 68,69,
14	- Turbidity	0.01 -----------------any size	X	X	X		3,13,23,36,55
15	- Ph/Dop. Sc.	0.5 -------------------10mm			X		2,22
16	- Photo. Dis.	0.5 -------100		X	X		61
17	- PCS	0.002 ----3			X		8,21,52,58
18	- Time Trans.	0.5 -------------------1.2mm	X	X	X	Yes	35,42
19	Radioactiv.	0.01 ---------------------3mm	X	X		Yes	1,73
20	Sediment.	0.01 ---------------500			X		4,7,9,32,40, 50,71
21	Sieving	5 ---------any size			X		5,26,31
22	Silting	3 -----------		X	X		6
23	Ultrasound	0.2 -------------------3mm	X	X			49
24	Visual App.	1 ------------any size			X	Yes	10,11,16,27 54,56,59
25	Wear	(1ppm upwards, eg. 35μm +)		X	X	Yes	34

CHAPTER 7

Quantifying Debris

In this chapter we examine the instruments and techniques which can be used to determine the quantity of debris present in a certain volume of liquid. Such measurement is sometimes referred to as the determination of the 'concentration' of debris. Although there may be a lower limit of size because of the detection or capture method used, and although there may be an assessment of the distribution ratio of sizes, there is no indication of actual counts. Counting devices (which, obviously, can also give a measure of the quantity of debris present) are detailed in Chapter 8.

For many monitoring applications quantitative measurement may be all that is required. Such measuring gives a feel for the likely wear that is occurring. It is the old 'rubbing some engine oil between the fingers' idea, brought up to date. (Just like spectrum analysis in vibration monitoring has developed from the locomotive wheel-tapper's ability to hear the change in sound when a crack is present.)

Whilst the maximum size of the particulate debris in a system may be quite obvious, the minimum size is indefinable. It is, in effect, infinitesimal. However, that need not worry us, because what matters is the lower limit which actually demonstrates that excessive wear is occurring, or will occur. If membrane filter analysis is being used then any of the commonly available filter sizes are possible, such as 0·45 μm, 0·8 μm, 2 μm or even 5 μm.

Provided the measured mass of debris is always determined with the same bottom limit of size, say 0·8 μm, then any significant change in the trapped quantity, for example,

- weight per mL,
- intensity per mL, or
- shape of the size distribution,

will signal a debris change. The change may be due to wear or to ingested contaminant, but a change has occurred and will need further investigation.

Quantitative instruments and techniques can be roughly separated into these three categories shown above — weight, intensity and distribution. Whilst the 'weight devices' are reasonably few in number, there are many different types of 'intensity' and 'distribution' monitor.

QUANTITATIVE WEIGHT MEASUREMENT

A common name for this type of analysis is 'Gravimetric Analysis', from the Latin meaning 'heavy measure'. Quite simply, all that is being assessed is the weight of debris in a certain quantity of fluid. This cannot be done by just passing the fluid through a membrane which has been preweighed, because the membrane material is affected by the fluid itself (either positively by adding from the constituents of the oil, or negatively by removing plasticisers in the filter). The way that compensation is obtained is by means of a double membrane arrangement, where the fluid passes through one membrane first (and also leaves the debris deposited) and then through a second membrane. The membranes can either be preweighed or can be matched to have the same weight by the supplier; either way it is possible to cancel out the weight of the fluid effect. Thus

a. Using two separate membranes — the first, A and the second, B —

Weight of A before passing the fluid $= A_0$ mg
Weight of B before passing the fluid $= B_0$ mg

Weight of A after passing 100 mL of fluid $= A_{100}$ mg
Weight of B after passing 100 mL of fluid $= B_{100}$ mg

Then weight effect of fluid $= B_{100} - B_0$
Weight of Debris $= A_{100} - A_0 - (B_{100} - B_0)$

which equals the gravimetric level in mg/100 mL.

b. Using a matched pair of preweighed membranes — the first, A and the second, B, there is only need to weigh the membranes once, ie. after the 100 mL has passed through –

Weight of A after passing 100 mL of fluid $= A_{100}$ mg
Weight of B after passing 100 mL of fluid $= B_{100}$ mg

Then weight of Debris $= A_{100} - B_{100}$.

Although it is possible to put together glassware and filters to undertake a gravimetric analysis, the Millipore Corporation provide a specific kit for the process. Millipore and Waters, founded in 1954, pioneered the development of membrane filters and high-performance liquid chromatography — powerful technologies used for the qualitative separation of components in fluids. Their Fluid Sampling Kit, housed in a portable brief-case container, includes:

    A Quick Release Valve and plug for fitting into the system
    A stainless steel Sampler with a 3-way valve and monitor holder
    A bypass hose and adapter
    A pack of Matched-weight monitors
    A vacuum syringe with 2-way valve
    A remote sampling assembly
    A pair of forceps for handling the membrane filters
    Teflon tape for sealing the attachments to the system.

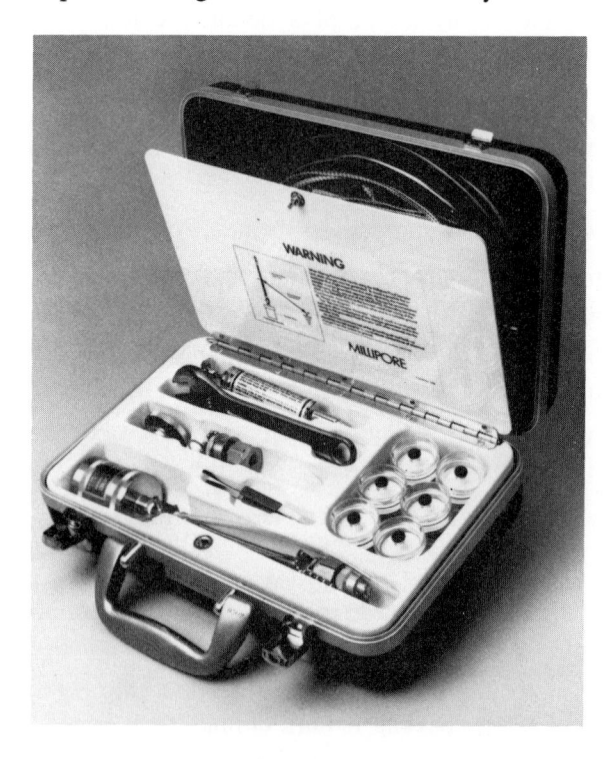

Fig. 7.1   *The Millipore Fluid Sampling Kit*

In addition a graduated flask is required and a suitable means for weighing to sufficient accuracy. The normal achieved accuracy can be as low as 0·02 mg/100 mL.

The sampling can be performed either attached to the system whilst it is running (although if the pressure is above 7 bar, a pressure reducer may also be needed), or as an off-line sampler. The Millipore kit for the on-line/in-line arrangement is shown in Fig. 7.1 and the bench-top arrangement for the off-line condition is shown in Fig. 7.2.

There are many advantages in making an analysis directly coupled to the system. This is the on-line technique which has been fully discussed on page 107. Although there are these advantages, it should be added that there are still certain precautions to take, in order to be sure that the result will be truly representative of the system debris.

For the arrangement shown in Fig. 7.1, the process is:

> the Sampler assembly is plugged into the Quick Release Valve socket (Sampling Point). After allowing a small quantity of fluid to

Fig. 7.2 The Millipore Fluid Sampler used off-line

pass to waste (to remove extraneous debris remaining in both the machine sampling point and the valve), the flow direction is changed on the valve to allow fluid through the matched-weight twin membrane monitor. The quantity of fluid passing through, perhaps 100 mL, is determined from the graduated measuring flask; when the right amount is reached the valve is switched off.

On releasing the Sampler assembly from the socket, the residual oil within the Sampler is removed as much as possible by, firstly, pouring or spraying a small quantity of precleaned solvent through the monitor. Then, using the hand syringe, either by sucking or pumping, dry air is forced through. The matched weight membrane filters in the monitor are then separated and individually weighed — the difference giving the weight of debris for the quantity of fluid which has passed through the monitor, as mentioned earlier.

The off-line arrangement shown in Fig. 7.2, requires a known quantity of the fluid in a sample container (or reservoir) to by forced through the Matched-weight monitor by means of the vacuum syringe or some other vacuum pump. (Only with very low viscosity fluids and small amounts of debris is it possible for the fluid to pass through under gravity alone.)

Even with a vacuum pump this can be very tedious if the fluid is extremely viscous or if its filterability property is not good. The oil, itself, actually blocks up the membrane very rapidly and a considerable vacuum needs to be applied for a long time in order to get sufficient sample through. There are two alternatives. The first is to allow a smaller quantity of oil through the membrane, but this gives a poorer accuracy. The second is to use a larger pore size membrane, such as 2 μm, but this means that a non-standard result is obtained.

The twin membrane method has an international Standard ie. ISO 4405, which quotes the result in terms similar to:

"x mg debris / 100 mL of oil"

This being related to a 0·8 μm membrane filter.

Conversion to ppm (parts per million)
The gravimetric level of mg/100 mL or g/cm^3 can be conveniently converted to parts per million (ppm) if the general density (g/cm^3) of the particles is known. This may not be particularly obvious where both wear debris (metallic) and ingested debris (silicaceous) is present; some

estimate would have to be made of the relative proportion of each. An approximate estimate could be made using the following densities:

Gold	19·3
Lead	11·3
Brass	8·5
Ferrous metal	7·5
Bronze	7·5
Light alloy, aluminium, rust,	2·6
Silica	2·6
Coal	1·5
Nylon	1·1

[**EXAMPLE:** If the gravimetric level is '12 mg' (per 100 mL), and there is about 60:40 volume ratio of ferrous metal:silica, then the ppm can be calculated as follows:

The weight ratio is calculated from $60 \times 7·5 : 40 \times 2·6$

ie. $\qquad\qquad\qquad$ 450:104 or 81·2:18·8

Therefore in the 100 mL sample fluid
$\qquad$there is 9·74 mg of ferrous and 2·26 mg of silica

The volume ppm is obtained by dividing the mg/L by the density of the particle.

Thus $\quad$ ppm ferrous = $9·74 \times 10/7·5$ = 13 ppm

$\qquad$ppm silica $\;$ = $2·26 \times 10/2·6$ = 8·7 ppm.

A much shorter route would be to use the direct equations —

The ppm ferrous = $\dfrac{60 \times 10 \times 12}{60 \times 7·5 + 40 \times 2·6}$ = 13 ppm

The ppm silica = $40/60 \times 13$ ppm $\qquad$ = 8·7 ppm.]

QUANTITATIVE INTENSITY MEASUREMENT

'Intensity' devices vary enormously, from the sublime to the ridiculous! But within that scope a large number of excellent methods have been built into reliable instruments, and are used extensively in industry and laboratory.

Before looking at the instruments it is important to appreciate what is meant by 'intensity'. Basically the concept is the determination of the quantity of debris present by means of a grading detector (apart from weight, which has already been discussed in the earlier section). This detection could be optical, magnetic, inductive, reflective, dielectric, piezo-electric, destructive, a change in pressure or temperature or flow, etc. But, from that sensor, a change is monitored and displayed as a quantitative increase (or decrease) in debris.

To separate the instruments into groups could be done in a variety of ways — cost, complexity, technique, success, ruggedness, on-line/off-line, range of particle size detected, etc. There is such interaction and overlap, that whatever grouping is chosen may be a little misleading. However, in order to be able to appreciate and compare the full spectrum, the instruments which follow are listed under generally comparable headings.

Seven groups of instruments and techniques

      Group 1 — Membrane filter examination (visual)
      Group 2 — Electrical (inductive, magnetic, dielectric)
      Group 3 — Nephelometric
      Group 4 — Silting/Filter blockage
      Group 5 — Wear of thin film
      Group 6 — Piezo-electric
      Group 7 — Photometric dispersion.

Group 1 — Membrane filter examination:

CCK 4 (Hydrotechnik)	Small microscope
Checker Kit (U.A.S.)	Small microscope
Conpar (Howden Wade)	Multi-purpose microscope
FAS-CC100 (Fairey Arlon)	Dual microscope
Oilab Kit	Small microscope
Patch Test (Millipore)	Direct visual
Patch Test (Parker)	Small microscope

Each of these kits is designed to compare the debris on a membrane filter with a standard membrane which may be supplied by the company. The content of each of the kits is described in Appendix 1. All the devices rely on careful visual identification and assessment by eye.

Each kit provides a means for acquiring the membrane filter from a system or sample of oil. Some of the kits, as with gravimetric analysis, use a simple set of glassware similar to Fig. 7.2; but only one membrane filter is required in the arrangement. After, say, 100 mL of the thoroughly mixed sample has passed through, a volatile precleaned solvent is sprayed into the top container to flush down any particles on the sides, and remove the remaining oil on the membrane. This has to be done very gently in order to avoid floating all the particles of debris into one spot — they should be allowed to settle uniformly over the whole surface of the membrane filter. This technique is fully explained in Chapter 5.

An alternative method of obtaining the membrane filter sample is where it is possible to attach a sampler to a pressure line in the system. This is the purpose of the kit shown in Fig. 7.1. Again only one membrane filter is used and Millipore, for example, supply these as standard items, either as 'wafer' membranes or sealed within the contamination monitor shown. This is also available with the Howden Wade condition monitoring system ie. the Conpar arrangement.

The main difference between the different kits, apart from price (which may of course be the deciding factor), is the accuracy in identifying particles by means of the microscopes supplied. A small hand-held microscope may be sufficient, where only the larger particles are likely to be the problem — where one is looking for fatigue chunks or excessive ingression — and where the overall contamination level is required. If a more detailed analysis is required, then the additional features of the Howden Wade Comparison Microscope (with its transmitted, reflected, polarised light and its optional ferrokinetic stage) may be necessary. (The Howden Wade kit and Millipore also include a fluid for making the filter media transparent so that the microscope transmitted light ability can be fully utilised.)

The simple idea of 'intensity' measurement is that of discerning a change in apparent darkness on a membrane when a fixed amount of sample fluid has passed through it. The more debris that has been trapped, the darker the appearance. This is the concept of the Patch Test supplied both by Parker and by Millipore. The kit is very similar to the Gravimetric Kit, but has in addition a small booklet, or set of comparative membranes, showing varying intensity plain photographs (shades). The booklet, the Patch Test Standards booklet, contains up to

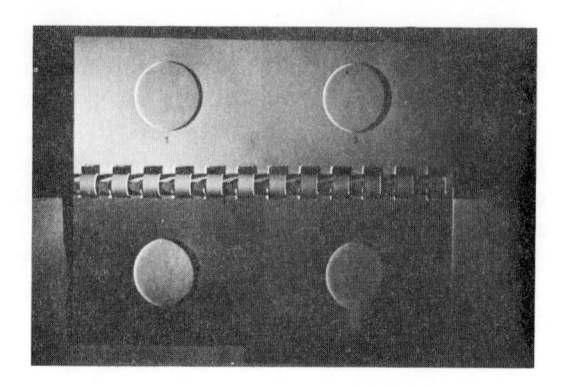

Fig. 7.3 The Patch Test booklet showing four intensities

four different intensities per general colour depending on the base colour of the oil (grey, tan and pink), as well as water and gross debris examples. Part of the booklet is shown in Fig. 7.3. Although originally designed for military hydraulic fluids (hence the pink colour), the method is now also used for lubricating oils, and in any industry. The Fairey Arlon microscope has built-in dot matrix 'patches' for comparison with the membrane filter obtained from the sampling.

The test proceeds with the use of a 5 μm pore size membrane and 100 mL of fluid sample. After drying, the membrane is placed under each cut out circle in the booklet until the comparison with the appropriate colour reference shade gives the closest resemblance. The intensities were originally classed 1, 3, 5 and unmarked (too extreme!) according to the U.S. Navy Standards Classes.

When comparison membranes or slides are used instead of the patch test booklet, they can take the form of actual slides, as with the Conpar, or be prints of intensities (or the multi-dot matrix), or photographs of real debris of certain size and quantity. The Conpar master slides are chosen by the purchaser to be relevant to his own situation, and they can be related to precise international standard levels of contaminant. The particles on the master slides have been counted and sized by eye under controlled conditions, and then given the appropriate rating.

Before leaving this section it is worth mentioning a novel Japanese device which has been used in robotic hydraulic fluid contamination control for some years. Two 13 mm membrane filters of 5 μm pore size are used for each test, one on top of the other (much like the gravimetric test method with the twin filter). After 7 mL of fluid has been passed through them, and the membranes dried, the lower one is used as a Control Filter and the top one, with the debris on, is the Sampling Filter. An infrared beam is directed on the Control Filter and the reflected rays sensed to provide a null reading; when the filter is replaced by the Sampling Filter the difference in absorption is the measure of particulate present. (The Control Filter is essential each time because of the different oil effects on the membrane causing a change in the infrared absorption independent of the debris.)

Group 2 — Electrical (inductive, magnetic, dielectric):

There is a double feature involved here. One is the method of separating or collecting the debris from a system and the other is the analysis of the debris; hence the two items mentioned in the list below (described in the way relating to their more usual applications).

CDM (Ranco Controls)	Magnetic/Hall effect
Debris Tester (Staveley NDT Technologies)	Sample/Inductance
FerroSCAN® (GasTOPS)	Magnetic/RF oscillator
Magnetic Chip Collector (MuirheadVactric Components)	Magnetic/Conductance
Oilcheck (UCC International)	Sample/Dielectric constant
PQ (Swansea Tribology Centre)	Magnetic or sample/Magnetic flux
QDM® (Vickers)	Magnetic/Inductance
WPA Model 56 (Tribometrics)	Sample/Magnetic flux

The debris being examined with these techniques is metallic. It is the change that the debris makes to an electrical or magnetic or inductive circuit which indicates its presence in a certain quantity. Not all metals will be detected necessarily, most likely are the magnetic or paramagnetic alloys, for example ferrous and nickel.

In the case of the electrical conductance sensors, the debris bridges a gap between two terminals or conducting wires and an instantaneous signal is generated. Any particle made of an electrically conducting metal will be detected provided it is trapped. Similarly most metals will influence the detection sensor of the dielectric constant instrument.

The other devices are able to sense a variation of signal depending on the quantity of magnetically sensitive debris which is present; either because they have used a magnetic force to attract the debris or because the sensor itself is magnetically orientated. The inductance, magnetic flux, Hall effect units are all sensitive to just the ferrous and nickel debris.

The Ranco Controls Continuous Debris Monitor was the result of a close cooperation between the Marchwood Laboratory (O. Lloyd and W. Hammond) of the Central Electricity Generating Board with, what they were then called, Gabriel Microwave Systems Limited. In 1984 preliminary results were published outlining the use of magnetoresistors (Bogue 1984). However, Mills 1985 describes the later developments using Hall Effect sensors instead of magnetoresistor sensors. It is worth noting the reasons for this change. The magnetoresistors were found to be:

Difficult to match pair thermally
Easily broken during assembly
Five times more sensitive to temperature than to magnetic field
Sensitive to temperature transients
Limited to a maximum temperature of 110°C.

On the other hand, the Hall Effect sensors had significant advantages

Only one required (not two)
More robust (no epoxy potting problems)
Twenty times more sensitive to magnetic field than to temperature
Less sensitive to temperature transients
Upper temperature limit 150°C
50% of the cost (at same price as a single magnetoresistor).

The CDM is designed to be built into a working system because it is continuous in action; debris is allowed to continue in the system after assessment. It is made as an intrinsically safe device and hence can been used underground.

The FerroSCAN® system is another on-line device; it first appeared on the market in 1989 when introduced by SENSYS. The method used is one of trapping the debris using an electromagnetic coil wound around the flow pipe. The field strength of the electromagnet is high enough to hold the ferromagnetic debris in the sensor region for a sufficient length of time to enable the change in frequency of an RF oscillator circuit to be measured. The change in frequency of the oscillator circuit is directly proportional to the mass of debris collected during the trapping period (typically 10 to 100 seconds). GasTOPS Ltd. acquired the technology from SENSYS in 1991 and has continued to develop the product as described by Faulkner & MacIsaac 1991.

The FerroSCAN® concept of passing the fluid through a cylindrical coil which is the inductive component of a RF oscillator, uses the following simple equation:

$$f = 1/(2\pi(LC)^{1/2})$$

where   f = oscillator resonance frequency
        L = inductance
        C = capacitance.

The presence of ferromagnetic debris increases the inductance and decreases the frequency.

GasTOPS Limited is also developing a particle counting sensor which will detect the presence of both ferromagnetic and non-ferromagnetic particles. This new technology has been chosen by Pratt & Whitney as the oil diagnostic system to be used on the new Advanced Tactical Fighter Engine (ATFE).

Whittington, et al (1991), from the Department of Electrical Engineering at the University of Edinburgh, describe a similar invention but, in their case, the particles are detected as they pass through the inductive coil (around 100 ms time band). They use a phase-locked loop (PLL) system to detect the important shifts in the frequency whilst, at the same time, correcting for any slow drifts in the overall frequency. They explain that whilst ferromagnetic debris decreases the frequency, as mentioned above, non-ferrous debris gives a corresponding rise in frequency.

The Vickers Systems QDM® is described in many papers (eg. Astridge 1987 with one example in Chapter 12). It is another device for building into a system and constantly monitoring for ferrous debris. In this case a magnetic plug attracts the ferrous debris to its surface and built-in circuitry, using the size of voltage pulses developed from an inductive sensor, indicates the magnitude of the debris. The sensor pulses are amplified and sent to a signal conditioner for classifying into two size categories (>100 μg and >800 μg). The sensor only needs inspection when an alarm count threshold is reached.

The Tribometrics Model 56 Wear Particle Analyzer also uses a filter trap, but this time off-line. The filter, through which a sample of oil is passed, is constructed so as to trap debris both by normal filtration and

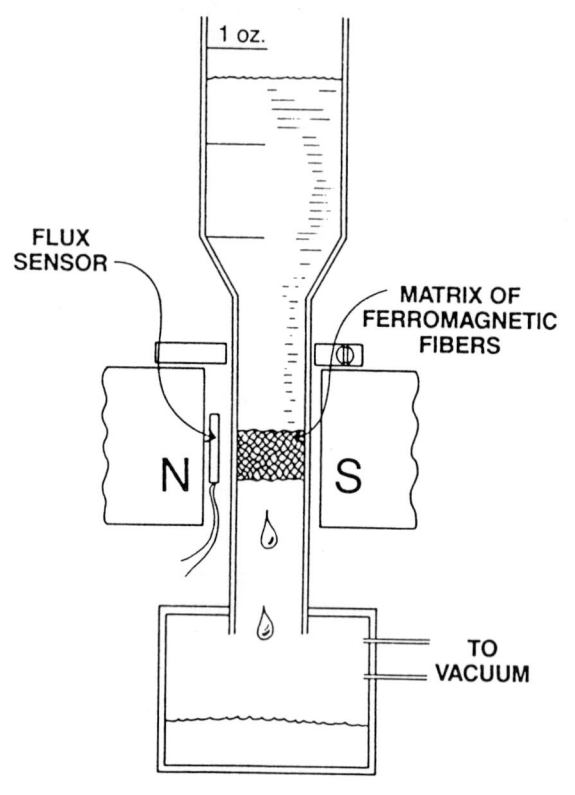

Fig. 7.4   Wear Particle Analyzer (diagrammatic)

also by magnetic capture. In other words all debris above a certain size will be held and all magnetic debris above a smaller size will also be held. The idea of the sensor is shown in Fig. 7.4.

In this device the presence of the magnetic debris is determined by means of a magnetic flux sensor.

The Staveley NDT Technologies (Inspection Instruments Division) Debris Tester Mk I was developed in conjunction with the National Coal Board for magnetic lube oil debris. It is an off-line device which rates the amount of magnetic debris placed within its sensor region by means of an inductive measurement coil detecting out-of-balance. The debris can later be inspected spectrometrically if required. Although a DTU

(Debris Test Unit) value is displayed, the instrument is usually best suited to trend investigations. The first models were launched in 1976. Examples of its use are given in Chapter 13.

The Particle Quantifier (PQ) produced by the Swansea Tribology Centre examines either a patch on which debris has been deposited, or looks at a 2 mL sample of oil direct. (The oil is put into a small plastic pot for insertion into the instrument.) Unlike the Debris Tester, the PQ uses magnetic flux as the detector of change. A PQ Index is displayed and a trend monitored.

The UCC International Oilcheck is a quicker simpler device for use on site with a small sample of oil being placed in the sensor well. The change in dielectric constant is used as the sensor technique. This means that it is an overall deterioration monitor, looking at both chemical changes (eg. water addition) and solid particulate (wear debris). It has been on the market since the late 1980's.

Other devices of this Group have had varied successes but are not included in the appendix. One such is a detector of any metal which conducts electricity, the debris being trapped on a special filter placed in the lubricating oil circuit. (In fact, all debris is trapped, but only the ferrous debris is detected). Whilst Miller and Rumberger 1972 and Beerbower 1975 describe a similar idea of filtering out debris, they use a coarse grid composed of weave/weft in insulator/conductor strands. The particular device referred to uses a grid element which is a small solid sheet of glass fibre perforated with small holes (acting as a filter); the element is clad with a printed circuit consisting of two interposed grids. When sufficient metal debris builds up on the horizontal element face, to bridge the gap between the two grids of the printed circuit, the electrical circuit is completed. The filter perforation size is around 800 μm diameter but the gap between the grids can be chosen to suit the particular application. The element needs removing and cleaning after each completion of the circuit.

Group 3 — Nephelometric:

    Analite (McVan Instruments)
    Climet CI-1500
    DAWN® (Wyatt Technology Corp.)
    Hach turbidimeter
    Monitek CLAM

These are automatic optical instruments which look at the change in intensity due to the presence of the debris. They, again, look at a total change and do not differentiate between particle sizes.

The range of instruments cover a variety of different methods for determination of turbidity. By discussing these the prospective user will have the right questions to ask when comparing one with another for a particular application.

The simplest technique is that of a single beam absorption photometer. All that needs to be measured is the attenuation of the light intensity caused through the denseness of the debris in the liquid. The change in light intensity from that reason, is described by Lambert's Law, which states that the logarithm of the transmission loss is proportional to the concentration of the particulate. This is applicable to dissolved as well as to undissolved solids.

However, the most common method, used for many years, is that of the detection of 90° scatter, as shown in Fig. 7.5 below.

The Hach Company have been making turbidimeters since 1957. (Note 'turbidimeter' is the familiar name for turbidity meter.) They have a large range of meters available suited to many specific applications. The original devices were based on the attenuation of light. Indeed, the Jackson Candle Turbidimeter determined the turbidity from the depth of a column of the liquid which just caused the light from a candle below to diffuse into a uniform glow.

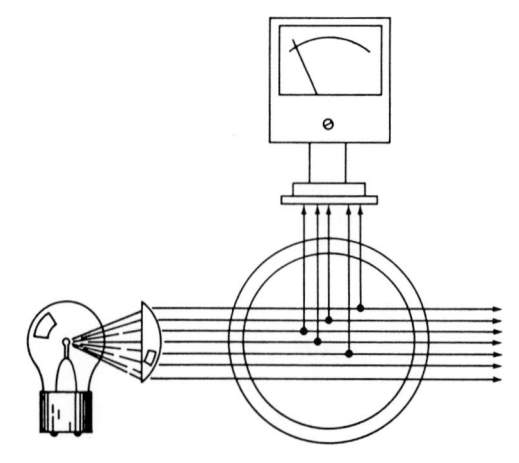

Fig. 7.5 *The 90° scatter measurement of turbidity (Hach)*

Today Hach use the 90° scatter, called nephelometry, because much smaller particles can be detected (using electronic photodetectors). The output from the meters is in NTU, the Nephelometric Turbidity Unit, with one NTU being originally equivalent to 1 ppm of suspended silica. (Formazin is the material used currently as a primary turbidity standard.)

Scatter occurs at all angles, and there is some discussion as to which angles give the best signal. Undoubtedly those devices which use multi-angle detection like the Wyatt Technology Corporation DAWN® turbidimeter tend to have a higher signal-to-noise ratio and therefore should have a better sensitivity. (The Wyatt Technology Corporation engineers were probably the first in the world to develop commercial laser light scattering instruments.) The Dawn uses a He-Ne laser light and 18 fixed detectors around the sensor cell.

The Climet Instruments CI-1500 also looks at multi-angle but in a different manner. In their case only the scatter which comes back at 180° is detected; but the sensor cell is so arranged that scatter at other back angles accumulates by reflection until received by the sensor (see Fig. 7.6).

The remarkable efficiency of such an arrangement enables mNTU's to be measured down to 0·01 μm particle size at 0·02 mNTU (see Kreikebaum 1990).

The Monitek® turbidity meters use a variety of scatter and transmittance methods including 90° and 45° back scatter. However, the ratioed

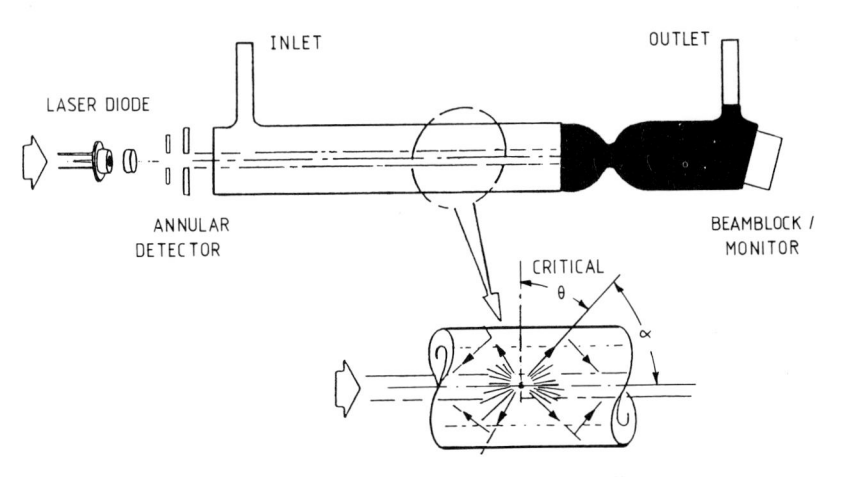

Fig. 7.6 The Climet Nephelometric Cell

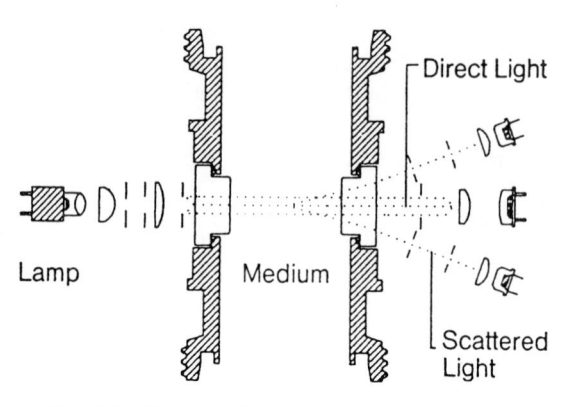

Fig. 7.7 Direct and near forward angle detection

light scatter and light transmittance is of particular interest. This is where the direct light and the scattered light at a small forward angle are both detected and compared (ratioed) (see Fig. 7.7). In this way the colour changes of the liquid or light source can be compensated for, and a real zero can be determined.

The turbidity level is not always comparable one unit to another because of the different light sources used. That does not matter if the unit is the sole means of monitoring. The McVan Analite hand-held portable turbidimeter uses a near infra-red light and back scatter (180°) for detection. It is a very compact unit because of the use of this type of scatter angle.

Group 4 — Silting / Filter Blockage:

BHR Group On-line Contaminant Monitor
Contam-Alert (Diagnetics)

These devices use a single clearance size or pore size, and detect the change in flow or pressure drop experienced by the oil as it flows through the clearance. Although a total level of debris above a certain size is determined, the output given varies considerably between each of these, depending on the use and assumptions taken into account.

Several silting devices are mentioned in a historical review by Raw & Hunt 1987, although most of them barely got further than the initial research models. Some, however, have succeeded, as exampled in this section. Fig. 7.8 shows the idea of the BHR Group monitor:

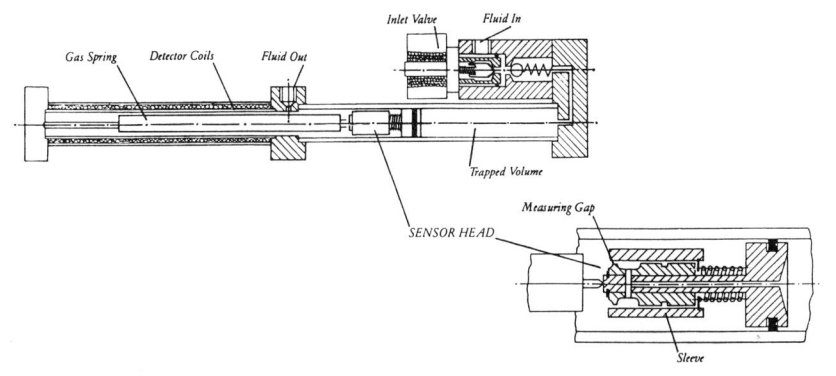

Fig. 7.8 *The BHR Group On-line Silting Contaminant Monitor*

The early successful work on the monitor is described by Heron & Hughes 1986. The monitor makes use of the phenomenon of silting of spool valves. System fluid is passed through a carefully controlled clearance between a specially designed spool and sleeve (piston and cylinder). Particles close to and greater in size than the clearance, become trapped and eventually block the clearance. The volume of fluid passing through such a clearance before blockage occurs serves as a reliable measure of fluid contamination level, largely unaffected by fluid pressure, temperature or viscosity variations. The monitor works cyclically by setting up a suitable clearance, measuring the volume required for blockage, clearing the accumulated contaminant and resetting the clearance.

The sensitivity of the BHR Group monitor depends on the clearance dimensions which may be altered to suit different applications. Although 3 μm was the proposed lower end of detection, it was found that a larger radial gap was acceptable as the smaller particles could bridge the gap. This eased the construction of the unit, which otherwise would have been extremely difficult to build.

If this method is classed as 'Filter Blockage', then it would be the type which uses a constant pressure upstream and the quantity of fluid to 'block' is measured.

It was found that a realistic debris size range of 3 μm to 20 μm could be reliably detected using a maximum flow of 20 mL/min. The monitor is continuous in that the silted cylinder/piston is automatically unblocked by the mechanical removal of the piston from the cylinder at the end of each cycle.

Although the BHR Group are aware of the possibility of relating the action of the monitor to the absolute gravimetric level (described earlier), they felt that the unit was best used as a direct comparator. That is, by observing a simple code number between 1 and 10 generated by the monitor, the increase in debris could be easily observed, and action taken when the numerical rise became significant. Repeatability was apparently good when operated on similar debris.

Heron & Hughes justify this apparently coarse means of monitoring from a table based on field studies undertaken by them and the National Engineering Laboratories in the early 1980's. The dramatic increase (or decrease) in field contaminant with condition of the machinery being operated, is described in terms of regularity of breakdown and the gravimetric level of the debris in the fluid. Part of the table is summarised below:

Condition (Breakdowns)	ISO Class 3 μm/5 μm/15 μm	Gravimetric level (mg/L) 3 μm/5 μm/15 μm
EXCELLENT (Less than one per 2 years)	16/13/9	0·5/0·1/0·3 ?
AVERAGE (1 per 2 years)	19/16/11	4·6/1·3/1·0
BORDERLINE (1 per year)	21/18/13	17/5/4
POOR (2 or more per year)	23/20/15	62/20/20

Table of relationship between reliability and contaminant in the fluid (from Heron and Hughes 1986).

(Note that an ISO code at 3 μm was expressed although no 3 μm class actually existed at that time. It served as a means of comparison.)

Diagnetics Inc. is totally committed to the use of contaminant monitoring as a predictive maintenance technique. Their feeling is that frequent monitoring and high fluid cleanliness control allow mean life between failures (MTBF) to be extended by five to ten times, hence their use of the wording "proactive maintenance" (discussed later in Chapter 14). They consider proactive maintenance to be uniquely different to predictive maintenance in that it extends machine life, not just predicts failure.

Their digital Contam-Alert, by deciding that the debris in the oil is of certain distribution, performs a calculation to display a predicted 'count' or 'Standard Contamination Class'. The device, like the **BHR** Group unit, also uses the filter blockage technique utilising a constant upstream pressure. The idea is shown in Fig. 7.9.

A cross-section of the dCA sensor is shown in Fig. 7.9. To initiate a test, the probe sleeve (2) is threaded by hand to a Minimess type sample test point on the fluid system. The piston (6) is gradually raised by the fluid (1) as it passes through the micro-screen (4). The resultant gradual blocking of the screen by debris causes a characteristic flow decay which is detected by the displacement transducer (7) and analysed by an attached hand computer (not shown). The manufacturer states that this decay profile is able to give an accurate indication of the contamination level.

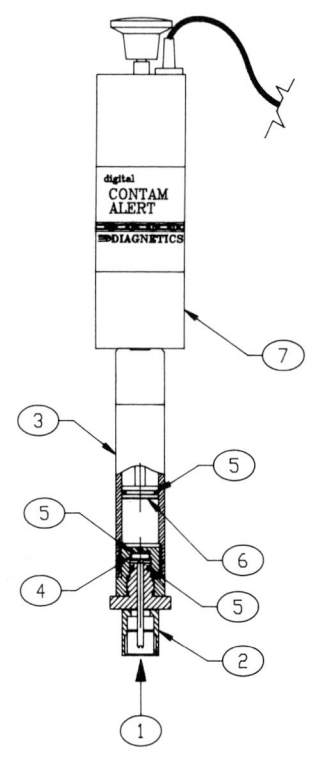

Fig. 7.9 *Cross-section of the digital Contam-Alert (dCA)*

The hand-held computer includes software which allows the test results to be compared to user-defined machine cleanliness targets. Results can be stored in its memory; extra prompt facilities are also included. The combined weight of sensor and computer is 1·5 kg.

Two devices of, sadly, historical interest (as they are no longer available) were the SDM 100 and the Silting Index kit. The SDM 100 was developed by engineers associated with the Central Electricity Generating Board at Marchwood, Southampton in the early 1980's and incorporated a number of novel features. Although filter blockage was again used to detect if any serious level of contaminant was present, the instrument had a secondary ability to retain debris for up to 25 days. The debris sandwiched between the filter strip and a cover strip, could then be analysed for its spectrometric content and hence enable precise identification of which metal components were shedding debris. The general arrangement of the SDM 100 is shown in Fig. 7.10. Its great advantage for large turbine monitoring was its ability to detect ALL debris including, on one occasion, salt!

The Silting Index kit was first proposed by Dwyer & Connelly in 1966 and it was primarily designed to look for particles between 0·5 μm and 5 μm using a 0·8 μm membrane filter inserted as shown in Fig. 7.11.

It required three timings to be made associated with the flow through the filter under a constant pressure. The pressure was applied by the force of gravity on a freely floating weight. No standards were associated with the

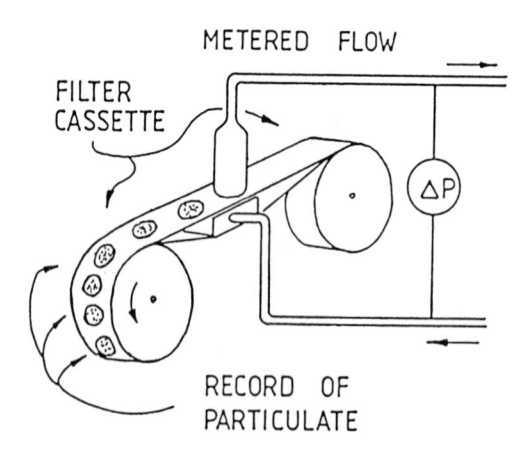

Fig. 7.10 The general arrangement of the SDM 100

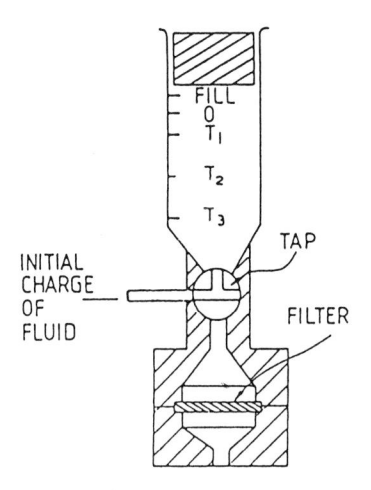

Fig. 7.11   The Silting Index arrangement

Silting Index number, it was merely used in a comparative sense. The Silting Index was calculated from

$$S = (T_3 - 2T_2)/(T_1)$$

where each 'T' related to the time at which the weight passed the engraved mark, starting the time at mark '0'.

Group 5 — Wear of thin film:
   WDM (Fulmer Wear Debris Monitor)

Fulmer Systems Limited had its roots in a foundation in the late 1940's (as Fulmer Research Institute). It has always been dedicated to providing a complete service to industry, including research and development of instrumentation where necessary. One example of that research and development expertise being put into practice is in the field of thin film wear.

The Wear Debris Monitor detects a change in particulate content in a fluid from the change in wear rate of a thin film when fluid at high velocity is projected onto it. Although it does not size the particles, the rate of change of the wear is dependent on the size, hardness and sharpness of the individual pieces of debris.

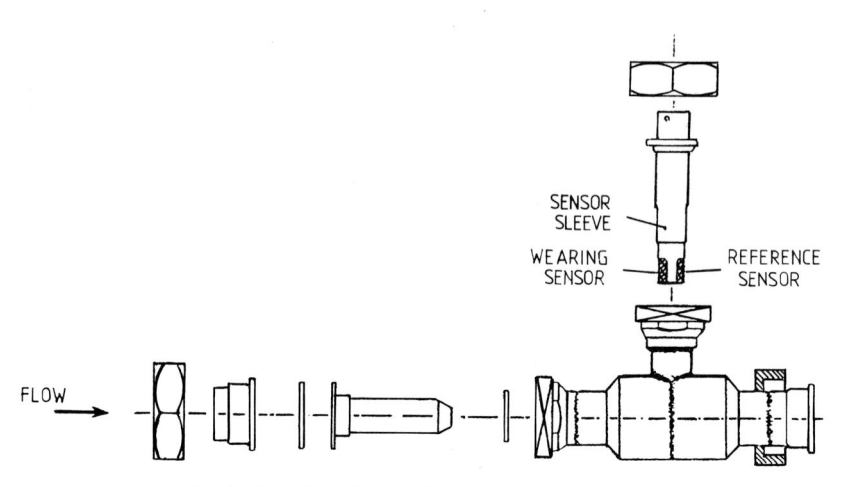

Fig. 7.12 *The Wear Debris Monitor sensor probe*

The sensor probe (see Fig. 7.12) consists of a cylindrical, ceramic substrate on which are deposited two thin films. The extremes of these films are attached to electrical connections. The probe is located in a housing through which a pressurised flow of the fluid passes. The flow is restricted through a 2 mm orifice which projects a narrow stream of the fluid onto the Wearing Sensor at a velocity around 25 m/s.

The second, Reference Sensor, is located on the reverse side of the probe (downstream), and is shielded from impingement from the fluid stream by the sensor body. This is used to compensate for variations in fluid temperature. The core of the probe also contains a temperature measuring resistor to measure fluid temperature.

When the wearing film is impinged by abrasive particles in the fluid stream, minute particles of the metal film are removed, thus changing the overall electrical resistance of the film. The rate of change of resistance is measured to indicate how much abrasive material is passing the probe. A typical example of change in resistance is shown in Fig. 7.13, showing the exceptionally long running-in period of a gas fuelled engine. (Other examples are given in Chapter 13.)

Group 6 — Piezo-electric:
    MPS-D Pure Alert (Micro Pure)

Since 1978 Micro Pure and its associated Monitek® have been leaders in the science of ultrasonics and liquid particulate measurement. The

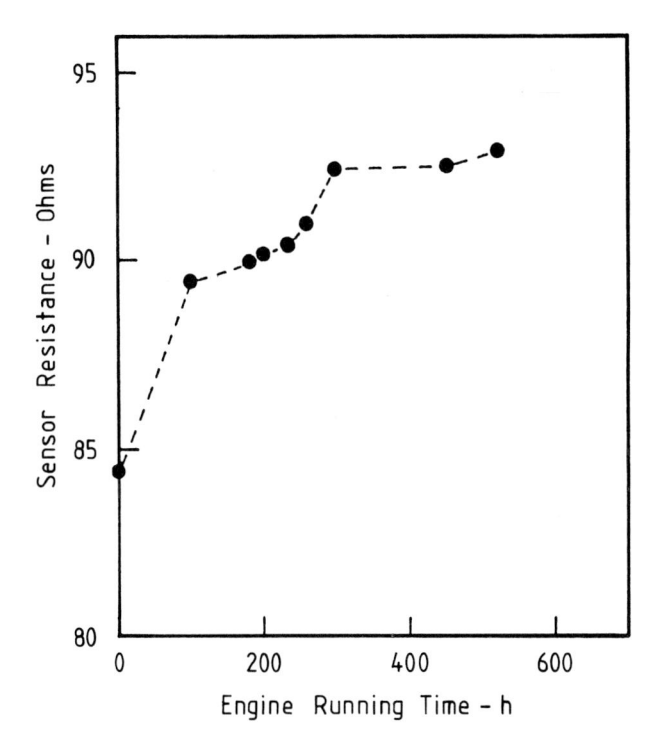

Fig. 7.13 Wear debris monitoring of the running-in of an engine

extensive research and development has resulted in the patented MPS-series Pure Alert technology using a focussed acoustical beam to sense contaminants in a fluid (as described on page 137). The contaminants may be oil or water droplets, bubbles or solid particles. The general arrangement of one of the range of probes is illustrated in Fig. 7.14. This probe can be installed and removed while the line is full and/or pressurised using the Monitek pipe adaptor.

The lens focuses the acoustic beam on a point at about 20 mm to 40 mm from the pipe wall or sensor tip. The measurement is made in the vicinity of this focal point to achieve maximum sensitivity to discontinuities in the fluid flow; it also operates in the maximum turbulent region rather than close to the wall. Particles as small as 0·8 μm can be detected.

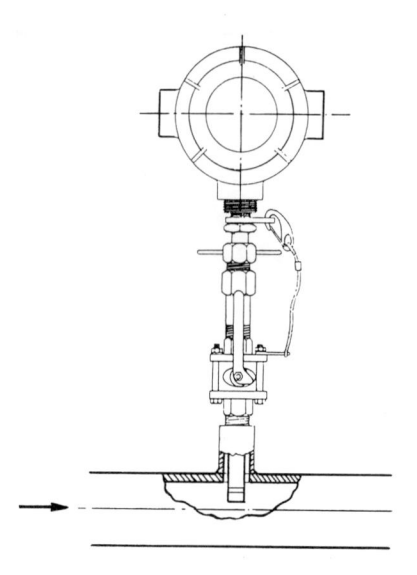

Fig. 7.14 *The fitting of a Micro Pure acoustic particle monitor*

The 5 MHz (or 15 MHz) transmitter also houses the receiver. When a short acoustic pulse is emitted, an electronically timed 'receiver window' determines the strength of the reflected echo. If it is low, then the reflection may have come solely from stray reflections from earlier pulses. If, however, it is significant (as in Fig. 6.23), then it will represent a particle of certain size.

Although individual particles (from submicron to several mm in size) can bounce back an echo from the projected ultrasound, the display indicates a total quantity from ppb to several percent content of particulate.

Another feature of the Micro Pure device is the ability to ignore particles below a certain size or, conversely, only measure particles below a size. Multiple sizing can also be undertaken to achieve distribution sizing.

Group 7 — Photometric dispersion:
PDA 2000 (Rank Brothers)

The photometric dispersion analyser PDA 2000, from Rank Brothers, is the result of development work undertaken at University College, London (Gregory & Nelson 1984). Although the instrument sounds very

Figure 1 Flow Cell (Schematic)

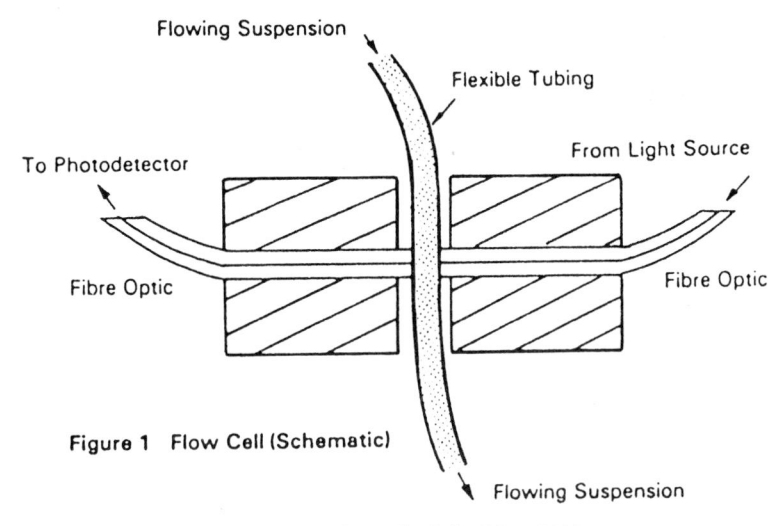

Fig. 7.15 The Cell of the PDA 2000

simple, with a high intensity light source and a collector to measure what passes through the liquid, in fact it is quite different to other devices (Fig. 7.15).

The output from the photo detector is converted to a voltage, which consists of a large dc component, together with a small fluctuating (ac) component. The dc is the simple part, being a measure of the average transmitted intensity of the light and is dependent on the turbidity of the suspension. The ac component, however, is more complex; it arises from random variations in the number of particles within the sample volume. Because the suspension flows through the cell, the actual sample number is constantly changing. It has been shown (Gregory & Nelson 1984) that the root mean square value of the fluctuating signal is related to the average number concentrations and the size of the suspended particles.

Very small concentrations (ppb) up to several percent can be detected.

QUANTITATIVE DISTRIBUTION MEASUREMENT

Although most of the advanced techniques mentioned in Chapter 6 would be able to give a distribution, they would not necessarily be actually used for the quantitative measurement or distribution analysis of debris or contaminant in fluids. Many with the distinct advantage of being able to 'count' are described in Chapter 8. The following types of instrument and technique are more appropriate in this chapter (being basically particle size distribution analysers — PSDA's), and are described here alphabetically under four different sensor methods:

USUAL NAME	SENSOR TYPE
Analysette 22® (Fritsch)	Fraunhofer Diffraction
Helos (Sympatec)	Fraunhofer Diffraction
Insitec EPCS	Fraunhofer Diffraction
LA-900 PSDA (Horiba)	Fraunhofer Diffraction
LS Series (COULTER®)	Fraunhofer Diffraction
Mastersizer X (Malvern)	Fraunhofer Diffraction
Microtrac (Leeds & Northrup)	Fraunhofer Diffraction
Analysette 20® (Fritsch)	Sedimentation
Applied Imaging Disc Centrifuge	Sedimentation
BI-DCP (Brookhaven)	Sedimentation
BI-XDC (Brookhaven)	Sedimentation
Bromley PSA	Sedimentation
CAPA-700 PSDA (Horiba)	Sedimentation
Microscan (Quantachrome)	Sedimentation
Sedigraph (Micromeritics®)	Sedimentation
Bahco Micro-particle classifier	Sieving (air elutriation)
Endecotts sieves	Sieving
Fritsch sieves	Sieving
CIS-1000 (Galai)	Time of Transition
PAR-TEC® 200 (LASENTEC®)	Time of Transition

A particle distribution is usually displayed as either a cumulative plot or a histogram. Because this applies to most of the instruments, an example of both is given here in Fig. 7.16.

Sensor 1 — Fraunhofer Diffraction

The idea of Fraunhofer diffraction has already been described on page 124. However, the large number of instruments which use this technique invariably incorporate ingenious extras to improve the performance. Others retain a lower cost with the more basic function using different collecting lens to capture each part of the size range. Others have such refined built-in checking that alignment of each new system can be done automatically (and can be automatically checked at regular intervals).

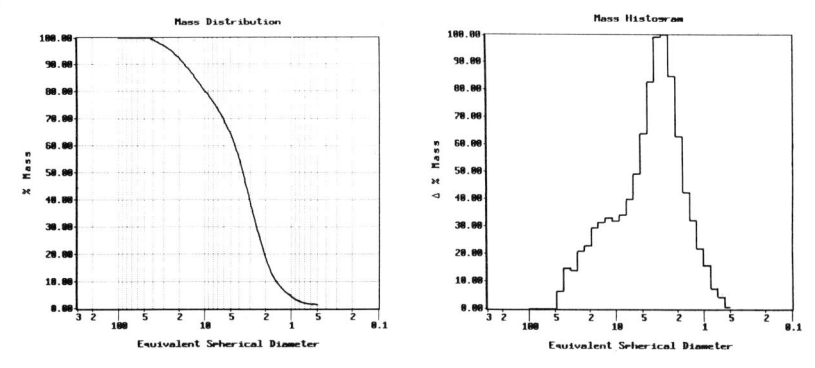

Fig. 7.16 Typical particle distribution plots (Microscan II)

The German firm of Fritsch has been established for several decades in the manufacture of high class laboratory instruments. The Analysette range of particle sizers is now well known; it started with the sieve shakers and moved into photosedimentation. However, the Analysette 22 Laser Particle Sizer uses the Fraunhofer diffraction technique with a new optical layout as shown in Fig. 7.17a & 7.17b. A highly stable spatial filter with a 25 μm aperture and optics produces a disturbance-free beam with a Gaussian intensity distribution. It is mounted very securely on the flange carrying the laser beam source. The diverging beam hits

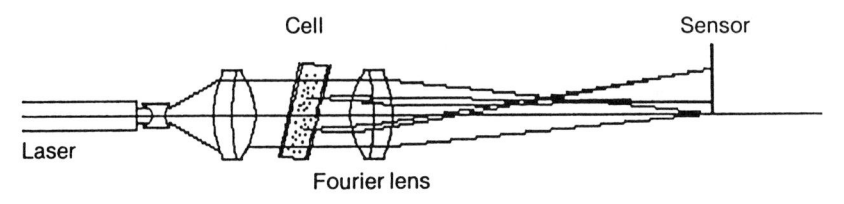

a. Particle sizer using a parallel beam

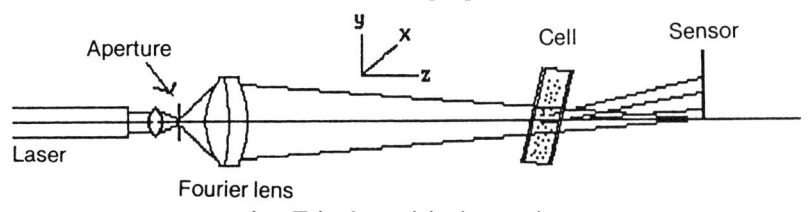

b. Fritsch particle sizer optics

Fig. 7.17 The Fritsch convergent ray principle compared with the more common parallel beam

169

an optical lens system which collects the beam in the centre of the multi-element radiation detector. The lens system and the sensor are fixed in a permanent position in relation to each other.

To alter the measurement range (which covers 0·16 μm to 1250 μm), only the measuring cell is moved between lens and sensor along the z-axis of the unit. Since no disassembly or assembly of the lens systems is required when the measurement range is altered (only the measuring cell is moved), no realignment of the optical system is therefore necessary. This greatly simplifies the mechanical construction.

The Horiba LA-900 does away with moving a system or of changing the optics by using an exceptionally large diameter lens. In this way the instrument is claimed to be able to cover the broad spectrum of 0·04 μm to 1000 μm very rapidly.

Sympatec GmbH, an advanced instrument firm based in Germany, grew from the Powder Technology department at the University of Clausthal and is primarily involved with marketing in those applications which use dry 'powders'. However, the size range they cover, now exceeds 0·1 μm to 2·5 mm which thus includes a number of different industries.

Particles may either be in the form of dry powders, a suspension, spray or emulsion. As with some of the other manufacturers, one highly important feature in the Helos is the ability to keep the particles fully in suspension during the testing. The purpose of the suspension measuring cell is to prepare, homogenise and disperse a suspension or emulsion, and keep it in circulation. An external ultrasonic generator promotes the dispersion in the suspension, and two agitators ensure homogenisation in the easy-to-fill liquid vessel.

Although the Helos full size range is given as 0·1 μm to 2·625 mm it is covered by a series of changes in the four standard central unit versions. The changes required are in the focal length of the lens, of which seven different ones are available. However, the analysis resolution is enhanced by the use of a multielement array of 31 semicircular rings, logarithmically dividing the size classes. On-line as well as off-line monitoring is possible.

The COULTER® LS series and the Leeds & Northrup Microtrac use a single lens system with an extensive, but very different, multi-collecting

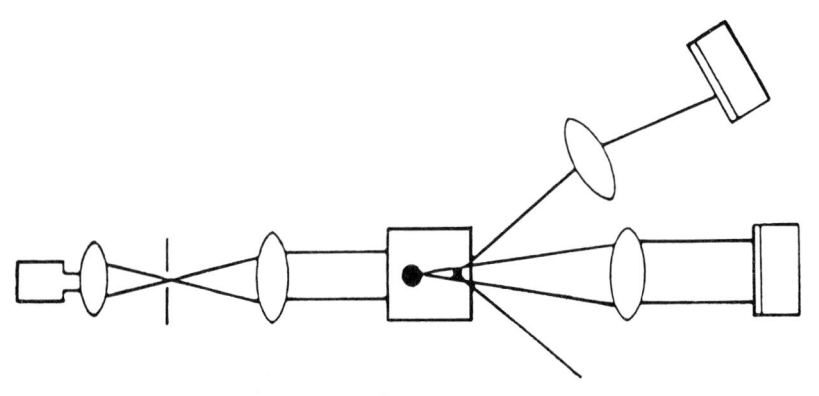

Fig. 7.18 *The COULTER® LS Series multiple optical trains*

system to extend the range, which means that the full instrument dynamic range of 0·1 μm to 700 μm or 800 μm can be determined simultaneously (Fig. 7.18).

The Malvern MasterSizer X claims a remarkable range of 0·1 μm to 2000 μm (2 mm) in a variety of configurations to suit most applications. The MasterSizer IP is designed for rugged use in an industrial situation, and includes special in-built checking features to ensure the instrument only works when the optics are fully aligned.

Insitec Inc are involved in measuring process systems in-line. Their EPCS-P (Ensemble Particle Concentration and Sizing for Process applications) optical system is contained in a rugged, compact, water or gas-cooled probe which can be extended several metres into the processing unit. The device is capable of operation at up to 1400°C from vacuum to 10 bar. Although designed for powder or spray processing systems, it can be adapted to monitoring particles in liquids.

Sensor 2 — Sedimentation

The process of sedimentation, as mentioned on page 134 involves two features. Firstly, there is the means by which the particles are caused to disperse (sediment), which may either be by gravity or by centrifugal force. The second variable feature is how the particles are detected as they disperse, eg. by X-ray beam or a light beam. These features are illustrated in Fig. 7.19.

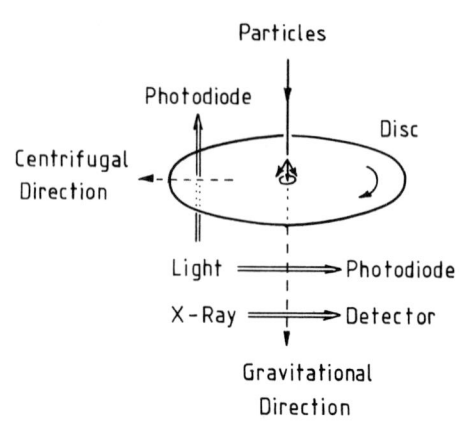

Fig. 7.19 *Different possibilities for sedimentation*

The list of instruments relate to these methods as shown:

Analysette 20® (Fritsch)	Gravity/Photodiodes
Applied Imaging Disc Centrifuge	Centrifugal/Light beam
BI-DCP (Brookhaven)	Centrifugal/White light
BI-XDC (Brookhaven)	Gravity + Centrifugal/X-ray
Bromley PSA	Gravity/Optical
CAPA-700 PSDA (Horiba)	Gravity + Centrifugal/LED
Microscan (Quantachrome)	Gravity/X-ray
Sedigraph (Micromeritics®)	Gravity/X-ray

Some of these instruments have been freshly designed with the latest technology, others have been on the market for many years, gradually developing and improving with customer feedback. The combined gravity plus centrifugal devices include both techniques in order to provide the user with the optimum method, one way or the other. Comments on some of the manufacturers, not mentioned in detail elsewhere, are included below. (Details on all the instruments are given in Appendix 1.)

Micromeritics offer a complete range of analytical instrumentation for particle characterisation, including measuring surface area, pore size, active site area, density and zeta potential. The Sedigraph has been a standard technique since its introduction in the early 1960's. The latest version, the model 5100 system maintains the accuracy associated with earlier models. Computer control with menu driven software makes the system easy to use and also provides rapid analysis time, improved data reduction and the ability to run up to 36 samples unattended.

The Scientific Instruments Division of Joyce-Loebl Limited was purchased by Applied Imaging International in mid 1991; hence the new name for a well established firm. The concept of the disc centrifuge was developed by ICI Paints Division during the 1950's and was production engineered into a commercial product by Joyce-Loebl around 1960. It is not surprising that over 100 application reports citing essential experimental data have been published on their disc centrifuge particle size analyser, particularly as it has been enhanced and refined through various models.

Bromley Instruments particle size analyser is a new venture aimed at the low cost market, but with extensive software included for a separate or customer's PC.

Quantachrome Corporation was established in the late 1960's with particle and powder characterisation as its major emphasis. The Microscan, based on sedimentation, has been available for a number of years, and the latest version Microscan II has many improvements, such as reduced size and weight, automatic system cleaning, and auto-initialization. An innovative fast scan mode enables a sample to be analysed for particle size distribution in as little as three minutes.

Sensor 3 — Sieving

Many sieves and sieving systems are available. They differ because of the way the manufacturers choose techniques to suit particular requirements, eg. accuracy and any required certification. However, in general, they follow definite standard sizes. There are many suitable shakers which will accept the standard sieves, and there are other elutriation devices.

Sieves are not the same as filters. Filters are designed to remove debris and in most cases keep the debris within the construction of the filter media. Large depth filters and good dirt capacity are thus valuable features of filters. A sieve is designed to hang on to the particles only until they are required for a further stage. In other words, the sieve is a separator into sizes.

Sieves are not the same as membrane filters. Membrane filters are usually exceptionally fragile, allowing the pass of only a small quantity of liquid once. A membrane filter will normally be supported by a coarser

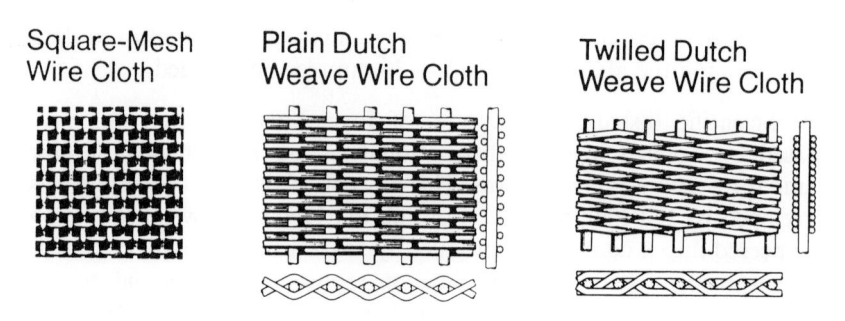

Square-Mesh Wire Cloth   Plain Dutch Weave Wire Cloth   Twilled Dutch Weave Wire Cloth

Fig. 7.20 Different mesh weaves top and side views (G. Bopp Ltd.)

mesh or sintered backing while the liquid is passing through. The sieve is designed for repeated use, and hence is more robust but that limits it at the lower end of the size range (to about 5 μm).

Sieve construction is usually either mesh or perforated plate. The mesh covers the bottom end of the size range and progresses well up into the higher range. Typical ranges would be as shown in the Endecotts description below. The mesh construction is plain for the larger sizes and twilled for the smaller. In fact in the smaller sizes, the weave and doubling effect mean that their is no direct means of seeing through the mesh (Fig. 7.20).

It must be remembered that the sizes of particles which pass through a sieve are those whose sizes are sized by sieve! Chapter 3 explains the importance of this statement when comparing size methods which other sizers use. Obviously particles with high aspect ratios may pass through or they may be trapped, depending on the amount of shaking involved and how much particulate is present; the odd shaped particles in smaller quantities will tend to pass through more easily than those in larger amounts.

Endecotts Limited is one of the largest manufacturers of test sieves in the world, going from strength to strength since its inauguration in the 1930's. Test sieves are made to every national and international specification. Sieves are available in brass or stainless steel in woven mesh and in tinned steel perforated plate. The standard size range covers from 20 μm (wire mesh) up to 125mm (perforated round or square hole). Other sizes can be made to order with 'micro plate' sieves down to 5 μm.

Size ranges for the different styles of sieve are as follows:

Micro-Plate	5 μm–75 μm
Wire mesh	20 μm–125 mm
Perforated plate (round hole)	1 mm–125 mm
Perforated plate (square hole)	4 mm–125 mm

The Endecotts shakers include mechanical and electromagnetic. Wet or dry sieving is possible.

Fritsch view sieving as a major part of their total particle size analysis equipment, as indicated below:

Method	Particle Size Range
Sedimentation (centrifugal)	0.05 μm–10 μm
Microscope	0.05 μm–1 mm
Optical Diffraction	0.16 μm–1160 μm
Sedimentation (gravitational)	0.5 μm–500 μm
Projection Microscope	1 μm–10 mm
Microsieving	5 μm–100 μm
Wet sieving	20 μm–200 μm
Dry sieving	63 μm–63 mm

The lower end of the Fritsch sieve range is covered by foil micro-sieves with coned shaped square holes. These enable a much larger open sieving area (eg. 4.3% at 5 μm) than normal meshes of the same pore size (eg. 0.7% at 5 μm).

They also produce some shakers for the sieves with a secondary oscillating movement to improve sieving efficiency.

Air elutriators have been used since the early 1900's, initially with mixed success, but since the mid century with a much greater acceptability. The Swedish firm of Bahco developed their Micro-particle Classifier as a means of assessing airborne particles; however, it is now used extensively for the evaluation of dry particulate and manufactured by the Dietert division of George Fischer Foundry Systems Inc. in the USA. The Bahco device uses both elutriation and centrifugation to separate the size fractions — the air elutriation tending to cause the particles to go to the centre and the centrifugal action causing the reverse. This means that size and mass are affected differently. Fig. 7.21 shows the cross section as described by Horowitz and Elrick 1986. The design of the air flow is complex, it being sucked in from the bottom past the adjustable

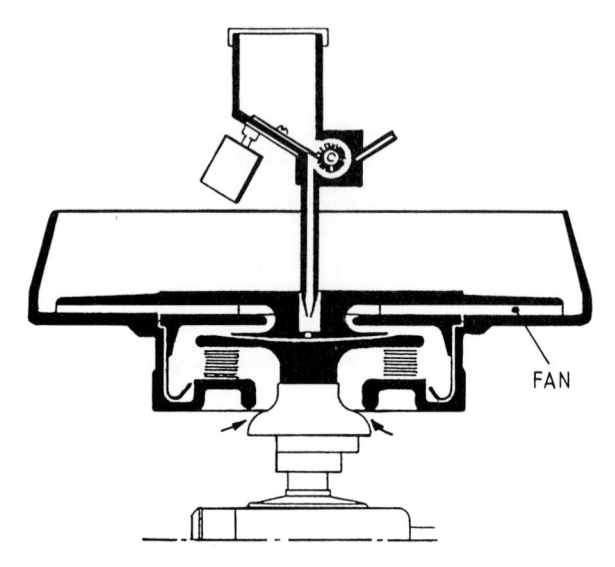

Fig. 7.21 The Bahco centrifugal air elutriator

throttle spacer, by the 80 vaned adjustable rotating fan. The air spiral, having both tangential and radial velocities imparts motion to the sample such that part of it is accelerated towards the periphery of the whirl (smaller particles) and the other part is moved towards the centre because of the friction between particle and air (larger particles).

There are certain similarities between this device and the California Measurement dry particle detector illustrated in Figs. 4.7 and 4.8.

Sensor 4 — Time of Transition

Two instruments, from Galai and LASENTEC®, use the time of transition measurement technique. This has the special feature of the rotation of the laser beam at high speed, negating any requirement for the liquid to be flowing, but yet it can work with a flowing stream as well. However, there is a significant difference between the two. The Galai uses a transmission geometry looking at the response ahead of the laser, whilst the LASENTEC® unit uses a back scattering technique. The LASENTEC® is thus able to determine concentrations at a much greater level because the beam does not need to penetrate the sample (or slurry). Conversely, the Galai will be more suitable for the cleaner fluids.

The Galai instrument is described in Chapter 11.

Laser Sensor Technology Inc, to give the full company name of LASENTEC®, was established in 1985 specifically in the field of the chemical process industry. Since 1988 the marketing of the PAR-TEC® and LAB-TEC® have extended their work dramatically. The PAR-TEC® was developed for industrial use, and many highly concentrated liquids have been successfully monitored, even up to 70% concentration. The idea of the instrument is illustrated in Fig. 7.22.

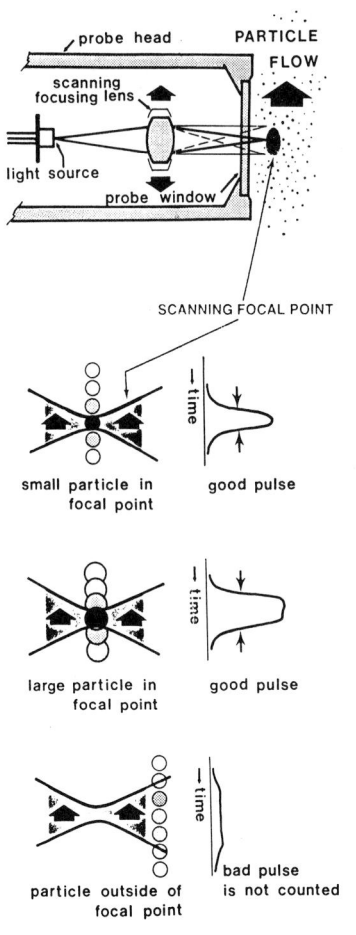

Fig. 7.22 Response due to particles in different positions in a scanning laser beam (Time of Transition) (LASENTEC®)

The laser beam is focussed to a very small beam spot with a very high light intensity at the focal point (greater than 1,000,000 watts per square inch). This means that particles as small as 1 μm can be detected. The system incorporates a unique discriminating circuit which only counts those particles which pass the focal point. The beam is scanned at a fast fixed velocity and the amount of time it takes for the beam to scan across a particular particle is a direct measure of that particle size. Size distribution is statistically determined by the number of pulses in the individual counting registers.

Particle size is determined by time (pulse width) not by the height of the pulse. This means that the technique is independent of the colour of the particles and the absorption coefficient of the suspension fluid. 38 separate size channels are displayed.

CHAPTER 8

Counting Debris — by 'size'

In the context of this book, a particle 'counter' can be described as an instrument which examines the particulate in a liquid sample and determines the number of particles greater than particular 'sizes'.

As has already been mentioned, the count/size value is not an absolute which can be compared one instrument type with another. 'Size' has to be defined in some way appropriate to the counter, which may mean a visible dimension (optical), a magnitude or an effect. These techniques are fully described in Chapter 3. The counters detailed in this chapter are real instruments which serve to illustrate the use of those three ideas of measurement. Just as the ability to measure varies from one device to another, so also does the accuracy. It is left to the reader to determine which type of counter is best suited to the situation in hand; a simple one may be quite adequate, but on the other hand, it may not be.

Basic information has kindly been provided by the manufacturers, and although comments have been added, it is left to the reader to undertake a comparative evaluation of the different techniques and models which he or she thinks may be appropriate. A brief company profile, the background and key features of each instrument, either known to the author or publicised by the manufacturer, are emphasised in order to help the reader make the choice. The nomenclature used to describe the methods is the preferred usage described in Chapter 6.

The counters are described in the following order of detection:

Visible examination (optical)
Magnitude examination
Effect examination

VISIBLE EXAMINATION TO DETERMINE PARTICLE COUNT

Visible examination as regards Optical — Fraunhofer Diffraction has already been described in Chapter 7 because that type is more closely associated with distribution analyses rather than direct counting. The more general Nephelometry/Turbidity and Photometric Dispersion similarly were also detailed in Chapter 7. Image analysers, however, can provide a very comprehensive range of counts but they are described later (see page 243). In this chapter we look at three other techniques — Light Obscuration, Light Scatter and Forward Reflectance.

Light Obscuration — General comments
The technique has been briefly described in Chapter 6. Because of its wide use in contamination control particulate assessments, it is important that its features are properly understood. I could have commenced this section explaining the advantages of this type of monitor, but instead I choose to set the scene as a problem; this is on purpose in order to understand why the various manufacturers include so many ingenious devices in their instruments.

The light obscuration devices are, in the main, scientific instruments. If they are handled well and are used for the correct applications, they will give highly accurate results. The important thing is for the operator to be fully aware of the instrument's capabilities and how it works.

If a distinct and clear shadow is to be cast by a particle as it passes a light beam, then the following must apply:

1. The beam must be of uniform intensity
2. The beam must have a clear sharp edged direction (collimated)
3. The liquid must be sufficiently transparent for the beam to pass through without unacceptable absorption
4. The particle must be of of sufficiently different refractive index to the liquid in which it resides
5. The particle must not cause significant diffraction to the beam
6. Only one particle may be in the beam at any one time [ie. no Optical Coincidence]
7. The particle must be face-on to the beam
8. If the particle is free to rotate (ie. No.7 does not apply), then a sufficient number of particles must give shadows for a statistical confidence that the average shadow size has been determined.

In addition there are certain electronic requirements which relate to what happens to the shadow being cast on the photodetector, such as:

9. The photodetector must be able to sense the shadow area
10. The photodetector must be able to sufficiently differentiate between one size of shadow and another [ie. Sensor Resolution]
11. The electronics must be able to accurately record each shadow before it disappears [ie. no Electronic Coincidence]
12. The electronics must be able to accurately record each shadow before the next one arrives
13. The electronic noise level must not mask the shadow detection signal.

Some of these features have specific names as shown in [square brackets] after the statement. There are also other physical aspects to be addressed; consider -

14. What happens if the size of the particle is close to, or larger than, the measuring cell?
15. What happens with air in the liquid?
16. What happens with water in oil?
17. What happens with a change in oil viscosity?
18. What happens with pressure pulses in on-line analysis?
19. What is the accuracy of the volume sensor?

The above indicates why the instrument is a scientific instrument and needs care in its use. The list does not mean it cannot work, on the contrary, it means that it is a precision instrument if designed correctly and used in the correct way.

Calibration is a primary task to be undertaken on all light obscuration counters. See page 289.

Light Obscuration — the instruments
We start with the instrument which is most often quoted with the standards relating to particulate in hydraulic fluids. HIAC (or HIgh ACcuracy Products) was founded in 1962 by Leon Carver in the United States; the firm pioneered the optical particle counting industry by developing products for aerospace hydraulics applications. ROYCO was another early leader, but primarily in the semiconductor clean room market. It was Pacific Scientific, however, who were to join the two together, purchasing HIAC in the early 1970's and ROYCO in 1980. Although both airborne and liquid borne counters are produced, only the latter are described here.

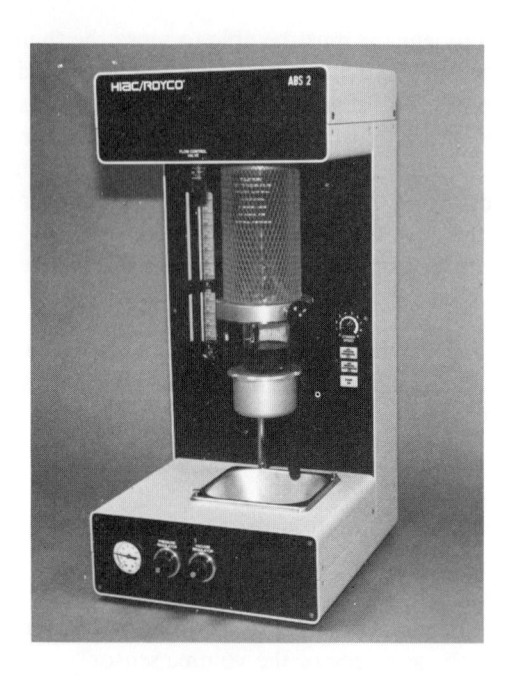

Fig. 8.1 The ABS HIAC/ROYCO particle counter

Originally, a basic means of transporting the sample fluid around the instrument was provided with a separately chosen sensor. Either on-line of off-line was possible provided the liquid was acceptable to the sensor as regards its viscosity and colour, and the particles were within the concentration limits of the electronics.

HIAC/ROYCO know about the limitations and hence have available a selection of sensors to cover the range of debris size and concentrations likely to be met — from 1 μm to 9 mm — and concentrations as high as 20,000 particles per mL. Uniform illumination of the sample cell yields resolution values better than 10%. The Automatic Bottle Sampler version is illustrated in Fig. 8.1.

Undoubtedly the great advantage of the HIAC/ROYCO is the favourable prestige and experience which has built up around it. It is a respected laboratory instrument which can be calibrated with either a distribution of particles (such as ACFTD — see index) or by using mono-sized spheres. The 'HIAC' is an accepted standard of measurement relating the measured shapes to Equivalent Spherical Diameters.

As with all optical devices, it does need care in the sense of ensuring that only appropriate fluids/particles are analysed. Where it is not known what level of particle concentration is present, it is essential to perform a preliminary membrane filter test to examine both the concentration and maximum likely size — and hence choose the sensor to fit. It may be necessary to dilute the sample with precleaned diluent in order to reduce the particle concentration or drop the viscosity of the sample, and hence provide the correct (calibrated) flow rate through the instrument. (An alternative method is to heat the liquid, but this may not be too consistent unless the test is undertaken very rapidly). Recently a sampler has been introduced that can handle viscous oils up to 2500cSt.

As an example of use, consider a typical bottle sample. Preliminary tests have shown that the fluid/particles are of the correct viscosity and numbers for one of the sensors. The sensor is fitted to the Automatic Bottle Sample (ABS) instrument, and the volume slider moved to an acceptable volume (eg. 15 mL). Appropriate sizes are chosen to cover the size range being measured. After thorough shaking of the bottle (by hand, preferably, but also possibly by insertion in an ultrasonic bath), it is placed on the stand, sealed and evacuated to remove air bubbles. The test cycle is then actioned with a certain pressure, and timed. (Provided the time is within the calibrated flow rate, the test proceeds, otherwise, a different pressure is applied to change the flow rate.) Three tests are conducted and the average taken. Although on the first occasion the whole test may take an hour of more to set up (with membrane filter tests and dilution checks), thereafter, and for future samples of the same type of fluid/particle combination, the tests should only take a few minutes — the faster the better in order to prevent particles settling in the bottle after the initial shaking. If the older type of magnetic stirrer is included, it may be necessary to remove this if wear debris is being analysed; the magnets tend to pull the ferromagnetic particles down, rather than distribute them in the fluid. The newer versions use an electromagnetic coil which can be switched off if not required.

Many laboratories and companies which provide wear checks, use an optical counter like the 'HIAC' as the ultimate test, after other tests, in order to give an accepted 'Standard' value for those who need to relate contamination levels. Others may use the older standard of microscope counting as the check.

Climet Instruments Company commenced the design and manufacture of liquid-borne and airborne particle counters in the late 1960's. High resolution counters have been a key feature of the company for longer than most.

Currently the Climet light obscuration sensors, like in the CI-1000, are Russell sensors. These are a considerable advance over earlier designs, because of the much improved uniform illumination and the fact that a slit of light is used rather than an overall collimated beam. This yields a better signal-to-noise ratio with a corresponding increase in accuracy when measuring small particles (down to 1 μm). Both ACFTD calibration and uniform latex spheres are possible.

Hydac has been a leader in the field of hydraulic measurements for some years. It operates a complete Oil Service with oil transfer equipment, laboratory analysis of samples including particle counting, sampling and on-site analysis kits. In 1991 it introduced the new RC 1000 contamination level portable instrument. This uses an optical gate of 1mm diameter which avoids the likelihood of blockage of the sensor measuring cell by larger particles. It is also said to be insensitive to vibration and run for 4000 hours without need of recalibration. The viscosity range is a stated impressive 15cSt to 1000cSt on-line. It can also work off-line with a pump attached to the system reservoir. Part of the reason for its being usable in an engineering environment is that the detected size range relates to 5 μm to 100 μm and the output is in standard values (ie. NAS 1638 or ISO 4406).

Malvern Instruments were one of the first companies to use Fraunhofer diffraction for measuring particle size distributions (in the late 1970's). More recently they have added light obscuration as a method for particle counting and sizing, with a white light single unit monitor for sizes 2 μm to 150 μm (ALPS 150H). Advanced signal processing allows liquids of widely differing viscosity (1cSt to 300cSt) to be sampled accurately without adjustment of the flow rate.

Particle Measuring Systems (PMS) was established in the United States in 1972 to make available state-of-the-art technology to particulate measuring problems. They have accumulated a wide range of experience in particle 'spectrometry' from small aerosol particles to much larger precipitations, covering fabrication of laboratory, industrial, aircraft and aerospace systems. Currently they state that they are the "only particle counter manufacturer which produces its own lasers and optical components to high standards, which allows for better quality assurance of all critical components".

The light obscuration sensor with a Liquid Batch Sampler (LBS 100), and two sensor designs based on light scatter (forward reflectance and right angle) enables PMS to cover the spectrum of sizing and sampling applications utilized by industries as diverse as semiconductor, pharmaceutical, aerospace, medical and hydraulic. The LBS 100, however, is the unit primarily used with hydraulic fluid and oil analysis. Using a sensing cell 500 μm × 1000 μm means that solid particle blockage of the sensor measuring cell is highly unlikely with the 1500 count/mL concentration. It will take sample bottles up to 1L in volume, and measure from 10 mL to 100 mL at a time with 90% accuracy (ie. less than 10% coincidence).

Partikel-Messtechnik (PMT) was founded in 1986. The company manufactures a considerable range of sensors and counter systems, and covers nearly the whole spectrum of particle characterisation with optical particle counting systems. This range includes the normal 1 μm minimum, but the maximum of 15 mm, with special sensors, is well above the usual upper size limit. They are also able to supply sensors for all three types of detection position, ie. off-line, on-line and in-line.

Rion, the Tokyo company, market air and liquid-borne particle counters covering a wide size range. The KL-20 and KL-22 counters use light scatter to cover sizes from 0·2 μm to 2 μm. However, the light obscuration technique KL-01 counter is designed for 5 μm to 150 μm counting in oils.

UCC International Limited has built upon its proven condition monitoring product range by introducing the CM20 Automatic Particle Counter. This easy to operate, tamper-proof instrument is designed ergonomically with robustness being a pre-requisite.

The CM20 principle uses pressure differential between two points on a system, which will cause flow. The idea of the device is shown in Fig. 8.2. It uses the light obscuration technique on-line which is a challenging concept with real fluid systems. By using a number of ingenious ideas incorporated in the design, a sample of 10 mL is taken, via a

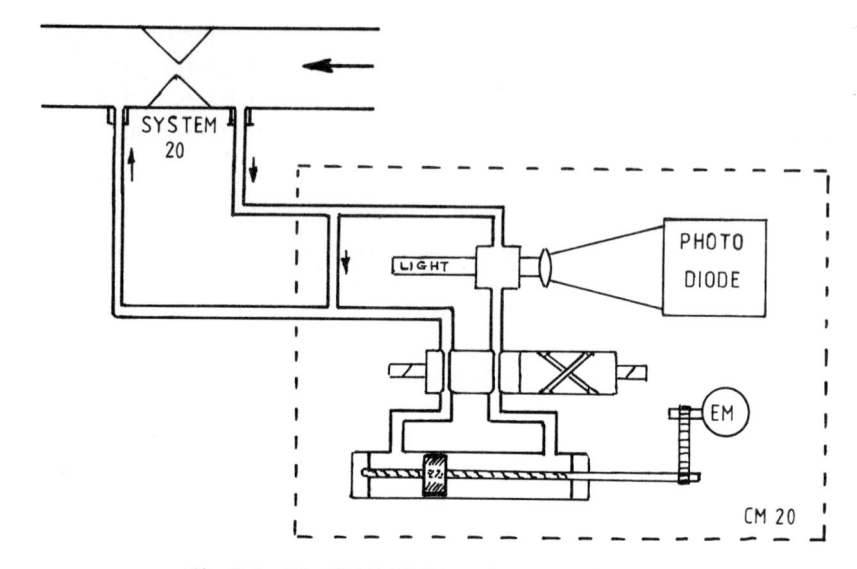

Fig. 8.2 The UCC CM20 on-line particle counter

tachometer controlled electric motor driving an axially moved piston (like a syringe pump) screwed backwards and forwards. A precise volume is sampled and kept at a constant flow rate no matter the viscosity of the fluid.

The CM20 can be attached to any number of monitor points, one at a time, recording up to 300 test results with its built-in microprocessor. 5 μm, 15 μm, 25 μm and 50 μm sizes are counted using the microprocessor controlled optical scanner. A 5 μm light slit within a much larger cell area is designed to help prevent saturation even at the highest contamination level (ISO 24). A maximum of four minutes is needed for each complete test where a hard copy of the results can be printed off with the CM20's built-in print facility.

Light Scatter — General comments

Light scatter instruments are able to count much smaller particles than the devices using the light obscuration technique. Although all use the reflection of a light or laser beam, the angle at which the reflection is detected varies considerably. This section deals with the scatter at or near right angles to the laser beam, or perhaps a little forward. The almost straight ahead (near forward reflection) is covered in the next section.

As with the light obscuration technique, there are many optical characteristics which require careful understanding if errors are to be avoided. For instance, bubbles are counted as well as particles and hence they need to be removed before counting commences.

Light Scatter — the instruments

Horiba Limited has been based in Japan for many years but its international expansion began with the 1970 establishment of Horiba Instruments Incorporated in California before rapidly moving into Europe and other parts of the world. It is involved with a very large range of scientific instruments for laboratory and industrial use. As regards debris and particle analysers, it has a turbidity monitor, X-ray Fluorescent analysers, particle size distribution analysers and liquid particle counters.

The light scatter (PLCA-520) particle counter uses a TEMoo He-Ne laser at 10 mw and 90° scatter. Particles in the range 0·2 μm to 10 μm can be counted in low viscosity clear fluids (not higher than 30cP) with counts up to 2000 particles per mL with a coincidence of less than 5%. Six simultaneous size readings may be recorded.

Two samplers are available. One is a typical bottle sampling unit with a required sample of between 200 mL and 1L, the other is an on-line suction sampler which can take samples of 50 mL by syringe pump. An automatic cleaning cycle is incorporated. The primary use of these two Horiba instruments is ultra pure water, organic solvents, strong acid/alkaline solutions and other chemicals.

The PMS partly-forward light scattering particle counters come in two distinct types. One looks at all the liquid flowing in a capillary (the 'volumetric' sensor with a diameter either 500 μm or 800 μm), and the other detects a small proportion of the pipe flow area (the 'in-situ' sensor).

The PMS volumetric sensor detects scatter which is forward of the laser beam. All the scattered light (around the light trap mask which absorbs the direct laser light) is parabolically reflected on to a single PHA (Pulse Height Analyser) which sizes according to the intensity. There is a wide range of size modules covering from 0·2 μm to 300 μm.

The in-situ sensor uses 90° scatter either side of the laser beam. Provided the particles are well mixed in the liquid under test, the results can be directly related to the flow knowing the proportion of the sample volume to the cross-sectional area of the pipe. Such testing has the

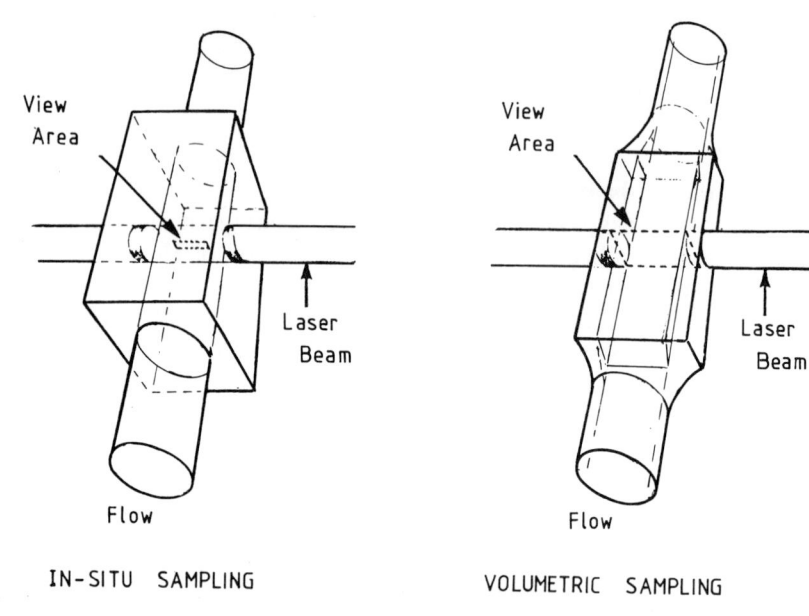

Fig. 8.3 The two types of PMS detector

added advantage of the less likelihood of the formation of bubbles because of the minimal pressure drop. (Fig. 8.30.)

Polytec GmbH is a high technology company, founded in 1967 and now based in southern Germany. It is engaged in the development and manufacture of electro-optical instrumentation.

The Polytec Particle Sizer, originally called the 'Karlsruhe' (the first development occurred in the Institute of Chemical Engineering at the University of Karlsruhe), again uses light scatter with 90° detection angle, as shown in Fig. 8.4.

The optical head of the instrument detects individual particles in a rectangular measuring volume through which the fluid is flowing. The light beam is scattered by the particles, the intensity of scatter at right angles being related to the particle diameter. In this instrument a white light source is used, and a relatively large aperture, to give a very smooth curve of proportionality. The near linear relationship then provides sufficient size resolution to allow a 128 channel analysis of the particle diameters over a dynamic range of 1:30.

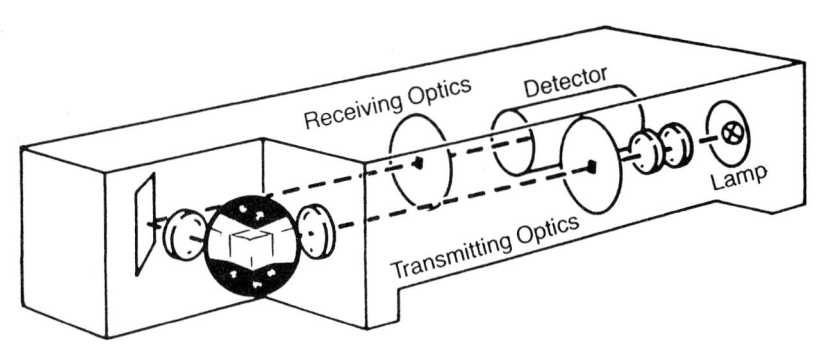

Fig. 8.4 The Polytec Particle Sizer Series HC

Maximum particle concentration over the ranges varies from 1000 to 100,000 per mL covering sizes from 1 μm to 125 μm. However, because the volume being measured (less than 500 μm diameter) is much smaller than the tube size, there is no problem with blockage from a larger sized particle; in addition there is no distortion of the particle stream as the technique is non-intrusive — on-line.

The Rion Company KL-20 and KL-22 particle counters both use laser scatter detected either at 70° or 90°. The laser beam, transformed into the shape of a rolled sheet passes through one side of the quartz glass cell; light then scatters sideways as the beam hits the particles and is collected on the other side of the cell at the angle to the inlet beam. The sizes covered are in the range 0·2 μm to 2 μm with a maximum concentration of 1200 particles/mL with 95% confidence (ie. maximum of 5% coincidence).

Light Forward Reflectance — General comments
This is a special type of reflectance (see page 123). The size of the particle at these angles of reflection is directly determined from the intensity of the reflected beam. In order to be able to count the particles, the level of concentration (the number of particles per volume) is limited.

Light Forward Reflectance — the instruments
The HIAC/ROYCO Submicron Liquid Sensor detects all particulate as small as 0·2 μm by using a He-Ne laser in a certain form. The laser is focussed with a cylindrical lens system to evenly illuminate the entire measuring volume of the sample cell. In this way all the fluid passes

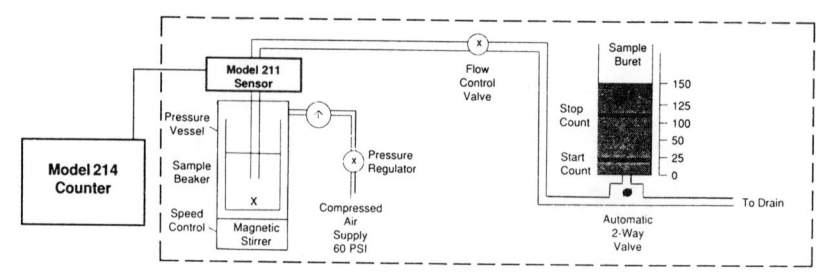

Model 250 Batch Sampler (Internal View)

Fig. 8.5   The Met One Liquid Batch Sampling System

through this volume. No particle can escape passage through the incident beam and all particles of the same diameter scatter light of the same intensity. Different sensors are available for different size ranges, the smallest currently is 0·2 µm to 2 µm.

Met One Inc was established in Oregon in the 1970's with the purpose of measurement technology advancement. They produce both airborne and liquid borne particle counters. The liquid borne analysers are based on the light forward reflectance. A laser diode source ensures a long life.

On-line and off-line (bottle) sampling is possible, with up to six different sizes monitored simultaneously. There are several sensor ranges from 0.5 µm–25 µm up to 5 µm–600 µm. Maximum counts vary from 3500 per mL to 10,500 per mL with not more than 10% coincidence.

The Liquid Batch Sampling System (the bottle sampler) is able to test between 10 mL and 120 mL in containers up to 500 mL. See Fig. 8.5.

A multi-channel, multi-function instrument is available to control an entire liquid monitoring process, as a centralised system. It is able to control multiple Met One liquid sensors during both on-line and off-line sampling with advanced data collection and manipulation features. An Intel 80386SX 16 MHz computer is used with a full suite of comparative programs and storage, enabling the user to quickly obtain standards and trends.

The Spectrex Corporation was founded in 1966 under the presidency of Mr. John M. Hoyte in California. From the start it was associated with particle counting in liquids, initially calling its particle counter the "Prototron". The Corporation has had many successful applications in oil

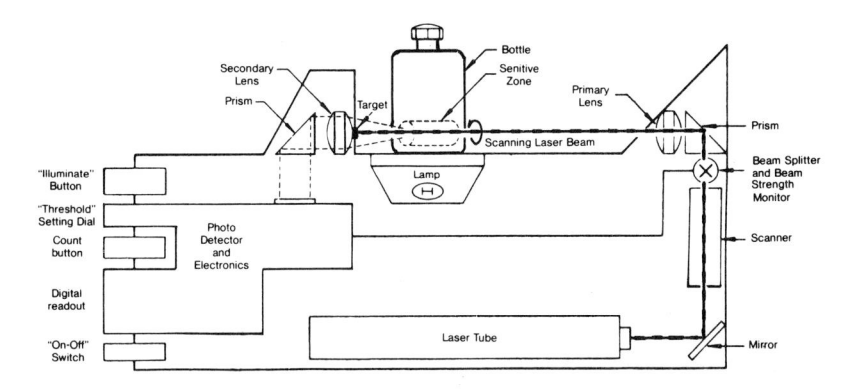

Fig. 8.6 The Spectrex Laser Particle Counter

laboratories, military bases and universities, as well as industrial applications in the aerospace, pharmaceutical, biomedical, semiconductor fields, to name but a few.

The key feature of the Spectrex Laser Particle Counter is the ability to analyse the liquid sample whilst it is still in its bottle! This avoids one of the hazardous operations associated with bottle sampling, ie. the removal of the bottle cap and insertion of a probe. Fig. 8.6 shows the schematic of the method.

A revolving laser beam passes through the transparent walls of the bottle, and, using the principle of near-angle light scattering, the unit counts and sizes the particles in the suspension in a 20 mm long central cylindrical zone (the "sensitive zone" of volume 1 mL). Only the light pulses generated from that zone alone are counted by the photo detectors.

The whole process is very simple provided a few precautions are taken. The bottle chosen (50 mm to 200 mm inside diameter) may be of normal quality as long as no severe optical discontinuities are present (or if they are, they are not in the line of the laser). Care should be taken to ensure the dust is removed from the outer surface of the bottle in case it should deflect the laser. The bottle needs to be agitated to thoroughly distribute the particles, but not too much as to cause aeration as the laser may count the air bubbles as particles. One method to achieve this is to agitate by hand, followed by agitation in an ultrasonic bath — the ultrasonic action is an efficient means of removing the bubbles and at the same time keep the particles in suspension.

A unique feature of the unit is that the calibration standard, ie. a bottle with a certain concentration of particles in it, can be used time and time again, unlike the normal flow path optical counters. (Note that in most cases the standard will have a certain shelf life which must not be exceeded.)

An opacity meter is provided as an accessory to make allowances for opaque liquids, and still give the correct particle count. However, as with other optical devices, very dark fluids should be avoided.

In addition to the off-line bottle sampling analysis, on-line sampling across a transparent pipe or tube is also possible. Again the presence of air bubbles must be avoided.

The basic particle counter is connected to any MSDOS compatible computer. Spectrex provides friendly software to give rapid (30 second) sizing and counting of in-situ samples, settling scans and automode scans to study agglomeration, dissolution and particle count changes with time. Fig. 8.7 shows a typical print-out.

MAGNITUDE EXAMINATION TO DETERMINE PARTICLE COUNT

The idea of magnitude examination is where, instead of looking directly at the particle, a secondary feature is monitored which is changed due to the presence of particles.

Two ideas of magnitude effects are discussed here. One is the effect that the particle magnitude has on the impedance between two terminals, when the particle gets in the electrical path. This is called the Electrical Sensing Zone method (ESZ). The second is the effect the particle has on the flow of the liquid through a filter when the particle partly blocks the filter. This is the Filter Blockage technique.

Electrical Sensing Zone — General comments
The idea of the electrical sensing zone came into being in the late 1940's, early 1950's. It is frequently termed the 'Coulter Principle' or the 'Electrozone', however, as more than one company use the method, the preferred term is as chosen in this book — the Electrical Sensing Zone.

The idea has been introduced in Chapter 6, see page 116, but there are features which need closer attention because of the totally different technique from optical methods. Initially the particles and liquid must have the following relationship:

SAMPLE FROM VAT 69

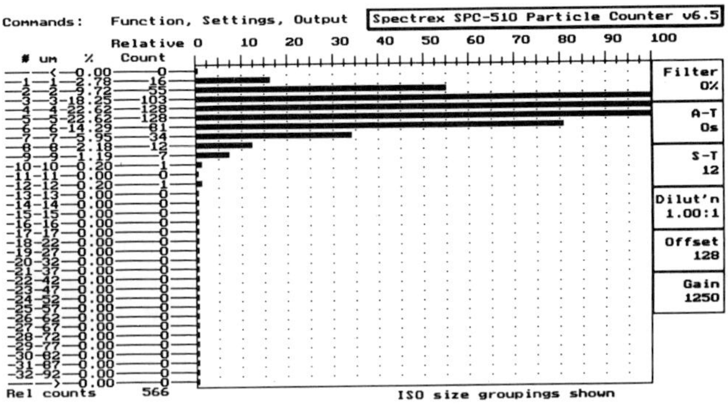

```
Commands:  Function, Settings, Output  [Spectrex SPC-510 Particle Counter v6.5]
           Relative 0   10   20   30   40   50   60   70   80   90  100
  # um   %  Count                                                      Filter
                                                                       0%
                                                                       A-T
                                                                       0s
                                                                       S-T
                                                                       12
                                                                       Dilut'n
                                                                       1.00:1
                                                                       Offset
                                                                       128
                                                                       Gain
                                                                       1250
 Rel counts  566              ISO size groupings shown
```

		Total counts	Counts	Surface area	Volume	Mass/bin
Bin	Size	/cc	percent	percent	percent	ppm
---	-----	------	------	------	------	--------
	<	0	0.00%	0.00%	0.00%	0.00
1	1um	16	2.78%	0.12%	0.02%	0.00
2	2um	55	9.72%	1.75%	0.62%	0.00
3	3um	103	18.25%	7.39%	3.94%	0.00
4	4um	128	22.62%	16.27%	11.58%	0.01
5	5um	128	22.62%	25.43%	22.61%	0.02
6	6um	81	14.29%	23.13%	24.68%	0.02
7	7um	34	5.95%	13.12%	16.33%	0.01
8	8um	12	2.18%	6.28%	8.94%	0.01
9	9um	7	1.19%	4.34%	6.94%	0.00
10	10um	1	0.20%	0.89%	1.59%	0.00
11	11um	0	0.00%	0.00%	0.00%	0.00
12	12um	1	0.20%	1.28%	2.74%	0.00
13	13um	0	0.00%	0.00%	0.00%	0.00
14	14um	0	0.00%	0.00%	0.00%	0.00
15	15um	0	0.00%	0.00%	0.00%	0.00
16	16um	0	0.00%	0.00%	0.00%	0.00
17	17um	0	0.00%	0.00%	0.00%	0.00
18	22um	0	0.00%	0.00%	0.00%	0.00
19	27um	0	0.00%	0.00%	0.00%	0.00
20	32um	0	0.00%	0.00%	0.00%	0.00
21	37um	0	0.00%	0.00%	0.00%	0.00
22	42um	0	0.00%	0.00%	0.00%	0.00
23	47um	0	0.00%	0.00%	0.00%	0.00
24	52um	0	0.00%	0.00%	0.00%	0.00
25	57um	0	0.00%	0.00%	0.00%	0.00
26	62um	0	0.00%	0.00%	0.00%	0.00
27	67um	0	0.00%	0.00%	0.00%	0.00
28	72um	0	0.00%	0.00%	0.00%	0.00
29	77um	0	0.00%	0.00%	0.00%	0.00
30	82um	0	0.00%	0.00%	0.00%	0.00
31	87um	0	0.00%	0.00%	0.00%	0.00
32	92um	0	0.00%	0.00%	0.00%	0.00
	>	0	0.00%	0.00%	0.00%	0.00

```
Total counts:         566/cc
Total suspended
       solids:        0.07ppm (mg/liter)
Spec. gravity:        1.80
Mean size:            4.40um
Standard dev:         1.68um
```

Sample taken on Fri Feb 21 1992 at 13:04:30

Fig. 8.7 Sample print-out (Spectrex SPC-510)

The liquid must be more (or less) conductive than the particles.

In essence, this usually means that the liquid which is chosen is considered a conductor, and the particle is considered a non-conductor. However, in practice, the two may not need to be as extreme as this comment may suggest, because of compensating effects.

Consider the operation of the process (Fig. 8.8):

As with all liquid monitors, the liquid (called an electrolyte solution, because of its conducting properties) needs the particles fully dispersed within it. The exact position of the electrodes (the two electrical poles between which the impedance is to be measured) is not critical. The aperture (the measuring cell) must be of certain shape and size to permit the free flow of the fluid and particles but, at the same time, not be too large to significantly reduce the impedance change which each particle will cause.

Usually the particles are much smaller than the aperture size, and it is found that when they are between about 1·5% to 80% of the aperture diameter, the optimum results are achieved, ie. the full impedance volume is detected.

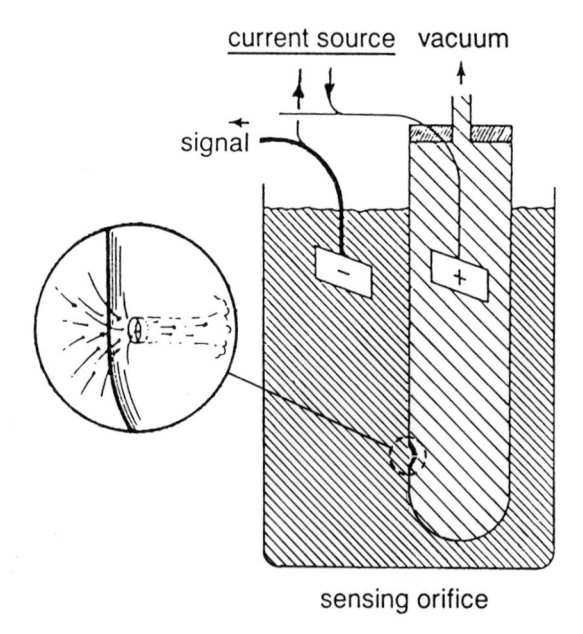

Fig. 8.8 *The Electrical Sensing Zone operation (ELZONE®)*

If the conductivity of the liquid is likely to change, ie. due to temperature instability, then results could be erratic. However, the design of the instrument should not allow this, eg. by built-in tracking.

Fluids which are considered non-conducting can be partially changed by mixing or dilution with an electrolyte, but this can be prone to error unless the diluent is fully miscible and thoroughly precleaned. Another method would be to filter out the debris and then resuspend it in an electrolyte solution. These techniques are complicated for regular use, but can be valuable for the occasional assessment.

It is usually considered that, unless the debris has a coefficient of conductance extremely close to that of the liquid, all particles will be correctly counted and sized by the Electrical Sensing Zone technique.

Electrical Sensing Zone — the instruments

The Coulter Corporation commenced in the late 1950's with the introduction of the first COULTER COUNTER® based on the ESZ technique. Since then not only has the company become worldwide, it has also brought onto the market other types of particle analyser covering a wide range of size and use. The COULTER COUNTER® models alone have been publicised in well over 6500 learned papers in medical, pharmaceutical, laboratory, university and industrial applications.

As regards the Coulter ESZ units, the models include advanced techniques such as the patented 'constant current' feature ensuring maximum accuracy by eliminating calibration drift due to changes in electrolyte conductivity and temperature. Another feature is the use of a mercury siphon to provide a reproducible and accurate method of sampling from a suspension with an accuracy better than ± 0.5% on the 0.5 mL volume position. An integral aperture image monitor is available which allows the operator to have a visual impression of the measuring zone to give confidence that the aperture is clean; remote focussing is included.

For the majority of applications a single aperture is probably all that is required. However, if there is a need to extend the range on the Coulter instruments, then a different cell can be fitted within a few seconds. Calibration is simple and needs only to be undertaken infrequently. A variety of analyses can be performed on the basic data, and displayed according to requirements.

Particle Data started in the early 1960's with the marketing of the EL-ZONE® particle counter based on the ESZ principle. Special features on the Elzone are such as the patented standard 'reference' pulse to ensure exact operational conditions and calibration validation at all times. Another is the patented variable logarithmic converter which gives six size ranges from 2·5:1 to 25:1 by particle diameter to give maximum accuracy and resolution over narrow or wide range distributions.

The electronic orifice monitoring and push button plunger on the Elzone enable rapid clearance, should a particle block the measuring cell. External interference is reduced by the advanced grounding network and sample area shielding. Latest models include high degrees of automation and 'Distributional Coincidence Correction' to alleviate dilution/concentration problems.

Filter blockage — General comments
For a long time the filter blockage idea was restricted to very rough estimates of debris in a liquid. Using an ordinary oil filter, with a pressure drop sensor, the operator of a system became aware of exceptional debris generation when the the sensor went into the red more rapidly than usual. Attempts were made to make the method more scientific with precision equipment (eg. see Raw & Hunt 1987 who refer to Dwyer & Connelly 1966, Swank 1966, Kasten 1967, Childs & Krulish 1969). However, all the early devices failed for one reason or another, mainly because of an inability to be able to function in a continuous mode. Today there are many devices which use the basic technique, and some, like the two below, which can give a remarkably accurate count of particles at specific sizes with a truly continuous operation. It is considered that provided the mechanics and pressure sensor are accurate enough, individual particle counting can actually be achieved in the same way as with light obscuration.

The size to be counted can only be that associated with the pore size of the filter. The greater the number of filters of different size which are fitted, the greater the number of size counts which can be determined. The current art uses woven mesh as the filter media because it is accurate in size with a high degree of consistency. Mesh is also suitable as a back flushable media, probably due to its depth and smooth sides which allow easy entry and exit of particles. (See the comments on sieves on page 174 for examples of the different types of mesh.)

The restrictions on the accuracy and range of the filter blockage technique are,

1. the physical limitations of woven mesh,
2. the maintenance of the pore size and number of pores,
3. the accuracy of the pressure sensor for the range required and
4. the accuracy of the measurements of the flow or volume passed.

In other words the mechanical components (pump, mesh, etc.) and the sensors (pressure, temperature, etc.) must be to the standard required for the particular levels of debris being monitored. Normally this is easily achieved with the basic design. A rare case where extra software may have to be included is where during the time of the test (typically 30 seconds) the temperature is changing significantly, causing variations to occur with volume and pressure.

The advantages of being a magnitude effect type of counter mean that the filter blockage technique is independent of colour and refractive index differences of particle or fluid. It is also able to cope with very high levels of debris or low levels. Water mixed in with oil makes no difference, and, unless the levels are considerable (like 50%) the presence of air in the oil makes little difference.

Fig. 8.9 shows the basic circuit diagram to cope with a single mesh. (Additional valves or pressure sensors are required for the real models where more than one mesh is incorporated.)

The position of the directional control valve would be central at switch-on, in order to determine the pressure drop through the circuit (P_0) and to ensure that the viscosity of the fluid is within the working range of the instrument. It would then move one way for the first flow direction before moving right across for the second one after 30 seconds, say, or at a certain pressure drop increase. This reversal of flow immediately acts as a back-flushing process on the debris which has been trapped on the opposite side.

The actual method of determining the particle count from the filter blockage technique can be seen from the following example using simple figures. (It should be noted that the number of pores in real instruments will be in the order of 10's or 100's of thousands per mesh sensor.)

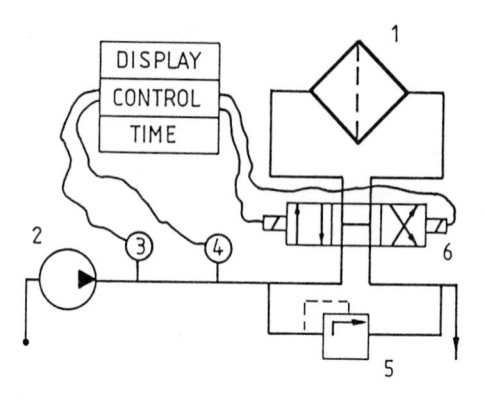

1 - FILTER
2 - FLOW CONTROLLER
3 - TEMPERATURE DETECTOR
4 - PRESSURE DETECTOR
5 - RELIEF VALVE
6 - DIRECTIONAL CONTROL VALVE

Fig. 8.9 A Filter Blockage single mesh circuit

Let the mesh area have 480 pores in it of size 15 μm
Let the flow through the mesh be 50 mL/min
Let the instrument have a pressure drop, P_0, of ... ... 0.5bar.
Let the initial pressure drop through the mesh, P_1, be ... 1·2bar.
Let the pressure drop after 12s of the liquid passing
through the mesh, P_2, be... ... ... 1·9bar.

Then the amount the filter is blocked is

$$(P_1-P_0)/(P_2-P_0) = 0\cdot 5$$

i.e. 0·5 × 480 = 240 pores have become blocked with 240 particles greater than 15 μm in size in the flow of 50 × 12/60 = 10 mL.

Therefore the count of debris is 240 particles > 15 μm in 10 mL.

It will probably come to mind that the particle is never the actual shape of the pore in the mesh. It can, therefore, never completely block a specific hole. In addition, it is possible for two small particles to get together and fit into a larger hole; conversely, a larger particle could lie across the mesh partially blocking several holes. These are the very arguments which can be directed at light obscuration counters! However, as with the older optical technique, statistics come to the rescue, and the final result is highly acceptable for a real system with real particles.

Filter blockage — the instruments

The original development for the two instruments described was undertaken at the University of Bath Fluid Power Centre in the early 1980's (see original patent Hunt & Bowns 1983/1987).

Lindley Flowtech Limited (formerly Lindley Flow Technology Limited) are part of the Lindley Group which has been involved with fluid systems from its commencement in the late 1940's. It was the first licencee for the filter blockage monitor, and brought the Fluid Condition Monitor (FCM) onto the market in March 1986.

The two key sizes of 5 μm and 15 μm (as associated with hydraulic standards) are the normal sizes for which counts are displayed. Two meshes of appropriate size are used, incorporated as standard. However, the very simple screw-out/screw-in change over to other sizes means that a wide range is available covering up to over 100 μm, the software being built-in for the different meshes.

Many different industrial applications have enabled Lindley Flowtech to sense the requirements of industry, and although the particle counting still applies, most situations initially have only required the indication of a 'go/no go' level or standard class.

Coulter Electronics Limited took on the monitor as the Liquid Contamination Monitor (LCM II) in 1987 as a totally new type of concept for their scientific range of particle analysis equipment. From the start the LCM II was designed to undertake three mesh size evaluations, 5 μm, 15 μm and 25 μm. In addition some indication of the viscosity of the fluid was presented, obtained from the pressure drop of the fluid though the 25 μm mesh.

Whilst the University of Bath had shown very good agreement between the filter blockage monitor and the HIAC, the Coulter evaluation of the technique (which they call 'mesh obscuration') revealed very close comparisons between three major types of particle counter when assessing ACFTD. A replot of their results is shown in Fig. 8.10.

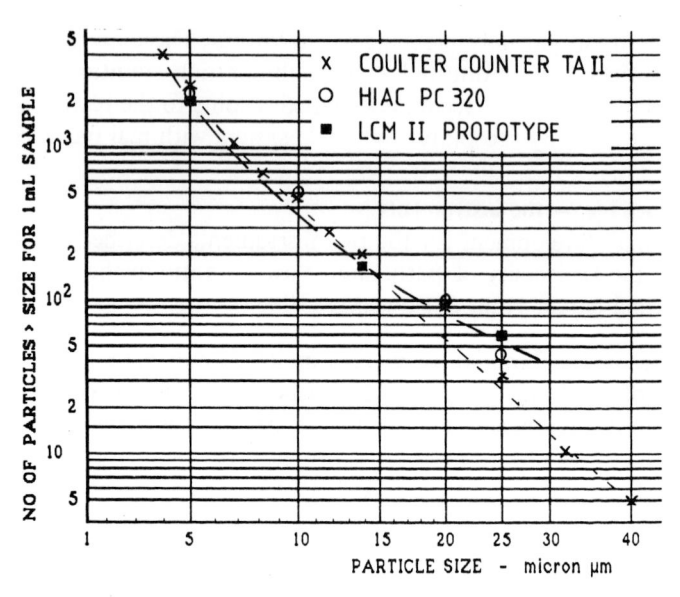

Fig. 8.10 *Comparison of 3 counters with 5mg/L of ACFTD (Miller 1988)*

(Another filter blockage instrument on the market is the Diagnetics digital Contam-Alert®. This uses a single filter element and is single action; it is described on page 161.)

EFFECT EXAMINATION TO DETERMINE PARTICLE COUNT

Although a number of effects are measured from particles, such as magnetic forces, electrostatic forces, wearing action, gamma rays, etc. these are not usually incorporated into particle counters. The bombardment by molecules, however, is used to measure sub micron (and up to 5 μm) particles. This technique is called Photon Correlation Spectroscopy and is normally associated with the reverse magnitude effect of the presence of molecules being shown up because solid particles can be seen moving under the impact from the molecules. Several instruments use this method.

Photon Correlation Spectroscopy — General comments
Photon Correlation Spectroscopy is sometimes called "Dynamic Light Scattering due to Brownian Motion". It is only suitable for the fine

particles, but in this analysis it is extremely effective. The behaviour of the particles is random but the movement (or rate of diffusion) is inversely proportional to the particle size. The technique of the analyser is that of measuring these random motions.

All the monitors use a laser light beam and measure scattered light in one way or another; probably 90° scatter is the most common. However, the scattered light is fluctuating in intensity, and in some way these intensity fluctuations need to be measured on a time scale and converted into size or molecular weight data.

Although 90° scatter alone can provide acceptable results in many applications, there may be the occasional combination of particle/medium which requires other angles to give the optimum determination of particle features as described further down in this section.

The actual analysis of the 90° scattered light is done by determining an intensity autocorrelation function $C\tau$, formed by averaging the product of the pulses in small time intervals of the time between intervals. For a monodisperse suspension of rigid, globular particles $C\tau$ decays exponentially, as

$$C\tau = A. \exp\{-2D_t . K^2 . \tau\} + B$$

where A and B are constants, K is a special constant calculated from the laser wavelength, the refractive index of the scattering liquid and the scattering angle, and D_t is the translational diffusion coefficient, ie. (from the Stokes-Einstein equation):

$$D_t = k.T/(3 \mu \pi d)$$

where k is the Boltzmann's constant, T is the temperature (K) and μ is the viscosity of the suspending liquid. Thus firstly the $C\tau$ is measured and D_t determined from the first equation. The diameter of the particle d is then obtained from the second equation.

Photon Correlation Spectroscopy — the instruments
The Brookhaven Instruments Corporation state that they have been manufacturing and installing laser light scattering instruments in the USA longer than anyone else, in the case of photon correlation spectroscopy. It is not surprising that their BI-90 instruments, based on this technique, include a wide range of features.

The COULTER® N4 series uses either 90° scatter or, in the N4MD, automated multiple angle measurements. This latter arrangement, which is also used in the Malvern Instruments System 4700c, enables a much wider range of sample sizes and concentrations to be evaluated because

> the low angles detect larger particle sub-populations which are masked by smaller particles at higher angles,
>
> the low angle detection also picks up extra signals from small particles or weakly scattering particles,
>
> the scatter angle can be changed to improve the signal-to-noise ratios by apparently slowing or speeding-up the diffusion process,
>
> the low angle measurements can minimize the effects of rotational diffusion and 'flexing' time constants when measuring non-spherical or elongated particles.

Besides the System 4700c mentioned above, Malvern Instruments also supply a fixed angle device, the AutoSizer IIc. This is more compact and is used for routine particle size (0·003 μm to 3 μm) and diffusion coefficient measurements; because it is fast (as short an analysis time as 10 seconds) and easy to use it relates with quality control and plant monitoring applications.

Leeds & Northrup have been involved in fine particle analysis for many years. Their laboratory facilities, using their own equipment, have provided analyses to outside customers since 1984. Their new Microtrac® Series 9200 analysers have continued the Fraunhofer analysers that they previously manufactured, but now include the UPA model which is based on Brownian Motion. (UPA stands for Ultrafine Particle Analyzer.)

The technique used in the UPA does not depend on measuring Doppler shift alone, which can be prone to poor signal levels. Instead a reference signal enables a combination of reflected and back-scattered light to create a high-level signal strong enough to be fed directly to a reliable solid-state photodetector without need of amplification (see Fig. 8.11). (Reflection is actually achieved from the interface between the Y-shaped waveguide used and the solvent in the cell.) This allows a very wide range of sample concentrations with minimum dilution.

The Otsuka Electronics Company has its headquarters in Osaka in Japan. It was established in May 1970 as a company with a high profile in optical technology, and today it has a strong research orientated emphasis with over half its employees engaged in research and development.

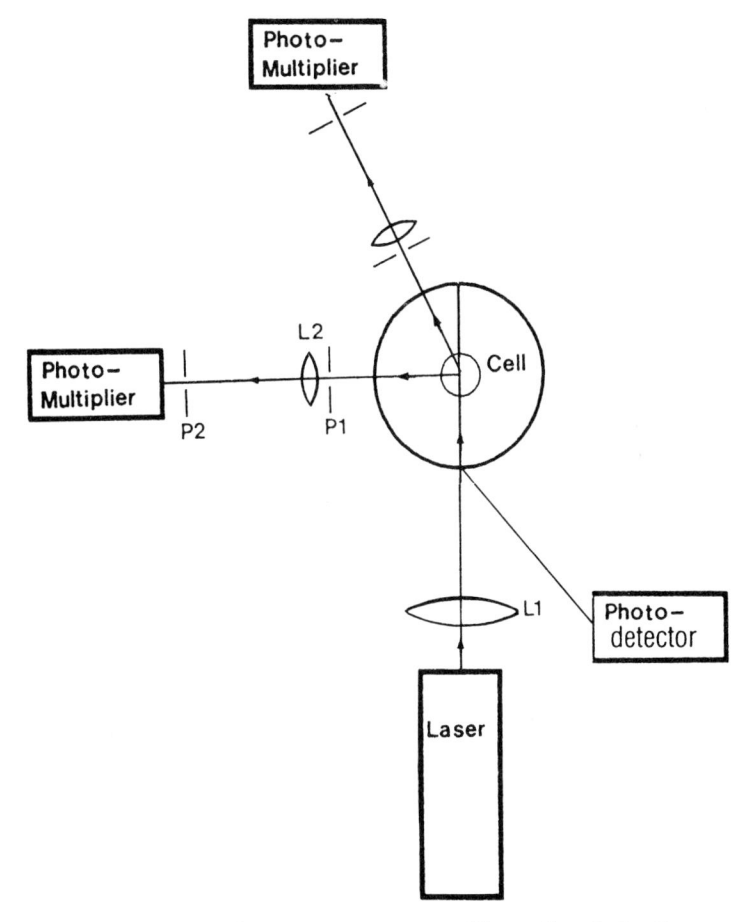

Fig. 8.11 Three types of scatter measurement in Photon Correlation Spectroscopy

In the LPA-3000/3100 particle sizer a total range of 0·003 μm to 100 μm is covered in one instrument — the range of ·003 μm to 5 μm by Photon Correlation Spectroscopy, and 3 μm to 100 μm by natural sedimentation. Separate dynamic and static light scattering instruments are also available for measurements of the mainly sub micron range (0·003 μm to 5 μm), molecular weight and zeta potential.

CHAPTER 9

Chemical Identification of Debris

Origins are more than fascinating. They convey the essential meaning as to why something exists, and, hence, what procedure or action should now be taken. Debris may be detected, but if any sensible action is to be taken, the source of that debris must be known.

There are different ways of determining the origin of particles. Some just give a hint of a possibility, or a probability if there is some past experience. Others give a highly significant indication. And still others, although probably the most expensive, can identify the source with no doubt at all.

Spectrometric analysis of debris has many advantages in the determination of origins. A spectrometer determines what chemical elements are present in the debris; it may also determine their relative proportions and possibly the molecular content. The elemental content is useful but it is not everything; for instance, the presence of the carbon element does not necessarily imply that the particle includes steel; the debris may include, for instance, carbon from combustion products. Molecular content combined with the ratio of elements is likely to give much more of a lead into the origin, as we shall see. We shall study the chemical/elemental analysis first, with an in-depth discussion on those techniques which are more related to the wear debris analysis; (a brief comment will also be given on many of the other methods). This is followed, in the next chapter, by a look at the more physical appearance methods of examination, including finally, a review of several other quite different ways of defining debris, each providing distinctive clues for determining the source.

DETERMINATION OF ELEMENTAL CONTENT

Notation
Elemental analysis has a number of similar names, eg.

> Spectrometric Analysis
> Spectrometry

Spectrophotometry
Spectroscopy

which, basically, all mean the same, except that when 'photo' is included within the name the spectrometry must include some type of photometer. A photometer is an instrument for determining the luminous intensity of sources of light. In elemental spectroscopy, the photometer determines the intensity of different light frequencies, or spectra, each one associated with a different element when burning in a flame or furnace or 'torch' as it is sometimes called.

Elemental analysis, in describing the individual processes, has an even greater number of names which for reference purposes are usually referred to by initials; the reason being obvious when it is realised how complex the actual process names are!

AA	Atomic Absorption
AAS	Atomic Absorption Spectroscopy
ADM	Acid Dilution Method
AE	Atomic Emission
AES	i. Atomic Emission Spectroscopy (as in this book)
	ii. Arc Emission Spectrometry
	iii. Auger Electron Spectroscopy
AFM	Atomic Force Microscope
AFS	Atomic Fluorescence Spectroscopy
ARDE	Ashing Rotating Disc Electrode
ATR	Attenuated Total Reflection
CE	Capillary Electrophoresis
CI	Chemical Ionization
DAM	Dry Ash Method
DCP	Direct Current Plasma
DDM	Direct Dilution Method
DSC	Differential Scanning Calorimetry
EDS	Energy Dispersive Spectroscopy
EDX	Energy Dispersive X-ray Spectrometer
EDXRF	Energy Dispersive X-ray Fluorescence
EELS	Electron Energy Loss Spectrometry
EI	Electron Ionization
EM	Electron Microscope
EPMA	Electron Probe Microanalysis
ES	Emission Spectroscopy

ESR	Electron Spin Resonance
FAS	Fluid Analysis Spectrometry
FE	Field Emission
FT-IR	Fourier Transform Infra-Red
GC-MS	Gas Chromatography Mass Spectrometer
GFAAS	Graphite Furnace AAS
HPLC	High Performance Liquid Chromatography
IC	Ion Chromatography
ICP	Inductively Coupled Plasma
ICP-MS	ICP Mass Spectrometer
ICP-OES	ICP-Optical Emission Spectrometer
IR	Infra-Red
LC-MS	Liquid Chromatography Mass Spectrometer
LPDS	Large Particle Detection System
MFM	Magnetic Force Microscope
MIP-MS	Microwave Induced Plasma Mass Spectrometer
MS	Mass Spectrometer
MS-MS	Double Mass Spectrometry
NDIR	Non-Dispersive Infra-Red
NIR	Near Infra-Red
NMR	Nuclear Magnetic Resonance
OES	Optical Emission Spectrometer
PAS	Photo Acoustic Spectrometer
PCS	Photon Correlation Spectroscopy
PSTM	Photon Scanning Tunnelling Microscope
PY-MS	Pyrolysis Mass Spectrometry
RDE	Rotating Disc Electrode [Rotrode]
RS	Raman Spectroscopy
SAM	Scanning Acoustic Microscope
SEAM	Scanning Electron Acoustic Microscopy
SEM	Scanning Electron Microscope
SEM-IPS	Scanning Electron Microscope Image Processing System
Seq.	Sequential
SFC	Supercritical Fluid Chromatograph
Sim.	Simultaneous
SIMS	Secondary Ion Mass Spectrometry
SNOM	Scanning Near-field Optical Microscope
SOA	Spectrometric Oil Analysis
SOAP	Spectrometric Oil Analysis Program(me)
SPM	Scanning Probe Microscopy

STEM	Scanning Transmission Electron Microscope
STM	Scanning Tunnelling Microscope
TA	Thermal Analysis
TEM	Transmission Electron Microscope
TG	Thermogravimetry
TLC	Thin Layer Chromatograph
UV-Vis	Ultra-Violet and Visible range
WDS	Wavelength Dispersive Spectroscopy
WDX	Wavelength Dispersive X-ray Spectrometer
XPS	X-Ray Photoelectron Spectroscopy
XRD	X-Ray Diffraction
XRF	X-Ray Fluorescence

Fortunately not all these processes and methods are appropriate to wear debris analysis; however, the initials do appear in connected literature and so it is important to have some idea of the subject under discussion.

Additional initials are also given to various components within the processes such as

APP	Analogue Pulse Processor
HCL	Hollow Cathode Lamp
PCA	Principal Components Analysis
PMT	Photomultiplier Tube
UDX	Ultrasonic Nebulizer
VIOB	Video Input/Output Board.

And, again, many manufacturers have coined other special initials for their own 'unique' processes, but in these cases they normally explain what the initials stand for, eg. Baird — MOA — Multielement Oil Analyser.

It should be noted that in the above lists, the 'spectroscopy' does not always refer to chemical element analysis. For instance, Photon Correlation Spectroscopy looks at a light fluctuation spectrum formed by scattered light, which conveys the SIZE of the particle, not the element content. Size determination is discussed in detail in Chapters 7 and 8. In this chapter we are concerned with chemical analysis and the identification of the origin of the particle.

Important Elements
The possible source of particles has already been discussed in Chapter 3. From that discussion one would have seen that the type of particle can

vary significantly. The elemental content also can be significantly different — the following is a very general suggestion of detectable debris, with the likely major elemental content associated with each:

Ferrous wear and steel	Fe, C, Cr, Mo, Mn, Ni
Non ferrous wear	Cu, Sn, Zn, Al, Pb
Quarry dust	Si, O, Al, Ca, Mg
Base oil	C, H, O
Additives	Zn, Mo, S, Cu

As a detailed example we could take a typical alloy in an advanced aero-engine. Czarnecki et al 1991 quote one alloy as having the elements, with percentage weight, of

Mo — 5, V — 2, Cr — 4, Mn — 0.01, Fe — 83, Ni — 0.01, W — 6, others — 0·01.

Looking at the analysis in the other direction, Hipkin & Heimann 1978 suggest an almost direct single element relationship with specific parts and problems in their chemical works:

Element	*Source*
Aluminium	Pistons, bearings
Boron	In-leak of coolant
Chromium	Rings
Copper	Bearings
Iron	Rolling element bearings, cylinders, gears
Lead	Bearings, fuel additive
Magnesium	Transmission
Silicon	Airborne dust
Tin	Journal bearings

Gulla & McCadden 1987 presented an even more detailed chart of element related components, covering a much greater field of operation with 20 elements. The table below is based on their work:

Element	*Source*
Aluminium	Spacers, shims, washers, pistons on reciprocating engines, cases on accessories, bearing cages in planetary gear, crankcases in reciprocating engines, some bearing surfaces.
Antimony	Bearing alloys, grease.
Barium	Oil additive, grease, water (leaks).

Boron	Seals, airborne dust, water, coolants.
Calcium	Oil additive, grease, some bearings.
Chromium	Plating metal (replacing silver in many newer engines), seals, bearing cages, piston rings and cylinder walls on reciprocating engines, chromate corrosion inhibitors (coolant leaks).
Copper	Main or rod bearing thrust bearings, wrist pin bushes, oil coolers, gears, valves, turbocharger bushes, washers, copper radiators (coolant leaks). Note that copper is usually present in the form of an alloy, either bronze or brass, and hence is determinable by the additional presence of tin (bronze) and zinc (brass).
Iron	Cylinder walls, valve guides, rocker arms, piston rings, ball and roller bearings, bearing races, spring gears, safety wire, lock washers, locking nuts, locking pins, bolts. Not surprisingly, this is the most common wear metal.
Lead	Bearing metal (usually in addition to high copper or aluminium), seals, solder, paints, greases. (Could be confused by the lead in leaded fuels.)
Manganese	Valves, blowers, exhaust and intake systems.
Magnesium	Aircraft engine cases for accessories, component housings, marine equipment (affected by water), oil additive.
Molybdenum	Piston rings (some diesels), electric motors, oil additive.
Nickel	Bearing metal, valve train metal, turbine blades.
Phosphorus	Oil additive, coolant leak.
Silicon	Airborne dust, seals, antifoaming additive (some oils).
Silver	Bearing cages (silver plating), puddle pumps, gear teeth, shafts, bearings in some reciprocating engines.
Sodium	Coolant leaks, grease, marine equipment (affected by water).
Tin	Bearing metal and thrust metal bushes, wrist and piston pins, pistons, rings, oil seals, solder.
Titanium	Bearing hub wear, compressors blades and discs (aero-engines).
Zinc	Brass components, neoprene seals, grease, coolant leaks, oil additive.

It must be understood that when debris is being taken from an oil and analysed, it is quite possible it is the OIL rather than the DEBRIS which is being identified. The main constituents of the oil may be carbon, hydrogen and oxygen, but in addition each manufacturer includes certain additives to improve the working property (as mentioned on page 20). It is unlikely that the oil supplier would be willing to divulge what these additives are, except in the case of zinc (which has an adverse effect on silver plating), so it must always be born in mind that small non-debris elemental traces may be in the oil, as indicated in the table above.

There are two ways out of this oil/debris dilemma:
> The first is to filter out the debris and thoroughly clean off the oil with an evaporative solvent so that only the debris is analysed; trifluoro-trichloroethane may be used for most mineral oils, or if a water based oil is being used, isopropyl alcohol (IPA, propan-2-ol). (If special precautions are taken, petroleum ether can be used for mineral oils.)
>
> The second is to determine the element content of the as-supplied oil, by analysing fresh oil which has been fine filtered before it has been used in any machinery. The base element spectrum obtained is then subtracted from the used oil analysis.

The first, 'debris only' analysis, is less likely to be confused and does allow the investigator to also get a visual impression of the debris under a microscope; however, the second method, which looks at the oil, is much simpler and quicker.

Atomic Number

Because the various techniques are restricted to different ranges of elements, it is worth bearing in mind the values of the Atomic Numbers associated with the most common wear and debris elements:

Symbol	Element	Atomic Number
H	Hydrogen	1
B	Boron	5
C	Carbon	6
O	Oxygen	8
Na	Sodium	11
Mg	Magnesium	12
Al	Aluminium	13

Si	Silicon	14
S	Sulphur	16
Ca	Calcium	20
Ti	Titanium	22
V	Vanadium	23
Cr	Chromium	24
Mn	Manganese	25
Fe	Iron	26
Co	Cobalt	27
Ni	Nickel	28
Cu	Copper	29
Zn	Zinc	30
Zr	Zirconium	40
Mo	Molybdenum	42
Ag	Silver	47
Cd	Cadmium	48
Sn	Tin	50
W	Tungsten	74
Pt	Platinum	78
Au	Gold	79
Hg	Mercury	80
Pb	Lead	82
U	Uranium	92

Size

The size of debris decides the technique to be used.

Spectrometers are sometimes classified according to their resolving power as regards depth of analysis. Time and time again, investigators have failed to fully determine the presence of debris because they have used an inadequate technique to detect it. It seems that having paid a high price for a highly efficient and advanced piece of instrumentation, the investigator (wrongly) assumes that it will do everything. For instance, much spectrometric analysis is only capable of determining the full quantitative value of particles below a few micron in size, as indicated by Jantzen & Buck 1983. They investigated the ability of Atomic Absorption Spectrometry to detect iron particles in an oil, both with the conventional vaporization of the particles and with a direct 'dilution only' method. They compared the result with an absolute determination by using a colorimetric process. It had already been found that 5 μm was the upper limit in size for the direct 'dilution only' AAS method if substantial errors were to be avoided (eg. Saba et al 1981) but Jantzen & Buck were to find that with the incineration method (vaporization) that a 50% error was present even for particles of size 2 μm, dropping to a 90% error at about 11 μm. They conclude that AAS is only fully suitable for ferrous wear particles at the 1 μm size and smaller, but for this it is extremely accurate.

The percentages and sizes quoted in the previous paragraph show the reason for the varied claims made by manufacturers and suppliers. There may be a hint of an element for a particle around 10 μm, but not its full elemental magnitude. If precise quantitative elemental analysis is really needed, then a much better accuracy is also required.

Other Spectrometric techniques, however, can cope with the larger particles associated with serious wear, as will be shown below in the different discussions relating to them. Instead of vaporization, a process of dissolving the particle into a solution is possible (in effect, reducing its size). With other techniques even that is not necessary.

Lukas and Anderson 1991 have outlined eight features in spectroscopy which adversely affect the true detection of large particles. Although these are discussed later, in various places, it is helpful to appreciate how many and varied are the problems, covering not only the instrument but also the way the sample is obtained and what type of debris is actually being observed. Their eight suggestions were along the lines of:

1. An unrepresentative sample
    as against being thoroughly mixed.
2. Inadequate excitation
    as against a longer dwell in the heat source.
3. Liquid viscosity too great
    as against being diluted prior to test.
4. Morphology too bulky
    as against long thin particles.
5. Different chemical composition of oil
    eg. hydrocarbon and ester oils.
6. Inadequate probability of being detected in sample
    eg. ICP only a few %, RDE 50% and ARDE 100%.
7. Inadequate transport (method of introduction)
    eg. an aerosol may have maximum droplet size of 5 μm.
8. Inability to provide true particle standards
    eg. 20 μm may be the upper size used.

SPECTROMETRIC TECHNIQUES

For our purposes we can consider spectroscopy under four headings, each relating to the basic technique used for generating the elemental signal from the atom or ion —

Atomic Absorption,
Atomic Emission,
Atomic Fluorescence and
Mass Spectrometry.

I have chosen to class 'fluorescence' as a separate category, although in essence it is an 'emission' technique; the other five emission methods involve a torch or spark. All the techniques explained in this section are used to a certain degree in debris analysis. The list is alphabetical as regards the four types mentioned above, with the more advanced techniques following the simpler ones.

a. AAS   Atomic Absorption
b. IR / FT-IR   Atomic Absorption
c. AES   Atomic Emission
d. RDE   Atomic Emission
e. ARDE   Atomic Emission
f. ICP   Atomic Emission
                    (mention also made of ICP-AFS)
g. DCP   Atomic Emission
h. XRF & EDXRF & WDXRF Atomic Fluorescence
i. XRD   Atomic Fluorescence
j. EDX & EDS with SEM   Atomic Fluorescence
k. MS & ICP-MS   Mass Spectrometry

a. AA — Atomic Absorption
(Eg. GBC, Hitachi, Perkin Elmer, Varian)

This is perhaps the earliest of the spectrometers for debris analysis, coming into general use in the early 1960's (an emission spectrometer had been used as early as 1941 for qualitative data on diesel engine oils, reported by Gulla & McCadden 1987). Atomic Absorption spectrometry is still considered a very precise method with clearly defined spectra relating to the individual elements. It is, however, very slow, taking up to 5 minutes per element analysed for the furnace AA, and it is only reliable for the debris particles at around 1 μm and below. Where a flame is used instead of the furnace a much quicker analysis is possible.

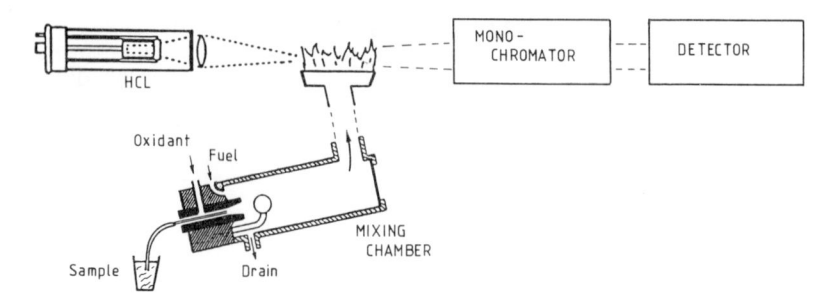

Fig. 9.1 Atomic Absorption Spectroscopy

Cumming 1990 reports its use on JT9D engines where only 0·2–0·4 ppm ferrous debris is being detected, 0·6 ppm being a significant change.

The technique gets its name from a process where the element atoms absorb energy in the form of light of a specific wavelength. It involves the

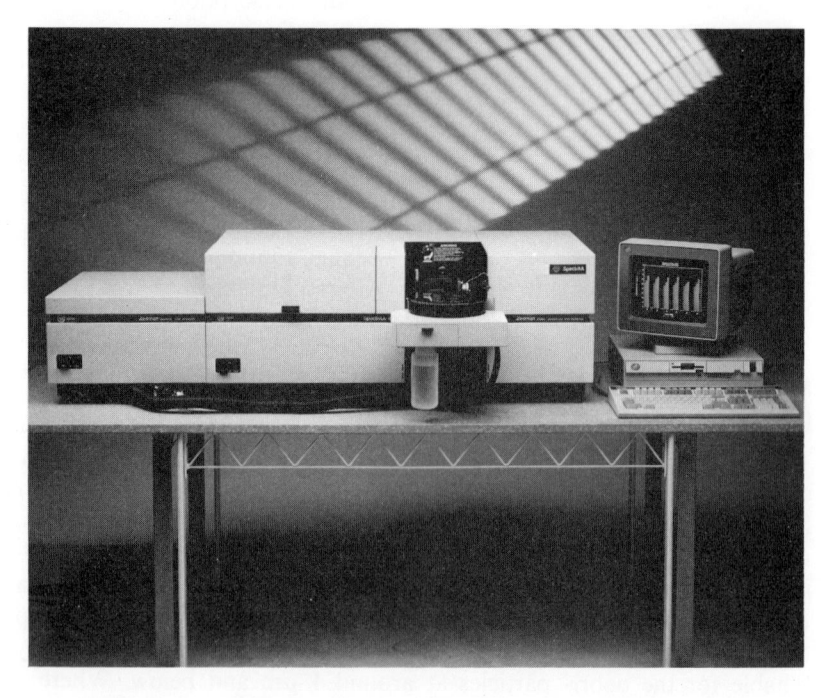

Fig. 9.2 An atomic absorption spectrometer (Varian SpectrAA)

burning of the diluted oil sample when sprayed (aspirated) into a high temperature atmosphere. A monochromator identifies the spectra associated with specific elements. Fig. 9.1 shows a typical arrangement.

The nebulizer and mixing chamber, in this example, involve the sucking out of the oil sample, mixing it with the required diluent and adding the necessary 'gas'. Because of the high temperatures required, an oxyacetylene, or air-acetylene, flame is usually used.

A typical diluent for mineral oil is methyl isobutyl ketone, in the proportion of, say, 0.3 mL of oil to 100 mL of diluent.

Up to 42 individual elements have been investigated in this manner, but because the light source (eg. an HCL) usually has to be changed for each new element, the time taken is considerable if more than one element is being investigated.

FOR A PARTICLE OF 6 ELEMENTS

AAS - ATOMIC ABSORPTION

PARTICLE SIZE (μm) FOR ELEMENT ANALYSIS	RANGE OVER WHICH ELEMENTS DETERMINED
up to 1μm	to ppm - (Flame) ppb - (Furnace)
TIME TO DETECT ONE ELEMENT	TIME TO DETERMINE PROPORTION OF 6
40sec (Flame) 3min (Furnace)	240sec (Flame) 18min (Furnace)
INITIAL COST OF INSTRUMENT (£)	COST TO ANALYSE ONE ELEMENT (£)
7k - 30k	very low

DESTRUCTIVE OF PARTICLE? ....Yes

COMMENTS: Precise and interference free. Tandem System (Flame & Furnace) available at £40k. Also 4-elements simultaneously.

The Non-Flame or Furnace Atomic Absorption can be used for very small quantities of debris. If the debris is still within an oil, the some oil has to be removed first by evaporation. The sample is then placed in an inert gas furnace, or electrically heated graphite tube, to become fully dry, decomposed and, finally, atomised. The light changes in the beam are detected as with the conventional Flame AA. Because in the Furnace AA the light beam passes through the tube where the atomization takes place, sensitivity and detection limits are much improved over the Flame AA. Fig. 9.2 shows the Varian SpectrAA Furnace AAS.

b. IR/FT-IR — Infra-Red
(Eg. Perkin Elmer)

Infra-red spectrometry for oil analysis is usually performed by FT-IR. Both dispersive infra-red instruments and FT-IR both give the same data but produce it in different ways. FT-IR is not used for elemental analysis, but is invaluable as a means of checking on oils and additives. It can be used as a comparative technique where the spectra associated with a particular type of oil are stored in a computer, and the differences can be highlighted. Sample preparation involves pipetting a few drops of the oil between two Sodium Chloride discs which are positioned in the infra-red beam (Fig. 9.3). The beam intensity is modified according to the characteristic vibrational modes of the molecules present. Each substance present in the oil can, therefore, be identified by the spectrum it produces.

An alternative technique to using sodium chloride discs is the use of Attenuated Total Reflectance (ATR). In this technique, an IR beam is passed through a crystal (rigidly mounted on the base of the tray) and penetrates one or two micron into the oil before being returning to the

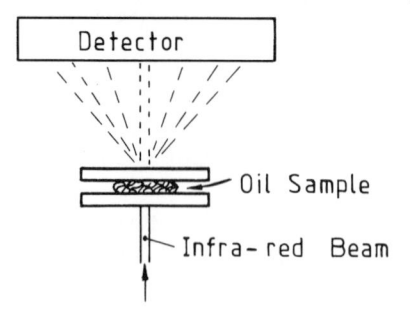

Fig. 9.3 Infra-Red Spectroscopy (Perkin Elmer)

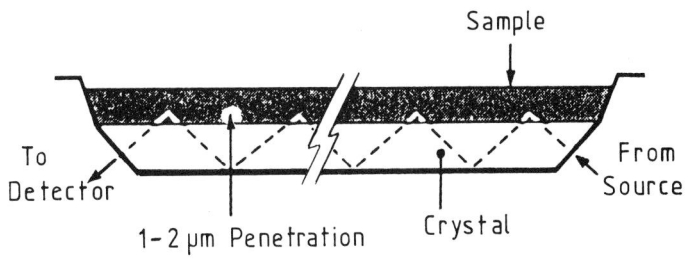

Fig. 9.4   Crystal sample tray for horizontal ATR

crystal. Several reflections occur to give a total path of about 30 μm before the energy exits from the crystal. Here the cell is much easier to clean, and dirty viscous samples can be handled more readily. A typical sample tray is shown in Fig. 9.4.

The full theory behind FT-IR is given by Perkins 1987, associated with the Perkin-Elmer Corporation. He also outlines some of the major advan-

FOR A PARTICLE OF 6 ELEMENTS

IR / FT-IR - INFRA-RED

PARTICLE SIZE (μm) FOR ELEMENT ANALYSIS	RANGE OVER WHICH ELEMENTS DETERMINED
N.A.	N.A.
TIME TO DETECT ONE ELEMENT	TIME TO DETERMINE PROPORTION OF 6
few seconds	few seconds
INITIAL COST OF INSTRUMENT (£)	COST TO ANALYSE ONE ELEMENT (£)
10k - 20k	very low

DESTRUCTIVE OF PARTICLE? ....No

COMMENTS: Used for oil analysis.

tages of FT-IR, in comparison with traditional dispersive instruments, such as

> Higher signal-to-noise ratio
> Greater throughput of energy (ie. more rapid analysis)
> Increased wavenumber accuracy
> Constant resolution
> Polarization effects reduced
> No stepping phenomenon (no sharp discontinuities in the spectrum).

c. AE — Atomic Emission (AES)
(Eg. Baird, Perkin Elmer, Spectro Analytical)

Atomic Emission spectroscopy, like AAS, also requires the wavelengths in a light spectrum to be determined. However, in this case there is no external light source. The light is generated when excited atoms or ions, which are unstable, return to their stable configuration. In order to cause the excitation, a flame (eg. Nitrous Oxide) is required to give an intense temperature around 3000 K. At this temperature, a characteristic spectrum is generated consisting mainly of spectral lines that vary in intensity in direct proportion to the quantity of the element present. (Fibre optics can be used to sequentially examine the light.) However, it should be noted that this temperature, like that used in the Flame AA, is not high and lacks full effectiveness with a certain amount of interference (Fig. 9.5).

It is for this reason that several developments of AES have resulted in such techniques as the Arc/Spark methods, eg. RDE, which give a higher temperature and can thus be more suitable for particle analysis. ICP also dramatically improves the situation. AES, in its basic form, is cheaper than ICP and is quite suitable for some elements such as Ca, Mg and Zn, however, the extensions of the technique are more common, particularly those which are able to examine many more elements

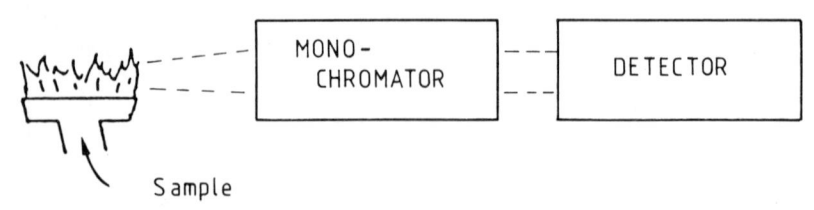

Fig. 9.5   The basic idea of Atomic Emission Spectroscopy

FOR A PARTICLE OF 6 ELEMENTS

AES - ATOMIC EMISSION

PARTICLE SIZE (μm) FOR ELEMENT ANALYSIS	RANGE OVER WHICH ELEMENTS DETERMINED
up to 5μm	variable
TIME TO DETECT ONE ELEMENT	TIME TO DETERMINE PROPORTION OF 6
40sec total	240sec total
INITIAL COST OF INSTRUMENT (£)	COST TO ANALYSE ONE ELEMENT (£)
10k - 30k	very low

DESTRUCTIVE OF PARTICLE? ....Yes

COMMENTS: Up to 64 elements, from traces to alloys, can be determined sequentially. Slow and may lack accuracy.

at one time. Cumming 1990 mentions its use for the JT8D engine which normally generates much more wear than the JT9D; in this case 10ppm to 20 ppm are being detected.

d. RDE — Rotating Disc Electrode (ROTRODE)
(Eg. Baird, Spectro Inc.)

The ROTRODE spectrometer is a 'burning' or arc device with optical emission detection. The process requires a few millilitres of oil to be placed in a tray from where a small proportion is lifted out on the edge of a rotating graphite wheel, like a water-wheel. At the top of the rotation a high intensity arc discharge, between two electrodes, burns the sample continuously. The burning causes light excitation with a wavelength associated with the elements present. This light is then analysed by means of a diffraction grating and photomultiplier tubes. The intensity of each wavelength is measured to determine the concentrations of the elements present in the sample.

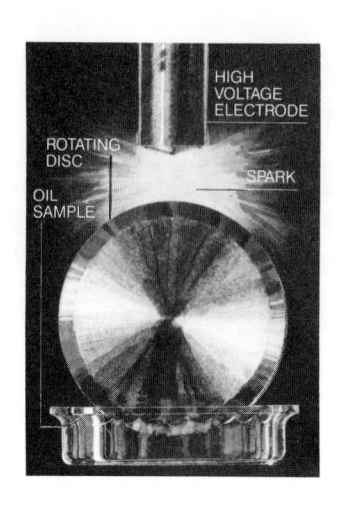

Fig. 9.6 The RDE technique (Baird)

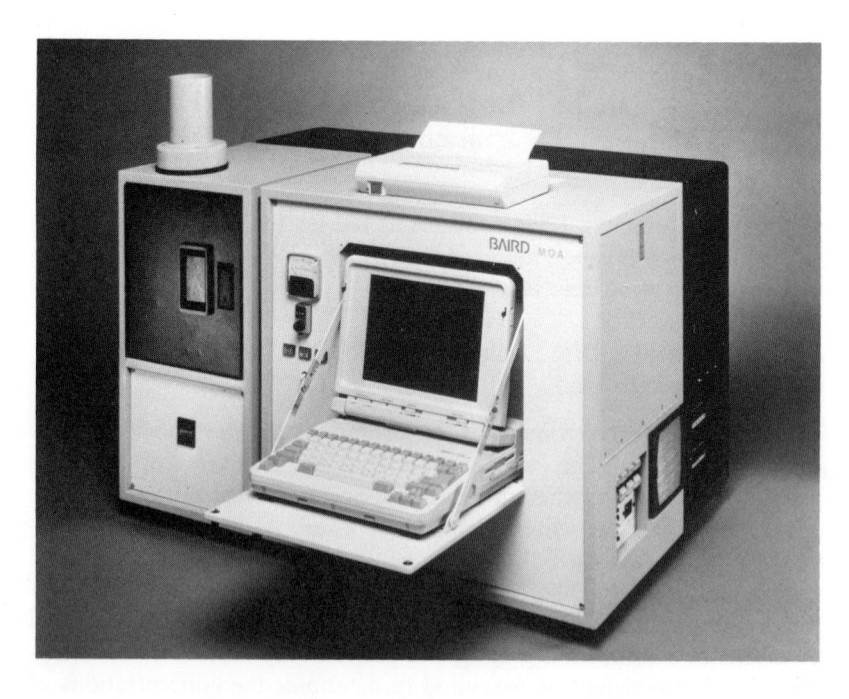

Fig. 9.7 The Baird Multi-element Oil Analyser

No preparation is required except to ensure full mixing of the particles in the oil sample. An analysis covering up to 30 elements can be completed in less than 60 seconds; this provides not only the elemental content but also the quantitative value of the elements.

Because the particles need to be drawn up on the wheel, it is only suitable for the smaller end of the particle size range. The particles also have to be burnt, which again excludes the larger particles — 8 μm is considered the maximum, although a smaller particle would give even better accuracy. Baird has extended the size range achieved with normal DC Arc source, by using what is called the High Repetition Rate AC Pulsed Arc source in combination with the Large Particle Detection System (LPDS); the actual value of maximum size can be altered depending upon the characteristics of the equipment from which the oil samples were taken, typical sizes being between 10 μm and 50 μm (This

FOR A PARTICLE OF 6 ELEMENTS

RDE - ROTATING DISC ELECTRODE

PARTICLE SIZE (μm) FOR ELEMENT ANALYSIS	RANGE OVER WHICH ELEMENTS DETERMINED
up to 8μm possibly '15μm'	0.5 ppm to 1000 ppm
TIME TO DETECT ONE ELEMENT	TIME TO DETERMINE PROPORTION OF 6
30sec	30sec
INITIAL COST OF INSTRUMENT (£)	COST TO ANALYSE ONE ELEMENT (£)
40k	£0.55 per electrode
DESTRUCTIVE OF PARTICLE? ....Yes	

COMMENTS: Direct analysis of debris in oil

is explained in detail in Gulla & McCadden 1987 who indicate that the method had been initially successful on six elements, and others were also feasible.) Another way to improve the size range is to dissolve the particles in acid — the Acid Dilution Method (ADM) — but this can be long winded and open to extraneous particle complication.

e. ARDE — Ashing RDE
(Eg. Spectro Incorporated)

Although there have been improvements in the particle size range with the conventional ROTRODE method, as outlined above, there are still many wear debris particles which would not be detected at all (except by ADM). These are the large break-up particles rather than the simple 'wear' particles. In order to cover these, the ashing method was developed. (This development work is described by Yurko 1987 and Lukas & Anderson 1991.) It is able to detect particles up to 40 μm in size.

FOR A PARTICLE OF 6 ELEMENTS

ARDE - ASHING RDE	
PARTICLE SIZE (μm) FOR ELEMENT ANALYSIS 40μm perhaps	RANGE OVER WHICH ELEMENTS DETERMINED
TIME TO DETECT ONE ELEMENT	TIME TO DETERMINE PROPORTION OF 6
INITIAL COST OF INSTRUMENT (£)	COST TO ANALYSE ONE ELEMENT (£)
DESTRUCTIVE OF PARTICLE? ....Yes	

COMMENTS: As RDE but with the possibility of larger particles being detected.
Not yet commercially available (1991).

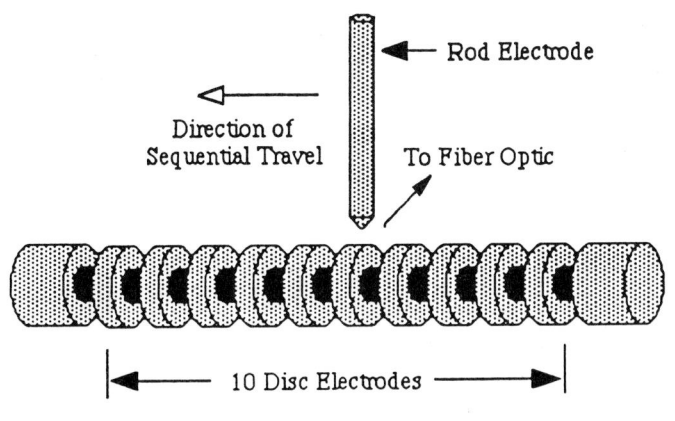

Fig. 9.8 The Ashing RDE

The technique involves the oil sample being directly deposited (by pipette) on a graphite electrode, heated electrically to remove the base oil, and then using the resultant 'ash' remaining on the electrode as the sample for testing. The ash is excited by a pulsed DC Arc with the resulting spectra simultaneously analysed by atomic emission.

Because all the debris in the (perhaps 30 μL) sample is analysed, in the ash form, the concentrations would be expected to be higher than the conventional RDE. This has been found to be true in cases where larger break-up particles, such as chromium, are present. Submicron particles give very good correlation with other methods, but 5 μm and greater see distinct variations (again, except with the ADM where there is good correlation). The technique is still relatively new. The idea developed uses a 10-disc electrode to avoid variations in the density of individual graphite electrodes; the spark probe electrode moves across the discs as they rotate at around 100 rev/min (Fig. 9.8)

f. ICP — Inductively Coupled Plasma
(Eg. Applied Research Laboratories (ARL), Baird, GBC, Jobin Yvon, Perkin-Elmer, Spectro 'Spectroil', Varian)

ICP became available commercially in the mid 1970's. It is an extension of AES. It is a much improved extension as regards speed of analysis and effectiveness throughout the elemental range, with a multi-element capability. (In some cases the absolute spectra may not be quite as clear as the

Fig. 9.9 An ICP spectrometer (Varian Liberty 100 ICP-OES)

AES; however, precision has been quoted as within 1% per element at concentrations greater than or equal to 50 times the limit of detection.) ICP is used extensively in laboratories for large number sample analysis. With mechanised sample placing and introduction to the machine, over 100 samples an hour have been reported, although this rate may cause a reduced accuracy, particularly if the oil viscosities differ greatly. The system is known for its good analytical sensitivity. Although values of ppb (parts per billion) are more the order of the day than are ppm (parts per million), it is possible to reach the percentage range using less sensitive wavelengths.

Fig. 9.9 shows the Varian ICP-OES designed for maximum analytical sensitivity and increased laboratory productivity.

Because linear calibrations over 5 orders of magnitude are typical, both wear and additive elements can be determined in a single analysis. Spectral interferences are irrelevant for oil analysis with the higher quality resolution spectrometers, allowing the accuracy to be as good as the precision (ie. better than 99%).

The ICP is an electrically-conducting ionized gas into which power from a radio-frequency (RF) source is transferred by inductive heating. Much higher temperatures (7000-8000 K or even 10,000 K) are achieved than in flames, which leads to good sensitivity for refractory elements such as silicon, aluminium, titanium, barium and phosphorus. The high temperatures of the plasma prevent chemical and ionization interferences. The sample is introduced into the source by nebulization. The larger drop sizes produced by the nebulizer are effectively filtered out by a spray chamber, so that particles greater than 10 μm are not analysed.

Most oils need to be diluted because of their viscosity, possibly at a simple 1:4 or 1:9 ratio with kerosine or xylene or a mixture of the two. The more advanced ICP instruments would perform this dilution automati-

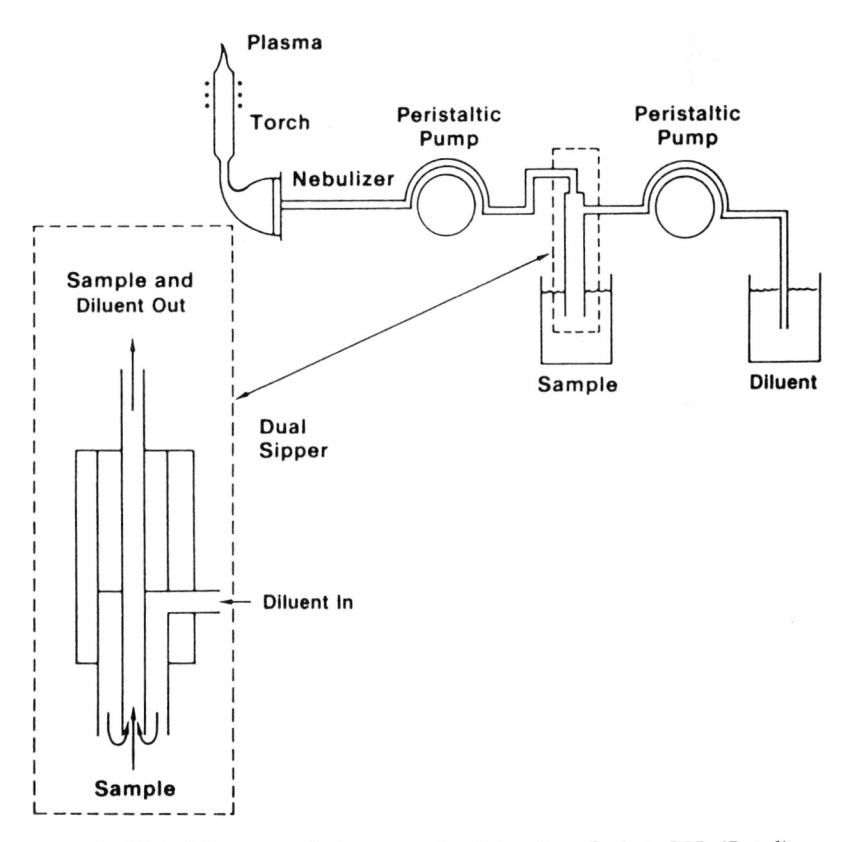

Fig. 9.10 The automatic dilution and nebulization of oils in ICP (Baird)

cally, as shown in the diagram below in Fig. 9.10, although at present it is more common to dilute off-line and run the samples automatically. Homogenisation of the sample may be required, say for marine diesel oils.

Generally the diluent will contain an internal standard element (or marker), say cobalt or zirconium, which is not present in the oil sample. Since the transport of the mixture into the plasma is dependent on the physical properties of the sample, monitoring of the internal standard element during the analysis compensates for widely different viscosities or varying levels of impurity. In other words, instead of running at a fixed rinse time for each sample, the optimum time is achieved for both analysis and rinsing, typically some 70–80 samples per hour in this way (Jobin Yvon 'Fast Oil Analysis').

After the dilution and nebulization, the nebulized sample at the high temperature of the plasma torch, provides atomic emission which is then analysed as with the basic AES.

FOR A PARTICLE OF 6 ELEMENTS

ICP - INDUCTIVELY COUPLED PLASMA	
PARTICLE SIZE (μm) FOR ELEMENT ANALYSIS	RANGE OVER WHICH ELEMENTS DETERMINED
up to 5μm, possibly larger	from ppb to high ppm
TIME TO DETECT ONE ELEMENT	TIME TO DETERMINE PROPORTION OF 6
30sec to 60sec	60sec with a polychromator
INITIAL COST OF INSTRUMENT (£)	COST TO ANALYSE ONE ELEMENT (£)
50k - 80k	medium (Argon gas cost)
DESTRUCTIVE OF PARTICLE? ....Yes	

COMMENTS: High sample throughput.
Large dynamic range (ppb - ppm).

The Spectro Incorporated Automatic Dilution System (ADS) includes a double mixing action (1:1 followed by the final mix) and a novel automatic tip cleaning action. The cleaning involves the use of a spiral action flow of diluent around the sample probe between the taking of each sample.

Some ICP instruments are designed with an additional autosampler specifically for wear metal analysis (eg. Baird Injection Dilution System — IDS, Jobin Yvon, Perkin-Elmer, Spectroil). Their design may include a means of withstanding organic solvents such as xylene and kerosine, and their software should give a rapid indication of the trend in wear particles.

A typical maximum size is 5 μm for approximately 100% recovery. Larger particles will give an indication but only to a partial degree.

Another type of ICP is the ICP-AFS. In this case the plasma torch is still generated but an additional set of combined excitation sensor modules are used to sense the fluorescent emission. The fluorescence can be used for quantitative as well as qualitative assessment of the elements.

A Hollow Cathode Lamp (HCL) radiates the plasma and the atomic fluorescence produced is collected and passed through an optical interference filter on to an inverted photomultiplier tube. Each element module is pulsed one at a time, but as the tubes are all in place the analysis is very rapid.

g. DCP — Direct Current Plasma
(Eg. Applied Research Laboratories (ARL))

Plasma Emission Spectrometry of another kind was also developed at the same time as the ICP. This is the Direct-Current Plasma emission spectrometry (DCP). Although a much lower temperature is developed, many of the analytical features are similar to ICP, eg. sensitivity for refractory elements, linearity and multielement capability. The source geometry is different with a Y-shaped plasma being struck between two graphite anodes and a tungsten cathode. See Fig. 9.11. The sample is introduced via a nebulizer which produces a coarser spray than the ICP; consequently particles of a larger size (between ICP and AA) may be analysed. Electrodes have to be changed which is inconvenient, but the capital and running costs are lower because of the much simpler DC supply required and the argon and electrode costs being slightly less. ARL make this instrument in the United States, supplied in the UK by Fison Instruments.

FOR A PARTICLE OF 6 ELEMENTS

DCP - DIRECT CURRENT PLASMA

PARTICLE SIZE (μm) FOR ELEMENT ANALYSIS	RANGE OVER WHICH ELEMENTS DETERMINED
up to 5μm, possibly 10μm	from ppb to low ppm
TIME TO DETECT ONE ELEMENT	TIME TO DETERMINE PROPORTION OF 6
60sec	60sec with a polychromator
INITIAL COST OF INSTRUMENT (£)	COST TO ANALYSE ONE ELEMENT (£)
25k - 50k	low (Argon gas cost)
DESTRUCTIVE OF PARTICLE? ....Yes	

COMMENTS: High sample throughput.
Large dynamic range (ppb - ppm).

h. XRF — X-ray Fluorescence (and EDXRF and WDXRF)

(Eg. Applied Research Laboratories (ARL), Kevex, Oxford Instruments — Industrial Analysis Group, Outokumpu, Seiko Instruments)

XRF is a common technique used to both identify and quantify chemical elements in a sample. It takes place at the atomic level, and could be described as similar to the effect produced from fluorescent paint when it is illuminated by light, only this time an X-ray is the 'light' source.

It has three distinct advantages over other spectrometric techniques, as far as wear debris analysis is concerned:

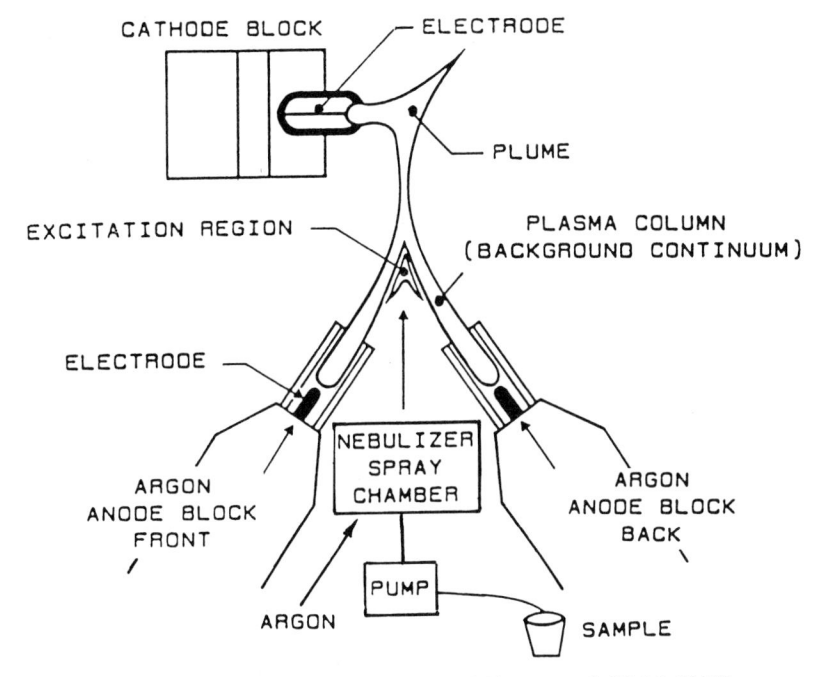

Fig. 9.11 The three electrode DCP (Thompson & Walsh 1989)

* Samples may be analysed in solid or liquid form, ie. oils, grease, thin films, powder or solid component pieces
* The elemental dynamic range is such that the simultaneous analysis of concentrations as low as ppm to high weight percentages may be undertaken on the same sample
* The technique is totally non-destructive.

The process involves putting a few millilitres sample of the oil in a special container and irradiating the container with X-rays. The radiation causes electrons in the atoms of the sample to move to higher energy levels which, when allowed to drop back to their original level, emit X-rays themselves at that precise energy level associated with the particular element. This energy emission is known as 'fluorescence'. The intensity of the radiation at that energy level is, at a first approximation, directly proportional to the concentration of the element in the sample. Thus, by collecting the emitted X-rays and analysing their intensities at the respective energies it is possible not only to detect the presence of an element but also to determine the quantity of each element present. Fig. 9.12 illustrates this point (the Oxford Instruments XR400 is regularly used for wear metals in oils).

The X-ray source is usually either an X-ray tube or a radio-isotope. In order to cover the required range of elements more than one source is used. In the case of the X-ray tube, different types of crystal are incorporated.

Whilst Uranium, at Atomic Number 92, is a common upper limit, the lower limit may vary between Boron (5), Sodium (11), Aluminium (13) or Potassium (19). For instance ARL cover Boron to Uranium with their X-ray tube crystal spectrometers, Oxford Instruments and Outokumpu devices have available four radio-isotope sources, namely Fe55, Cm244, Am241 and Cd109, to cover from Aluminium to Uranium, although only two are normally fitted in the supplied instrument. Fig. 9.13 shows the overall range.

The Cd109 is particularly useful because it is able to cover most of the steel-related range on one setting; a typical industrial bench-top model will probably utilise the Fe55 with the Cd109. To cover a multi-element analysis, of up to 80 elements, a better resolution detector is required, possibly either one based on a Si(Li) detector (Fig. 9.12) or crystal spectrometers.

The reason why the lower end varies is because X-rays from elements of atomic number below titanium are severely absorbed by air; helium is often used to purge away the air in the region between the sample and the detector (particularly suitable for oils and powders). Other techniques (more suitable for solid samples, eg. debris on a filter membrane) are those which use a vacuum (eg. ARL) or bring the two very close to-

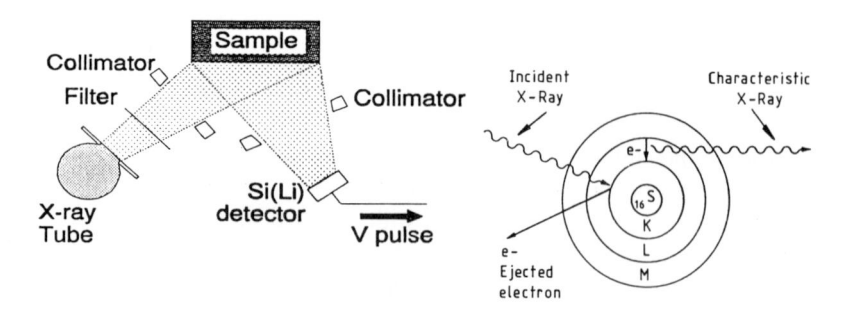

Fig. 9.12 XRF of a sulphur atom (Oxford Instruments)

```
        Fe-55
   _____    Cm-244  (K x-rays)  _____              _ _ _    Cm-244  (L x-rays)  _____
                              _____  Am-241  _____
              _____  Cd-109  (K x-rays)  _____           _ _ _        Cd-109  (L x-rays)
  I  I  I  I  I  I  I  I  I  I  I  I  I  I  I  I  I  I  I  I  I  I  I  I  I  I  I  I  I
  Al S  K  Ti Mn Ni Ga Se Rb Zr Tc Pd In Te Cs Ce Pm Gd Ho Yb Ta Os Au Pb At Ra Pa Pu
     Si Cl Ca V  Fe Cu Ge Br Sr Nb Ru Ag Sn I  Ba Pr Sm Tb Er Lu W  Ir Hg Bi Rn Ac U
        P  Ar Sc Cr Co Zn As Kr Y  Mo Rh Cd Sb Xe Ld Nd Eu Dy Tm Hf Re Pt Ti Po Fr Th Np
```

Fig. 9.13 *Range of elements covered by separate sources in XRF (Outokumpu)*

gether to minimize the problem (eg. Outokumpu). Fig. 9.14 shows the ARL arrangement. It should be noted, too, that the windows which separate between the sample and the detector will often effectively absorb the radiation of the lower atomic number elements.

Analysis time with the different XRF instruments can be between a few seconds to a few minutes; the presence of elements can be assessed in the seconds, but a longer time is needed to compute the quantitative proportions. Another difference between the instruments is the type of detector used, for instance the Seiko SEA 2001 XRF analyser and the Oxford Instruments XR 400 use specially developed Si(Li) semiconductor devices.

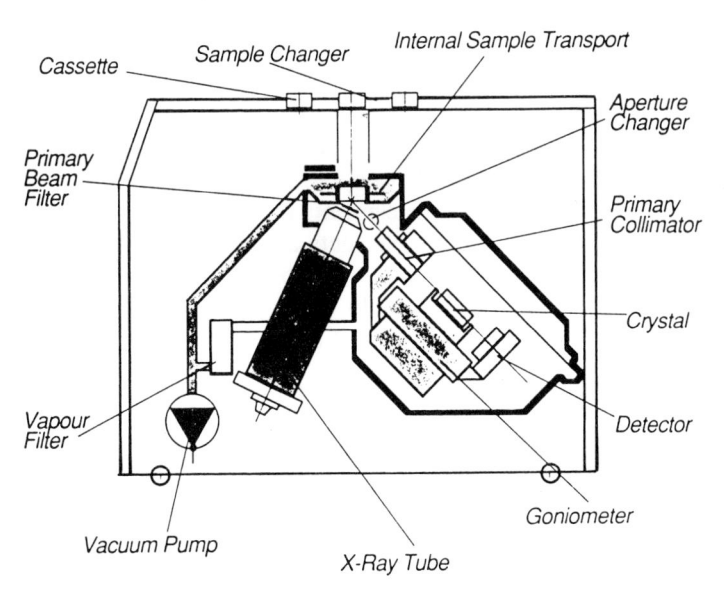

Fig. 9.14 *The ARL 8410 XRF Spectrometer (diagrammatic)*

231

FOR A PARTICLE OF 6 ELEMENTS

XRF - X-RAY FLUORESCENCE

PARTICLE SIZE (μm) FOR ELEMENT ANALYSIS	RANGE OVER WHICH ELEMENTS DETERMINED
any sample, rather than size	ppm up to 100%
TIME TO DETECT ONE ELEMENT	TIME TO DETERMINE PROPORTION OF 6
5sec	60sec
INITIAL COST OF INSTRUMENT (£)	COST TO ANALYSE ONE ELEMENT (£)
20k - 100k	very low

DESTRUCTIVE OF PARTICLE? ....No

COMMENTS: Elements Boron to Uranium. Not suitable for ppb. ppm only for transition elements, 100ppm upwards for low atomic nos. Oil or individual particles from oil.

One on-line wear debris application is that of aircraft in flight. Outokumpu in conjunction with Allison Division of General Motors have developed an XRF to detect the elements titanium, iron, silver and copper. The parts which use these elements are known to erode slowly in normal operation, but towards the end of their service life the wear rates accelerate rapidly. This should enable the aircraft users to programme the maintenance and overhaul to the optimum time.

Because of the multi-element facility, it is possible to 'tune' the microprocessor in an analyser to look for specific alloys. These could be aluminium alloys, stainless steels, nickel and cobalt alloys, tool steels, as well as brasses and bronzes. This availability allows a much improved means of identification and sourcing of the debris particles. However, it must be borne in mind that the X-rays only penetrate a narrow surface layer of the sample. While this is no problem for homogeneous liquids, this may present misleading results with suspended particles and hence would need to be used with caution in such circumstances. (The particles could be taken out of suspension and individually examined.)

The technique by which the X-rays are separated may either be energy dispersive (EDXRF) or wavelength dispersive (WDXRF); the latter allowing of sequential (seq.) and or simultaneous (sim.) analysis. The EDXRF is used for complete spectrum analysis (80 elements or more simultaneously including semi quantitative calculations) but the WDXRF, which could include inbuilt standards for comparison, usually has a better resolution.

The Kevex XRF includes many other features such as the ability to determine small or large areas, the determination of a variety of layers in a surface coating (including their depth), and it competes very strongly with an SEM.

(Note the very wide range of initial cost. The particular application must be discussed with each of the manufacturers/suppliers in order to obtain the most cost effective instrument.)

i. XRD — X-Ray Diffraction
(Eg. Outokumpu))

Whilst XRF looks for the content of the chemical elements in the sample, XRD can be used to estimate the content of minerals in the sample. The same type of X-ray as used for the XRF can be simultaneously appropriate for the XRD, as shown in Fig. 9.15 below. However, the intensity of a radio-isotope source would be too weak for this type of application

An X-ray diffraction tube and proportional counters are used for both the XRF and the XRD. The XRD would normally be less sensitive but

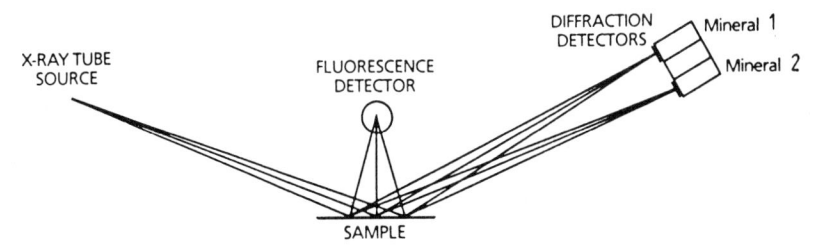

Fig. 9.15 *Combined XRF and XRD analysis (Outokumpu Courier 40)*

FOR A PARTICLE OF 6 ELEMENTS

XRD - X-RAY DIFFRACTION

PARTICLE SIZE (μm) FOR ELEMENT ANALYSIS	RANGE OVER WHICH ELEMENTS DETERMINED
any size	0.1% up to 100%
TIME TO DETECT ONE ELEMENT	TIME TO DETERMINE PROPORTION OF 6
5sec	60sec
INITIAL COST OF INSTRUMENT (£)	COST TO ANALYSE ONE ELEMENT (£)
100k	very low

DESTRUCTIVE OF PARTICLE? .... No

COMMENTS: Suitable for detection of crystaline structures and the mineral content in the slurry.

could detect 0.1% concentration covering such minerals as sylvinite, apatite and industrial minerals. The XRD would be used in such applications as a flotation plant with slurry measurements.

j. EDX & EDS with SEM — Energy Dispersive X-Ray Analysis
(Eg. Kevex)

This is a distinctive energy dispersive X-ray for direct connection to a specimen prepared for SEM evaluation. The electron beam striking the prepared specimen causes the emission of unique X-rays each associated with a chemical element. A common method is to use a semi-conductor X-ray detector for collecting the rays. The specific wavelengths are then displayed alongside the usual SEM photograph with suitable identification as to which refer to which parts of the photograph. 'Mapping' of individual elements is sometimes possible. This is particularly useful for illustrating the scatter or concentration of specific elements within the debris.

Whilst the method is excellent for qualitative identification, it is also able to give the quantitative levels for particles greater in size to 0·5 μm. The full range of elements from at least Boron (Atomic Weight 5) upwards can be detected.

The preparation of specimens for SEM requires mounting and coating which must not adversely affect the EDX. In addition one should be careful not to allow other particles in the vicinity to influence the readings. However, a distinct advantage is to be able to choose a particular

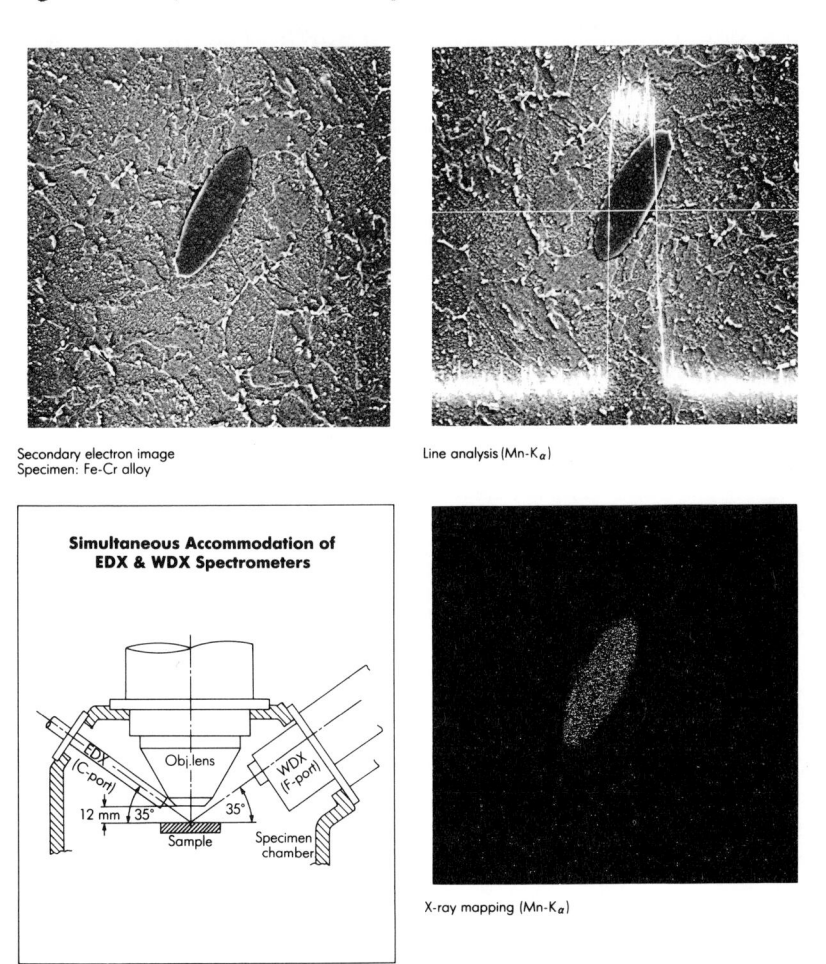

Secondary electron image
Specimen: Fe-Cr alloy

Line analysis (Mn-K$_\alpha$)

Simultaneous Accommodation of EDX & WDX Spectrometers

X-ray mapping (Mn-K$_\alpha$)

Fig. 9.16 *The EDX arrangement, with WDX, on an SEM (Hitachi)*

FOR A PARTICLE OF 6 ELEMENTS

EDX - ENERGY DISPERSIVE X-RAY

PARTICLE SIZE (μm) FOR ELEMENT ANALYSIS	RANGE OVER WHICH ELEMENTS DETERMINED
0.5μm up	
TIME TO DETECT ONE ELEMENT	TIME TO DETERMINE PROPORTION OF 6
100sec	100sec
INITIAL COST OF INSTRUMENT (£)	COST TO ANALYSE ONE ELEMENT (£)
30k - 65k	liquid nitrogen up to £1 per day
DESTRUCTIVE OF PARTICLE? .....Not usually	

COMMENTS: Highly selective.
An SEM costs around £30k - £150k.

area as small as 0·5 μm to analyse and be able to cover the complete specimen. Particles smaller than 0·5 μm will also be accounted for, but not with the same accuracy.

k. MS — Mass Spectrometry
(Eg. Carlo Erba, Hitachi Scientific Instruments, VG)

Mass spectrometry is another technique used with oils, both to determine elements and the organic products present. A small quantity of the oil is entered into the spectrometer by means of a pipette or syringe. On passing through the instrument the oil is ionised in a chamber so that ions are emitted with a variety of mass-to-charge ratio characteristics. The mass spectrum which is formed is examined by the detector on which the ions fall, ie. the ions impinging on the detector generate a signal proportional to the concentration of the ions, as shown in Fig. 9.17.

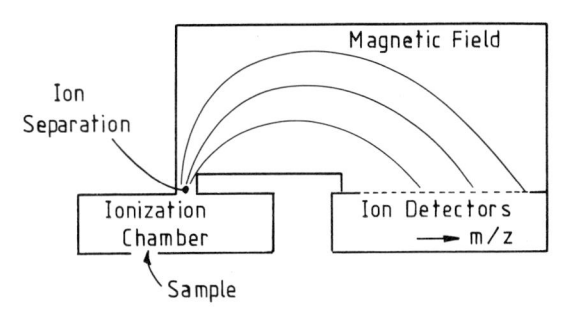

Fig. 9.17 Mass Spectrometry

Although early mass spectrometers required highly specialised operators, the advent of user-friendly software in personal computers has enabled the technique to be more readily useable. Industrial mass

FOR A PARTICLE OF 6 ELEMENTS

MS - MASS SPECTROMETRY

PARTICLE SIZE (μm) FOR ELEMENT ANALYSIS	RANGE OVER WHICH ELEMENTS DETERMINED
solution	ppb - 100%
TIME TO DETECT ONE ELEMENT	TIME TO DETERMINE PROPORTION OF 6
1sec	1sec (simultaneously)
INITIAL COST OF INSTRUMENT (£)	COST TO ANALYSE ONE ELEMENT (£)
20k - 1.5M (ICP 90k - 500k)	wide range
DESTRUCTIVE OF PARTICLE? ....Yes	

COMMENTS: Compounds, not particles.
ie. it looks at a solution.

spectrometers are now available as well as the better known laboratory instruments. Difficulties have been experienced with liquids with high levels of solids, in the sample introduction operation, but this is not a problem with the cleaner fluids.

The technique may be expensive both as regards initial cost and as regards operation and maintenance costs. However sample size can vary between less than 1 μg up to 1 mg (ppb to per cent levels), and simultaneous analysis is possible. Care has to be taken as regards ambiguities arising from the same combination ratio being achieved with different elements.

Microwave induced plasma (MIP-MS) is less costly than ICP-MS because of a lower gas consumption and the use of nitrogen, but ICP-MS is more common.

CHAPTER 10

Physical Identification of Debris

In Chapter 9 we considered identification of particles only as regards the CHEMICAL content. The methods described may be able to decide quantitatively (ie. how much of an element is present in comparison with others) as well as qualitatively (which elements are present). In some cases the compounds are evaluated. The evidence from such techniques is usually quite a good indicator of origin, but not necessarily completely reliable. We will now look at debris from a different viewpoint, that of the PHYSICAL characterisation which can add a further dimension in the identification. Physical, however, implies more than just visual; there are other features which are peculiar to particular particles, such as hardness, bulk, etc., which are all part of the particle's physical make-up.

We have already read of the various shapes which debris can take in Chapter 3. If we know that a particle looks like a

Sphere, Pebble, Chunk, Slab, Platelet, Flake, Curl, Spiral, Sliver, Roll, Strand or Fibre

then an immediate suggestion would come to mind as to where the particle originated. As discussed in Chapter 3, there may be more than one possibility, but at least we do get some positive pointers.

We will look first at the visual analysis of particles and debris, but before we do so, there is a preliminary action which must be taken — the debris must be suitably separated from the system oil or fluid. How that is done may help or hinder the analysis, and it will certainly have some influence on the findings. Each particle needs to be presented to the eye or image analyser with as little disturbing or extraneous aspects as possible.

DEBRIS SEPARATION

Separation can be achieved by basic filtration through a membrane filter, usually 0·8 μm pore size, but sometimes smaller at 0·45 μm, or larger at 2 μm. Two Millipore arrangements for obtaining a debris sample on a membrane are shown in Figs. 7.1 and 7.2; the kit (Fig. 7.1) provides fittings for attachment to a low pressure system on-line, similar to the Howden Wade Conpar system, and the off-line layout (Fig. 7.2) achieves the result from a bottle sample of oil.

The type of result is shown diagrammatically in Fig. 10.1. The lines on the membrane filter are not strictly necessary for identification purposes, however, they would be useful for counting particles per square or even making a quick estimate of the size of a particle, as the line is usually about 150 μm thick (but check this with a vernier or graticule before making too confident an assessment of the size). The disk itself is usually 47 mm diameter, but smaller ones (eg. 13 mm and 25 mm) and larger ones (eg. 90 mm) are also available from companies such as Millipore or Whatman LabSales; care must be taken to ensure the membrane media is compatible with the fluid in use. The membrane filter sampling is discussed further in Chapter 5.

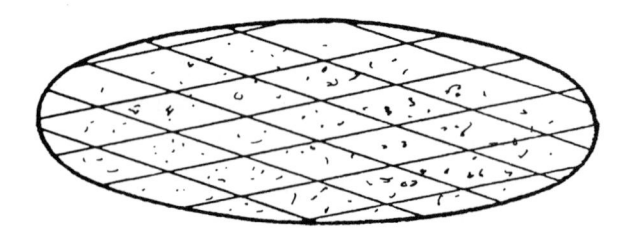

Fig. 10.1 A typical membrane filter sample with debris

Another way to separate the debris is by means of a magnetic field, such as that used in ferrography. Although the debris is usually drawn down in a pattern associated with its size, microscope analysis is quite usual afterwards. Because the debris is on a glass slide another possibility arises — that of heating the slide. This shows up certain metals very clearly, eg. at 330°C, medium and low alloy steel would show up as straw and blue in colour, respectively (see page 258).

One warning with magnetic induced debris — it may be biased towards the ferromagnetic content in the oil. True, other debris does show, but only a limited proportion of it.

Once the debris is on a slide or filter, the oil should be removed by gently passing precleaned solvent across the debris, or through the filter. It must be done gently else the smaller particles will be lost.

MICROSCOPE ANALYSIS

This is the visual examination of collected debris under a microscope.

Goldsmith 1986 has done much valuable work in this field of identification of debris through a microscope. He has several papers to his name, and the Conpar comparison microscope has been his work for many years. He stresses that there are a variety of ways of looking at a particle, or rather, there are a variety of ways of illuminating a particle in order to draw out its different features.

Most research microscopes would have both top illumination and bottom illumination, better known as 'reflected' and 'transmitted' light. (The top light is sometimes oblique to give shadows.) In addition some microscopes will have a removable lower polarising filter for use with the transmitted light. Another stage, associated with Goldsmith is the ferrokinetic stage which involves a small magnet which can be rotated below a sample membrane, while the debris is still moist, to show which particles are ferromagnetic (they move). These different techniques are shown diagrammatically in Fig. 10.2.

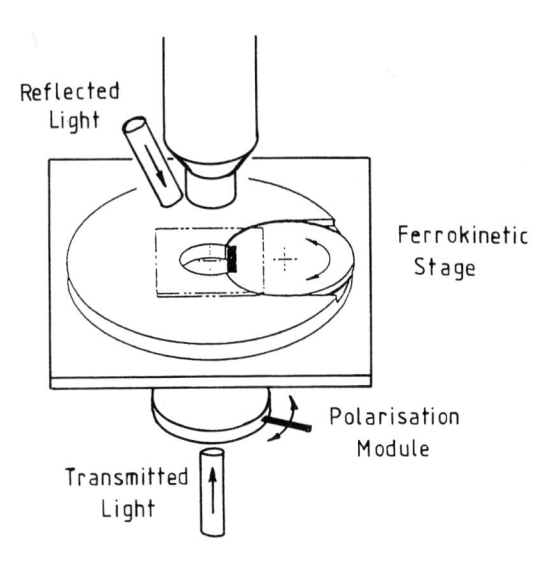

Fig. 10.2 *The various microscope particle assessment features*

Each one of these methods will help to decide the true identity of particles of known shape.

The Wear Particle Atlas (1982/91) points out a useful process for the evaluation of spheres. Smooth spheres can cause light from above to be reflected off to the side and not directly back into the objective lens. An objective lens that can accept light from a wide angle is therefore needed in order to show up such spheres. This type of lens has a high numerical aperture (NA) objective; the NA is the sine of the half-angle. For this sphere examination the NA should be 0·85 or higher. (Examples are also given in the Atlas of actual photographs where the high point of the sphere reflects back what it sees, eg. the lamp filament or an object placed in the light path.)

SEM ANALYSIS

This is a technique for visually examining particles at a very high magnification. Normally if an attempt is made to use an ordinary microscope at a high magnification, two problems exist. The distance from the objective to the particles is so small that damage may result to the debris; secondly, the depth of field is so minute that it is difficult to fully appreciate the true shape of a particle.

Fig. 10.3   The Scanning Electron Microscope (Quantimet S360FE)

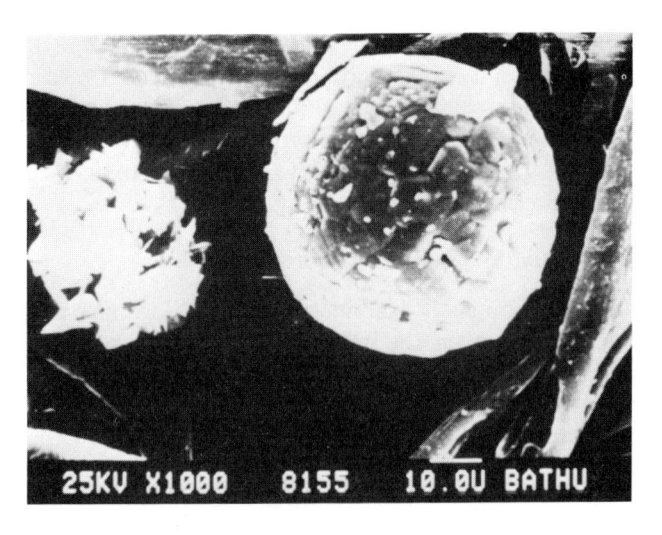

Fig. 10.4 *A typical SEM photograph (University of Bath)*

The Scanning Electron Microscope (SEM) is able to achieve a much greater enlargement and a large depth of field, allowing the complete particle to remain in focus. It does this because it uses a parallel beam of electrons rather than a light beam. It is focussed magnetically or electrostatically in a similar way to the optical microscope but to a far greater magnification. The example shown in Fig. 10.3 shows a highly advanced version called the Field Emission Scanning Electron Microscope produced by Leica Cambridge Limited; it can achieve an image resolution of 2 nm at 25 kV — up to three times better than conventional SEM's. A typical high quality SEM photograph of particles is shown in Fig. 10.4.

SEM instruments often have an attached EDX device for the spot elemental analysis of the minute particles detected in an SEM picture, as mentioned on page 234. This is of considerable value because particles of micron size can individually be analysed for shape, size and chemical content (see Figs. 9.16 and 11.11).

IMAGE ANALYSIS

This is the electronic visual examination of collected debris.

The image analyser attempts to do what the eye does when an image is presented to it. The human eye is designed with infinite care and is vastly superior in detecting minute features of interest, but the Image Analyser has the ability to provide precise dimensional measurements, and record the evidence. The human also is less consistent in the sense that, what one particular human sees today, he or she may completely miss tomorrow, particularly if their health or happiness has been affected in the interim period.

Most Image Analyser systems come in three parts —

1. A means of optically examining the debris (eg. microscope)
2. A device to convert the image to electronic form, eg. a TV camera, electron microscope (and a 'frame grabber')
3. Image processing hardware which converts the image to digital form, stores the image for re-display, and performs high-speed analysis.

There are many variations in the quality of item 1, and certainly the costs of a microscope can vary from a few hundred pounds to well over £10,000. The earlier section on microscope analysis identified some of the important features necessary in one if used for wear debris analysis, but ultimately, the one used will probably depend mostly on the finance available. The SEM is another provider of an image for analysis.

Item 2 and 3 also have wide differences. The 'frame grabber' comes in a variety of guises and can affect the final resolution as much as the digital video camera. A 'bare' frame grabber is very slow, and if hundreds of particles are to be analysed in different ways, a system not optimised for processing image data quickly will probably be a failure. Some frame grabbers will only hold 64 grey levels (as against the generally accepted top level of 256, currently). The resolution of the video camera or chip can have significant differences, and an almost limitless range of software is available for item 3. In comparing image analysers in depth, another critical question to be asked is how any particular model deals with a Fourier Transform; speed is often gained at the expense of the working resolution, eg. by the use of integer arithmetic rather than that of floating point arithmetic. Incidently the use of a colour video camera can triplicate the number of 'bits', eg. $3 \times 8 = 24$, but it needs to be a triple CCD camera in order to get sufficient resolution

Care must be taken to choose those features which are relevant to wear debris analysis, and a resolution which will accurately detect the size of the particles of interest. Because what is needed is a determination of the origin of a particle, then what is required is a means of identifying (and counting) the relevant particles on the image, then deciding what features of the particle are important, such as its shape characteristics, its colour and its size.

But, we have to be more precise. The operator must choose what shape characteristics are important to him. If it is fibres, then high aspect ratio particles must be identified. If it is engine component break-up, then the low aspect ratio particles with a rugged outline and a shiny surface. But there is no problem in making these choices if the correct software is chosen. (Overlapping particles can be 'separated' by the software so that all are counted, providing the program has been carefully written.) Actual methods of sizing and shaping can be chosen as described in Chapter 3.

Grey scale in a large range of ratios (typically 256) is usually available, enabling the operator to highlight, or drop out, particular parts of the image depending in their 'colour'. Real colour screens are able to also perform a semi-automatic differentiation with some 16.7 million variations!

Although the image analysis technique is automatic, it normally requires some operator interaction. By having a visual display of the particles which are being counted (by means of a drawn red perimeter around them, say, or even giving certain shapes or sizes a distinctive colour or grey scale), the operator can see whether the important particles are being assessed or whether too many are being neglected. He can then change the acceptability level for whatever parameters he is using. Fig. 10.5 shows a typical screen display.

Image analysis systems can come in special packages or even in single units specially made, but very common are the individual modules which can be attached to a customer's PC and microscope. Hard wired cards can be easily fitted into most PC's with very little trouble, and floppy discs usually have the necessary software, at a price. However, a word of caution; although these modules can be fitted very easily, their relevant operation may be much harder to implement. Image analysis is still a specialist technique, and it is rare to find users who gain full benefit from the programs without some external advice.

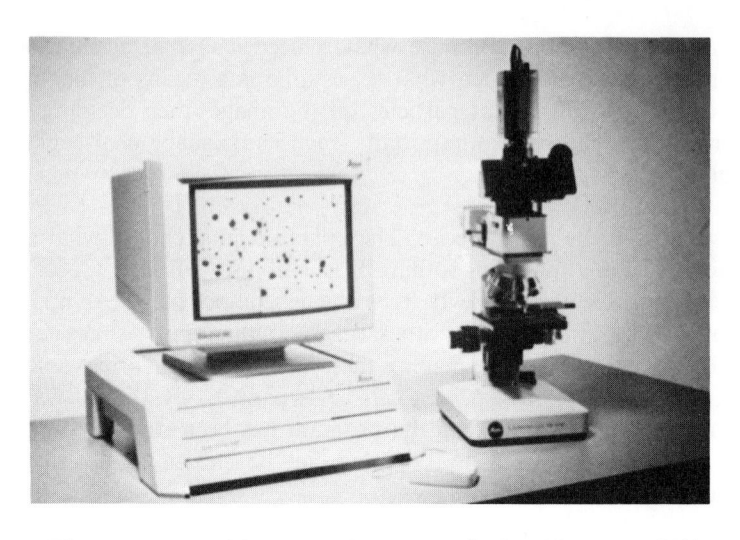

Fig. 10.5 Typical Image Analysis screen display (Quantimet 500)

Automated identification
Recent research in a number of establishments has highlighted the possibility of completely automating the identification of particles under a microscope, in image analysis. For instance, Anderson & Stecki 1990 describe a process of gradually grading particulate through a series of question and answer diagnostics. They recognise that this will differ depending on which method of particle capture has been used (eg. magnetic or filter), but basically it is suitable for all particle images. The types of question they ask are:

>   Is the particle reflective?
>   What is its basic colour?
>   What is its texture?
>   Does it look 'hard' or 'soft'?
>   If transparent, is it polarised?
>   Is it of fibre form?
>   What is its basic shape?
>   What is its basic size?

For each question a series of possible answers are projected, allowing the operator to make a definite reply and move to the next appropriate question.

This type of analysis is currently best operated by eye, but the work at the University College at Swansea and the Swansea Tribology Centre, as initially outlined by Albidewi, Luxmoore, Roylance, Wang & Price 1991, shows that automatic machine recognition is not far away. The particle is defined as regards Surface and Profile, and subdivided as shown in the chart below:

SURFACE		PROFILE	
Texture	Smooth	Shape	Regular
	Holed		Irregular
	Rough		Circular
	Striated		Elongated
	Serrated		
	Cracked		
	Pitted		
Colour	White, etc.	Edge	Smooth
			Rough
			Straight
			Serrated
			Curved
Thickness	1:1	Size	1-5 μm
	1:2		6-10
	1:10		11-25
	etc		etc
	1:41		100+

After the above, which gives the Particle Characterisation, additional information is programmed into the computer along the lines of what debris is expected because of the type of machine and the environment in which it is operating. This all building up into a sufficiently comprehensive analysis to give a first attempt at identifying the debris automatically. The procedures are currently embodied in an expert system package known as CASPA (Computer-Aided Systematic Particle Analysis). Roylance, et al 1992 includes some examples of particle identification using CASPA based on analytical ferrography.

It is likely that as the programming improves, and as more information is able to be fed in, that confidence in the identification will be sufficient without the need for any final human confirmation.

Another series of automated tests and identification is described by Uedelhoven, et al 1991. Their work was aimed at the debris generated in transmission lubricating oil. Initially they obtained the images from the debris deposited on a Ferrogram, and then, using image analysis, they identified both a change in shape and overall size as the debris generation became more serious.

Hand-held scanner

A recent valuable contribution to microscopic examination, is the hand-held scanner. This is a kind of portable video-microscope which is able to be pointed to a particular area, such as a membrane on which particles are resting, and transmit the results to a video screen or analyser. The Keyance Monitor Microscope and the Moritex Scopeman are two examples; the Scopeman is shown in Fig. 10.6.

Fig. 10.6   Hand held video-microscope (Moritex)

Counters based on a specific relevant debris feature, such as hardness, sharpness, bulk, size concentration, magnetic property, etc. can also be used successfully. We will mention four of these here, all completely different techniques, but full details on them are given in the chapters to which reference is made.

THIN LAYER WEAR

This is the effect particles have on machine wear.

The wear propensity of particulate is specifically detected by the Fulmer Wear Debris Monitor. This is a highly significant feature because a major problem in machinery is the wear caused by debris. In discussion on this instrument, it was pointed out on pages 139 and 163 that when a particle of suitable substance begins to be generated in quantity, then a rapidly increasing wear rate signal is given by the instrument. This is a useful identification feature because the thin films used in the technique are worn away not only at a rate proportional to the quantity of debris, but also proportional to the hardness and sharpness of the particles. True, there is a possibility that the increased resistance could be either due to quantity or to quality, but from experience this may be able to be resolved quite quickly.

FILTER BLOCKAGE

This is the effect particles have on blockage.

Bulk is detected by the Filter Blockage technique, ie. the blocking ability of particulate. This is another highly important feature of particles which can jam spool valves and fill up small control orifices. Normally refined and sophisticated mesh is used to detain the particles as the fluid passes through the filter. If the particulate is 'chunky', then the debris will block the mesh very easily because of the sharp edges and general bulk. If the particulate is spherical it is more likely to slip through. Fibres, too, may go through if they can be flexed into the shape of the tortuous fluid route; but if they become more serious to the mechanical system, they too, will block the pores. The hand-held Diagnetics Contam-Alert® uses the same principle but has used as its 'filter' a photoetched disc with sharp square holes in the thin membrane.

The continuous multi-size particle count feature of some filter blockage instruments enables a more detailed analysis to be undertaken. This is where two or three sizes are counted simultaneously or in rapid sequence. Typical sizes are 5 μm and 15 μm (see Coulter Electronics, which also includes 25 μm, and Lindley Flowtech, the first to use this

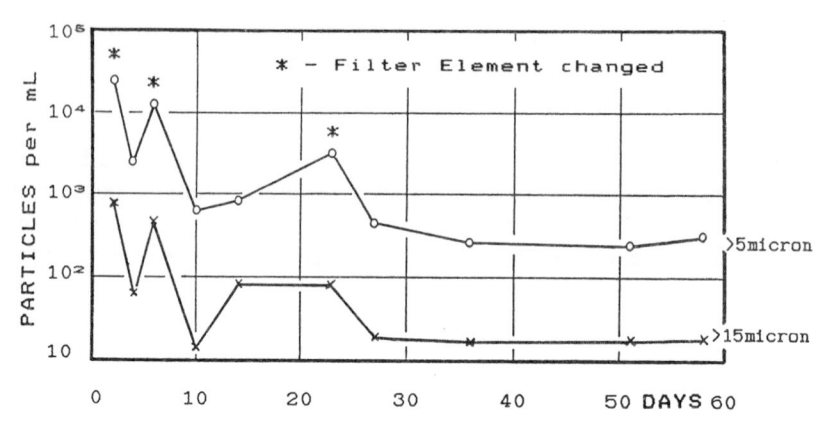

Fig. 10.7 Filter blockage detection of two important sizes

device, on page 199). Such sizes represent different features of debris and can thus be able, in addition, to differentiate between fine wear and dangerous wear; they give that vital indication of an increase in silt debris (5 μm) in comparison with an increase in silt debris (15 μm). Fig. 10.7 shows the two sizes varying during the flushing out operation on an hydraulic transmission; the comparison here is between the larger built-in debris, which was getting dislodged, and the finer silt.

FERRO-MAGNETISM

This is the isolation of ferrous particles.

Ferrous wear particles are usually identified by magnetic devices, although the magnetic properties of debris will vary from metal to metal. Where a monitor specifically observes the presence of ferro-magnetic material, it is highly suitable as a means of detecting the individual wear of a metallic component fabricated in such a material. In some applications this is ideal, such as with mild steel, or some other magnetic steel, being the only metal needing detection. In fact, it can be an advantage if the silicaceous ingested debris is to be neglected (as would be the case where clearances are large and particulate contamination is not a problem). Care has to be taken to ensure the components are, in fact, ferro-magnetic.

Note, though, that the non-ferrous debris may be present, either because the ferrous debris has dragged it bodily to the detection region, or because a certain amount of magnetic attraction has rubbed off from the ferrous onto the non-ferrous. With both these circumstances, it will be true that a greater quantity of ferrous debris has to be present first, if the non-ferrous is to be influenced.

Ultimately the best use of a device such as the Ferrograph mentioned on page 122 or the ferrous wear monitors on page 151, is in the visual analysis of the Ferrogram or the particles, allowing positive identification (as to whether they are metallic or silicaceous). However, the on-line or in-line monitors, have to be trusted to ONLY detect the ferromagnetic particles. Fig. 10.8 shows the change in size generation detected by the QDM when a fault began to develop on a helicopter gear box.

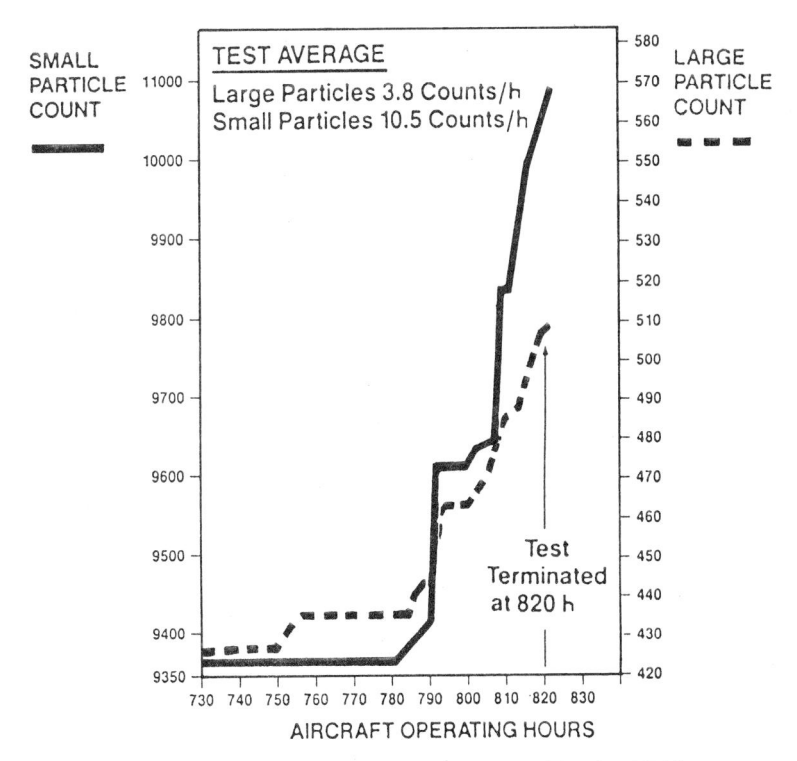

Fig. 10.8 Wear rate on a gear box detected by the QDM

THIN LAYER ACTIVATION

This is the detection of released irradiated particles.

If the reader has already read of the different techniques in Chapter 6, then the idea of Thin Layer Activation will have made an impression. In this technique a component, which is possibly going to wear, is irradiated prior to assembly; when it wears, the particles are released into the oil. Using a gamma-ray scintillation detector, the particles can either be detected in the oil (or filter), or the drop in intensity can be observed at the wearing part. In either case the change can only be due to wear of the irradiated component!

The two techniques are illustrated in an example in Fig. 10.9 below. The top picture shows how the radiation has been applied to a piston ring — the activated spot — and how a scintillation detector can be attached outside of the engine case observing the radiation being emitted. (The gamma rays penetrate the engine metal.) As wear occurs, the level of radiation is reduced. The second picture shows the alternative technique

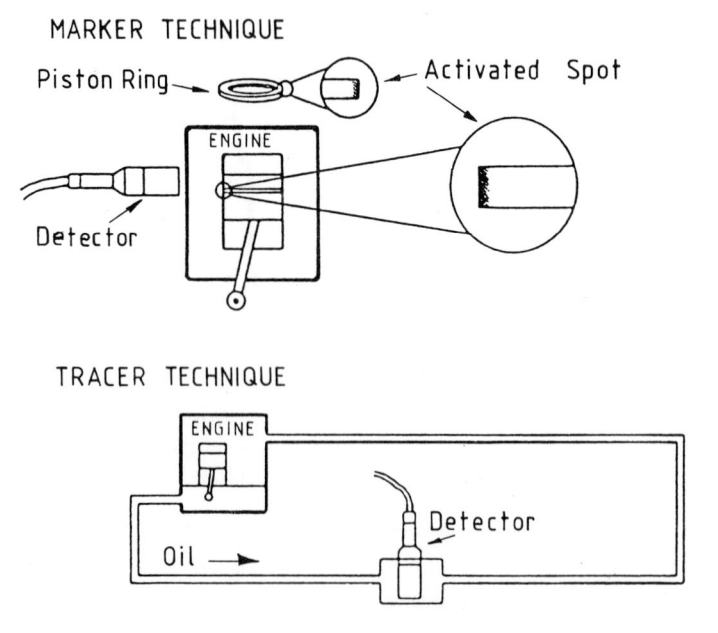

Fig. 10.9 *Wear measured using a NaI (Tl) gamma ray detector (Spi-wear®)*

using wear debris analysis. In this case the oil is monitored with a detector, and only when the component wears (the piston ring, in this example), does the detection system measure any traces of radioactivity.

Because the distribution of activity as a function of depth is known, measurements of activity loss precisely indicate surface loss. (The level of activity in the component is also reduced due to natural decay, but this rate is fully predictable and usually slower, and a correction can be applied.)

Another means of detecting the depth of wear is to produce multiple layers with different isotopes at different depths as explained by Schwabe & Asher 1987. Each isotope can be separately monitored.

Although this approach is more expensive than some competing techniques, the special advantages of thin layer activation provide benefits which are more than economical, particularly for critical and inaccessible components.

Safety must be taken into consideration, but this is usually not a problem because of the more than adequate precautions which are easily taken; the level of radioactivity is also very low. A typical radioactivity level is about $1 \cdot 0$ μCi for a monitoring life of four years.

CHAPTER 11

Multi-value Equipment

One would have liked to have had a much greater selection to present in this chapter. Perhaps, when the time comes for a second edition, there will be a few more devices that have moved into the multi-detection field. In wear debris analysis there is great need for identification as well as the usual quantitative/qualitative output of an automatic monitor.

Consider, for instance, the metal/non-metal identification. Simple portable instruments which are able to discern between ferrous and non-ferrous debris — with an equal amount of confidence — are highly desirable. It is too easy to have a ferrous detector and say that it will capture perhaps 15% non-ferrous as well. That is not good enough.

On the other hand there is the 'all particle' detector which responds to all solid particles of whatever structure. Such a particle detector is at least truthful; it does what it says and records the presence of both ferrous and non-ferrous debris without any bias. But it misses the key question —

> How much of the debris is wear debris, and
> how much is damaging contamination particulate?

(The reader will perhaps realise that that question is not quite sufficient, either. Some wear debris will be ferrous, and some will be non-ferrous; so a further division may be needed for identifying the particles. This has been discussed in the previous chapter.)

The individual instruments and techniques will be discussed fairly fully in this chapter, although one must be limited to a certain extent by what manufacturers have agreed to be presented on their behalf, or by what has already been published.

FERROGRAPHY

Ferrography was developed in the early 1970's, and wear debris analysis publications in the early 1980's were almost exclusively based on the technique. Not entirely, though, because the spectrometric oil analysis program (SOAP) which commenced in the United States in the 1960's, had built up a substantial following. However, SOAP required expensive laboratory equipment and can only really deal effectively with particles of 3 μm and under. At the other extreme, larger particles (>50 μm) were trapped by magnetic chip plugs, so ferrography was seen as the device which bridged the gap between the other two, with a range of about 1 μm to 200 μm. Since the original design, additional equipment has been produced by other manufacturers, such as the RPD from Swansea Tribology Centre; this will be explained after the section on ferrography. The Czechoslovakian firm of VD Amos have now commenced marketing a simpler ferrograph, the REO 1, available in the UK through Oilab Lubrication Limited.

Ferrography has been touched on briefly in Chapter 6 in connection with methods of detecting particles. A fuller description will now be given as regards its working and its ability to define a number of features from the trapped debris.

Making a Ferrogram
The basic ferrographic action is illustrated in Fig. 11.1.

The sample oil is diluted with a special fixer liquid, and about 4 mL is caused to flow down a precleaned prepared slide, over a length of some

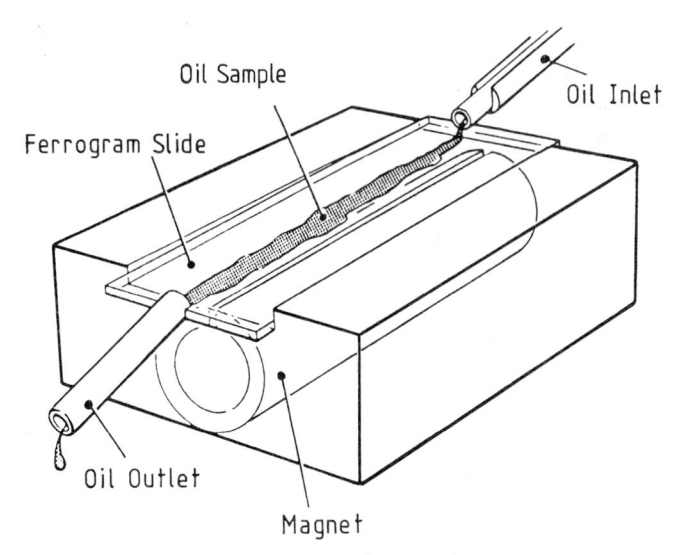

Fig. 11.1 The basic ferrographic action

55 mm. The slide experiences a gradually increasing magnetic flux because of the distance from the permanent magnet positioned beneath it. Because of the slightly 'sticky' nature of the pretreated slide, and the flux force, particles which respond to magnetism are drawn down on to the slide — the larger particles first and then gradually the smaller ones.

In addition to the ferromagnetic particles, other particles which have been in contact with the ferrous ones in the oil may have some ferrous rub-off which causes them to be attracted. Still other particles will just find that the tortuous path remaining on the slide (once a number of ferrous particles have been deposited) is too big a hurdle, and will get trapped. In practice, therefore, a small amount of non-ferrous debris will also find its way on to the slide, but usually in a random manner.

After the deposition of the particles, a second more volatile liquid is caused to flow down the slide to clean off the oil, and leave a dry specimen with the particles stuck to the glass slide. This slide is called a 'Ferrogram'.

The ferrous debris is not scattered over the slide. The flux lines cause a very distinct pattern of horizontal lines, particles being deposited in strings across the slide, as shown in Fig. 11.2.

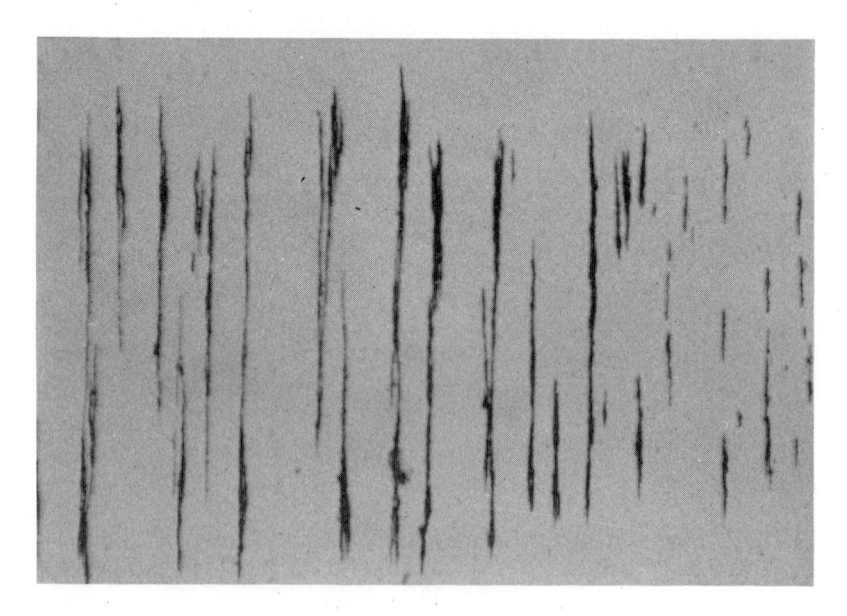

Fig. 11.2 A Ferrogram ferrous particle deposit from a diesel engine (George Smith)

This is a great help in particle identification. The large particles are well spread out first, and even the much smaller particles cover a good area for ease of examination.

Examination of the Ferrogram under the Ferroscope
There is, however, another very valuable addition to the identification provided by the ferrography system. This is the visual examination of the Ferrogram by the Ferroscope microscope. This microscope includes the usual transmitted and reflected light sources, a polarising filter and two colour filters (red and green). By using the filters in a consistent manner it is possible to build up an atlas of shapes, colours and sizes which relate to specific wear and non wear particles. The main users of ferrography have produced numerous colour photographs of a variety of wear conditions, which give a good start to this identification procedure. The complete Wear Particle Atlas 1982/1991 is an even greater asset for this analysis. (See also the Guide to Wear Particle Recognition from Swansea Tribology Centre.) Two typical photographs are reproduced in Fig. 11.3.

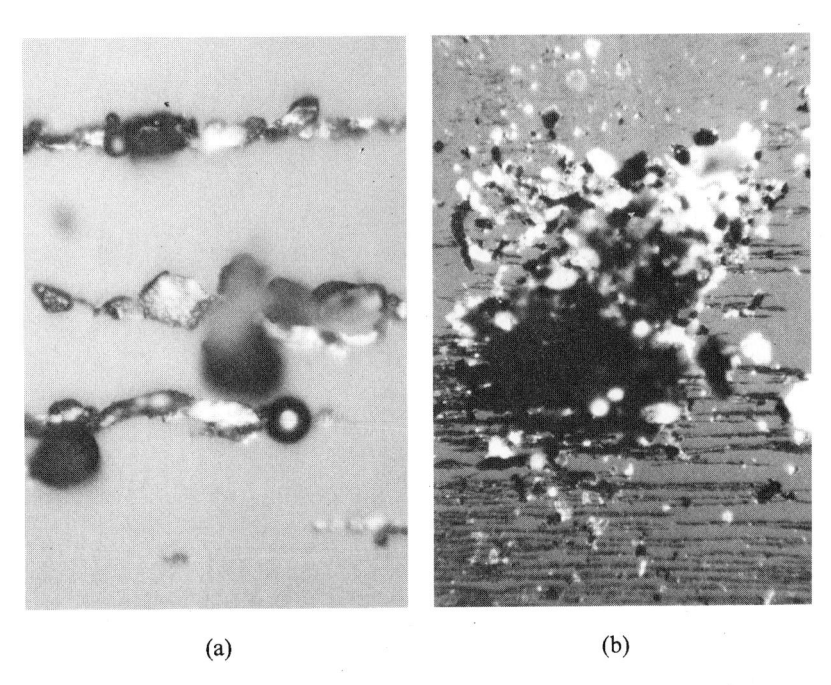

(a)　　　　　　　　　　　　(b)

Fig. 11.3　Ferrogram photographs *(George Smith, SGS Redwood)*

(a) shows a smooth 25 μm sphere and other ferrous wear from a gas compressor. (b) taken with polarised light, shows the random nature of silicaceous debris deposited on top of lines of ferrous wear.

The heating of the debris on a slide will convey further evidence of the nature of the particles. Although it is feasible to heat the slide in an oven, it is better practice to use a hot surface or 'hot-plate'. The temperature and time are both important, hence the need to have a precise technique. In the Wear Particle Atlas 1982/1991 it is suggested that the slide is placed on the hot surface, debris side up, for 90 seconds; with this time the following should be apparent:

at 20°C	RED/YELLOW	Copper alloy, Gold
at 330°C	Speckled BLUE/ORANGE	Probably oxidised Lead/Tin
	BLUE	Carbon and low alloy steels
	TAN	Cadmium
	STRAW/BRONZE	Medium alloy steels
at 400°C	LIGHT GREY	Carbon and low alloy steels
	LIGHT TAN	Titanium
	TAN/STRAW with some DEEP PURPLE	Molybdenum
	DEEP BRONZE with some speckled BLUE	Medium alloy steels
	Some YELLOWING	Stainless steel
at 480°C	BRONZE with BLUE	High nickel alloy
	TAN	Titanium, Zinc
	STRAW/BRONZE with some BLUE	Stainless steel
at 540°C	BLUE or BLUE/GREY	High nickel alloy
	DEEP TAN	Titanium
	TAN/BLUE	Zinc

at higher temperatures the glass will go unstable!

Note that aluminium, silver, chromium and magnesium will not noticeably change in colour over the above temperature range. Organic materials will char, contract and distort, or even vaporize at 330°C.

Fig. 11.4 is part of a Ferrographic Analysis report from SGS Redwood on an hydraulic oil sample. Use is made of different enlargements and heating to enable a realistic assessment of the debris to be made by the analyst. In this case the wear was not extensive, and the diagnosis suggested that there was no suggestion of a serious problem.

TYPES OF PARTICLES ON FERROGRAM

D.R. Readings, L = 6.3 S = 3.5 Sample volume to prepare Ferrogram = 3.0 (ml)

TRACES	FEW	MODERATE	HEAVY
Inorganic Crystalline Cutting Wear	Severe Wear Normal Rubbing Wear Dark Metallo-Oxides Spherical Particles		

REMARKS

Low magnification view of Entry Deposit showing strings of ferrous wear debris.

PHOTO 1 — Magnification = X50

PHOTO 2 — Magnification = X200

REMARKS

Higher magnification view of Entry Deposit showing strings of ferrous metal particles the largest of which is 35 microns in major dimension. Many of these particles show dark oxidised surfaces.

REMARKS

High magnification view of a string of Ferrous wear debris in the Entry deposit showing Free Metal Wear particles together with a number of darker Spheres approx. 3 micron in diameter.

Also several 'Dark metallo-oxides'.

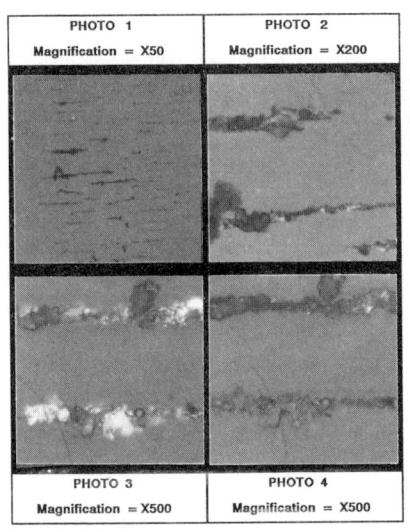

PHOTO 3 — Magnification = X500

PHOTO 4 — Magnification = X500

REMARKS

Similar view to photo 3 but after heat treating the ferrogram for 90 secs at 625 degrees F.

As you can see, all free metal particles have turned a 'straw' colour suggesting Medium Alloy Steel e.g. Cast Iron / Case hardened Steel.

DIAGNOSIS : NORMAL

The distribution of debris in this sample would appear to have increased slightly since the last sample (431615) but while a number of abnormal particles are still present the level does not suggest any serious problem.

Fig. 11.4 Part of a Ferrographic report on a hydraulic oil (SGS Redwood)

The SEM can also be used on the Ferrogram as discussed in Chapter 3 when describing 'spheres'. The excellent separation of the particles on the slide by ferrography, particularly enhances such analysis. Special preparation, as with all SEM analysis, is necessary in order to achieve the best results; for instance, the Ferrogram should be coated with a protective conducting thin film (usually sputtered or evaporated) as appropriate with the SEM manufacturer's instruction.

The Direct Reading Ferrograph

The intensity of the large particulate compared with the smaller can also be determined and given a numerical values. An automatic light density reading is obtained in the Direct Reading Ferrograph, illustrated in Fig. 11.5. In this case 1 mL of oil is syphoned up a tube with the two light paths 5 mm apart. It can give the proportion of large particles (D_L) against the proportion of small particles (D_{ds}). The large particles are typically the serious wear and fatigue (5 μm and above), whilst the small particles are the regular wear particles (around 1 μm to 2 μm). Density is given by a digital meter reading from 1 to 190; 100 representing a half blockage. A rise in either density value would indicate an increase in one or other of those two types of wear debris. The DR Ferrograph can obtain a reading in five minutes, once set up, but unlike the Analytical Ferrograph it does not retain the particles for examination.

It should be noted that some dark oils can be so dense that a reading as high as 190 could be achieved from the oil alone, without the presence of any debris. Care must, therefore, be taken to dilute the sample, if necessary, to bring the initial reading down to 100. An alternative is to use what

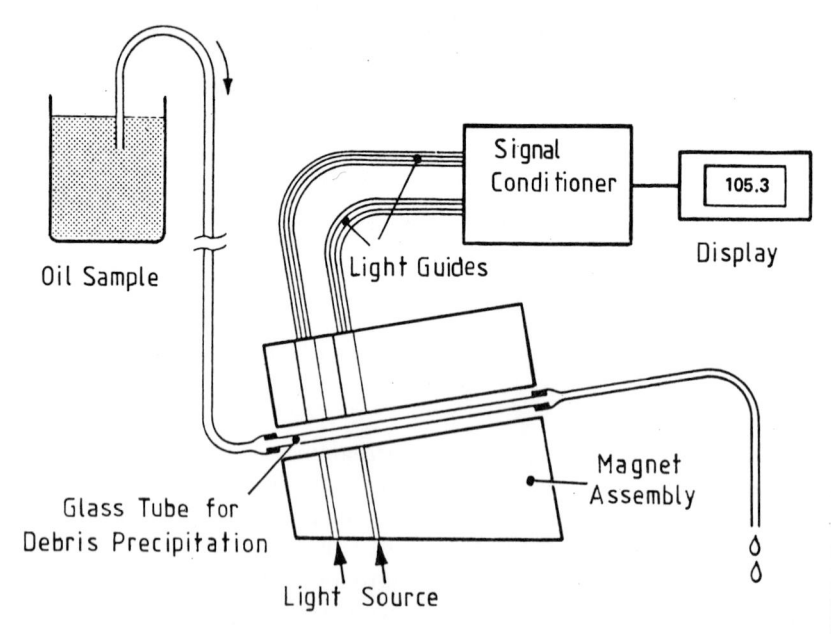

Fig. 11.5 The optical layout of the Direct Reading Ferrograph

is called the 'dark oil technique'. In this case the Ferrograph is zeroed using a clear fixer, before running all 1 mL of the dark oil sample through the tube. The dark oil is then flushed out with the clear fixer fluid. What is left in the tube is the clear fixer and the precipitate from the dark oil.

The Wear Particle Atlas 1982/91 explains an apparent quirk in the D_L and D_S readings:

> "If a sample contains only small particles, the readings at the L and S channels will be equal. Fig. 11.6 illustrates why this occurs. When particles enter the precipitator tube and are influenced by the magnetic field, vectors of motion can be used to describe their trajectory within the precipitator tube. The vector of motion in the flow direction will be equal for large and small particles. However, the vector of motion in the downward direction will be greater for large particles because the magnetic force is proportional to volume (diameter cubed), whereas flow resistance in the downward direction is proportional to surface area (diameter squared). Large particles move so rapidly to the bottom of the precipitator tube that all magnetic particles larger than about 5 μm are deposited at the first sensor location. Small particles, which migrate much more slowly to the bottom of the precipitator tube, penetrate much further

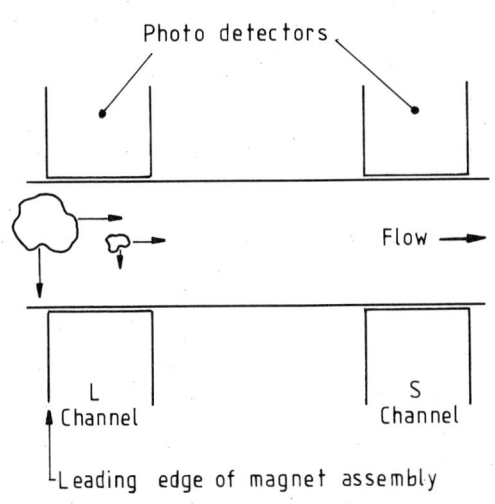

Fig. 11.6 Particle deposition in a DR Ferrograph (*Wear Particle Atlas 1982/91 page 151*)

along the tube being deposited. As a sample enters the precipitator tube, small particles are dispersed throughout the volume of the tube so that some small particles are near the bottom of the tube and will be pulled down over the first sensor location. Other small particles, however, which enter the tube higher up from the bottom will travel further along the tube before being deposited. Therefore, if the L and S readings are equal, only small particles are present, and a true indication of large particles is L minus S."

A 'Severity of Wear Index' can also be calculated to take into account the ratio of large to small particles. It has been quoted as

$$I_D = D_L^2 - D_S^2.$$

If the I_D increases, then obviously there is a greater quantity of serious wear debris being generated.

If the concentration of ferrous debris is too high, some overlapping of particles may occur. In this circumstance, the larger particles (even up to 200 μm) may be totally obscured by the smaller particles on top of them. A dilution of 100:1 or even 1000:1 may be necessary. Reciprocating engines are cited as being particularly prone to this problem.

Ferrography has achieved many successes over the years, however there has also been the occasional confusion. This can arise, as with all off-line devices, from the actual method of obtaining the original sample (see Chapters 5 and 14). It can also happen if the level of non-ferrous particulate is high; this is deposited randomly and can seriously affect the optical density readings.

Xu, Qiao & Huo 1989 from the Friction and Wear Research Laboratory in the Beijing Graduate School, China University of Mining Technology, describe the use of Ferrography in the Huo Lin He coal mine. The 75 ton capacity trucks are brought in for service every 250 hours. During this service a Ferrographic analysis is undertaken, but also samples are taken every 50 hours during the oil change. Both Direct Reading and Analytical Ferrography is used. However it is the microprocessor attachment to the DR Ferrograph which enables a clear trend to be presented in a new manner. By determining the optical density values between the D_L and D_S values over the full length of the ferrogram path, a pattern of debris can be obtained. The two examples given in Fig. 11.7 show the difference between an acceptable wear

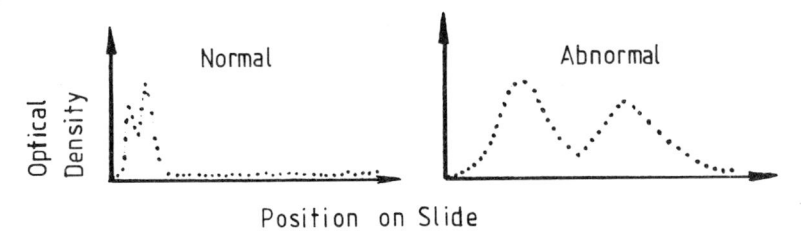

Fig. 11.7   Examples of the Quantitative Ferrogram (Xu, et al 1989)

pattern and an abnormal one. Although they report that the method has been "effectively used" in the mine they could not describe any precise achievement as yet.

Oilab Lubrication Limited now market the Czechoslovakian Ferrograph REO 1 which is able to produce a much simpler 'slide' which can be checked for density over its whole length using the Densimeter REO 2. This is thus able to give an indication in the same way as the examples in Fig. 11.7 although not in as many steps. The 'slide' which is used is actually an elastic foil.

THE RPD — ROTARY PARTICLE DEPOSITOR

As the title suggests, this device deposits the particles in a rotating flow field. It is a ferromagnetic particle device, requiring a magnetic force to draw down particles onto a substrate. Williams 1988 provides not only a description of the RPD but also thoroughly compares it with the Ferrograph. Fig. 11.8 is a cross sectional diagram of the unit.

The Sample Delivery Pipette is used to take a sample of the oil needing to be analysed. The pipette is then slowly depressed by hand to cause the sample liquid to flow onto the centre of the rotating substrate. The substrate, which is a flat square plate of glass is held securely to the RPD spindle by means of a suction rubber seal. Within the head, below the slide, are two circular magnets which cause two circular fields of magnetic flux. As the liquid slowly flows outwards on the rotating slide, ferromagnetic debris is deposited, large particles first and then the

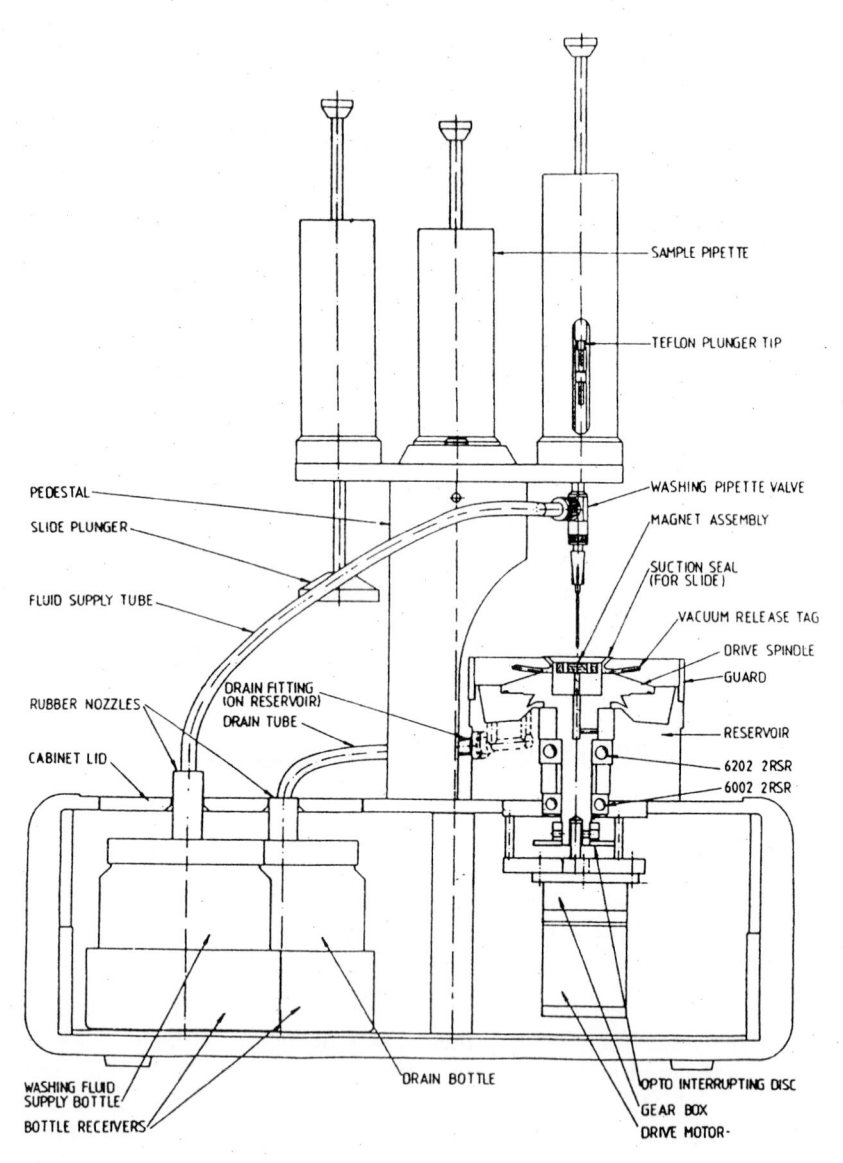

Fig. 11.8 The RPD

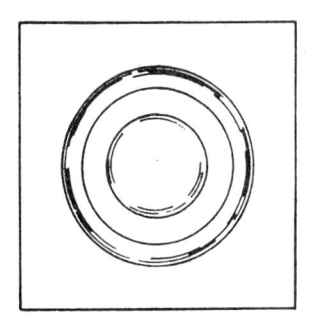

THREE CONCENTRIC RINGS

Fig. 11.9 The RPD slide with debris deposited

smaller particles, in two or, sometimes, three rings, as shown above in Fig. 11.9.

In comparing the RPD with the Ferrograph, Williams draws attention to the larger surface area available on the RPD slide compared with the Ferrograph. The RPD uses three concentric rings at radii about 5, 7 and 10 mm; in other words the initial ring is able to spread out the large particles well, and the much larger 10 mm radii ring is valuable for the very large number of small particles. The processing time for the RPD is claimed to be between 5–8 minutes for a complete cycle.

The analysis of the RPD slide can be undertaken in the same manner as the Ferrograph slide; although, if density measurement takes place, it should be noted that there is no direct comparison between the two techniques because of the different surface areas used.

THE GALAI CIS-1 AND CIS-1000 — TIME OF TRANSITION PARTICLE ANALYSER

Galai Production Limited, in Israel, started with an emphasis on computerized electro-optical inspection systems. However, it was in the late 1980's they came into world-wide prominence, in the particle analysis field, with their 'Time of Transition' particle monitors. These Galai instruments are able to operate as dual measuring devices. In simple terms this could be described as the measurement of both size and shape. Although the

COULTER COUNTER® has had volume size and shape ability for many years, the shape feature is only designed to look at particles which have become lodged in the orifice of the instrument, and have become stationary. The Galai instruments go very much further. They are able to photograph the moving particles and offer an image analysis feature on all the particles in view.

Before looking at the working of the 'shape' aspect, it is essential to examine the major feature of the instrument — the particle size analyser and counter. The Time of Transition theory which the instrument uses has been generally explained on pages 131 and 176, however, there are specific features on the Galai instruments which need describing in more detail. (Fuller specifications are given in Appendix 1.)

The primary model (CIS-1) is for laboratory use. It consists of four items, involving sample presentation, analysis, computing and image viewing.

The CIS-1 system's modular design enables airborne, liquid-borne or dry samples to be introduced for analysis through the wide variety of measurement cells available. A Galai Cell Module (GCM) can be firmly mounted inside the measurement area of the optic unit by means of a tightening screw. The measurement cells also give the user the flexibility to measure samples in a disposable plastic cuvette, in adjustable flow or on a microscope slide, depending on the application. Measurement in the cuvette is usually achieved using a magnetic agitator; however, when ferromagnetic debris is suspended, an alternative mechanical agitator is available. In cases where samples form aggregates, a sonication procedure may be necessary in sample preparation for the CIS-1. The automated on-line systems have an integral ultrasonic probe incorporated (see the APM description below).

The time of transition analysis is undertaken on the fluid in the Cell Module. Although the particles are not stationary, the rotating laser spot is moving so rapidly that the particle movement has little effect on the results. The strength of the time of transition technique is that it relates solely and directly to particle size rather than to any secondary feature such as movement, refractive index, viscosity, concentration, etc.

The computing is undertaken in a separate computer. This enables a choice of which size range to use and also how the results should be displayed. Up to 300 discrete size intervals are possible between the extremes of the size range chosen; this enables secondary peaks and fines to be detected which might otherwise be missed by other techniques. Up to 1000 consecutive analyses can be programmed which is ideal for trend analysis.

The image viewing involves the second aspect of the Galai instruments, based on Dynamic Shape Characterisation (DSC). Although the actual analysis is undertaken in the computer, the visual display enables the operator to see exactly what is happening and what is being measured. The method used for detecting the shape involves a video camera microscope, much as in normal image analysis. However, in this case the device is used concurrently with the time of transition analysis. Illumination of the particles is undertaken by means of a variety of sources although the synchronised strobe is the most usual. Acquired images (either one a second or 30 a second) are passed to a frame grabber for analysis. The software includes such features as rejection of out-of-focus particles, separation of overlapping particles and automatic lighting correction, as well as the usual operations.

Although the image analysis is performed using a black and white video, considerable detail can be observed down to 0.8 μm particle size. The final feature of the combined unit is the ability to combine the two aspects in the sense of determining the volume of particles by incorporating a shape factor in the analysis.

The on-line model, the CIS-1000, takes the flow from a bypass process line, and analyses it in a typical time of 30 seconds, before returning it to the line or some other part of the system. It can withstand up to 11 bar pressure and a maximum sample temperature of 100°C. (Other sample cells are available which can withstand a temperature of up to 250°C or 120 bar.) The CIS-1000 covers the larger range of 2 μm to 3.6 mm in three separate settings which can be remotely set by computer. No calibration is necessary.

Up to seven on-line sensors can be used in suitable bypass positions, and the signals relayed to a central processor unit. In this case the option of a sensor which only uses the time of transition technique means that a smaller cheaper unit can be used, weighing only 5 kg instead of the 11 kg of the dual unit.

Two systems based on the Galai CIS-1000 are particularly aimed at production monitoring. One is the Automatic Particle Monitoring (APM) system and the other the Automatic Sample Management (ASM) system. The APM is like a fully automatic, round-the-clock,

Fig. 11.10 Galai shape screen and histogram output

quality control laboratory at hand; it comprises automatic modules for sampling, sample preparation for measurement, sample feed to sensor, particle measurement by core sensor, data acquisition, real-time outputs and trend analysis, active control outputs and self-cleaning of the sensor. The ASM is a mobile unit for use on-line and fulfils a number of functions on the production floor for quality control, measurement and process evaluation; it also can be seen to replicate the work of a laboratory technician in the sense of preparing the sample for analysis.

The typical video display of particles, and a VDU histogram output are shown in Fig. 11.10.

OTHER MULTI-VALUE EQUIPMENT

The dual feature on the Galai instruments, highlights the idea of image analysis. Image analysis, with its direct connection to a wide range of

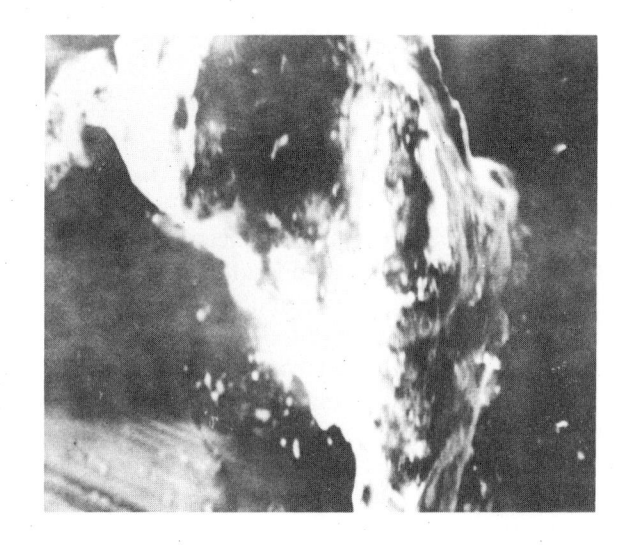

S.E.M.
x 300
(One Particle)

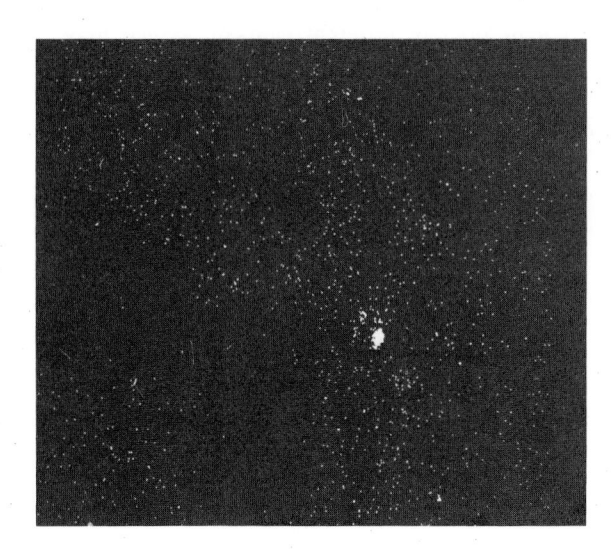

Cr X-ray
Note
Cr segregation

Fig. 11.11 Particle shape and elemental content from an SEM/EDX

computer software, opens up a vast set of possibilities for particle analysis. These features have been discussed in Chapter 3 as regards shape, size and density. In Chapter 10 more detailed analysis revealed the possibility of identifying the source of a particle from the shape profile. This is a multi-value technique, albeit off-line at the moment.

Image analysis on an SEM allows another feature to be assessed of vital importance to the analyst. This is the elemental content of the particle. Where an SEM is attached to an EDX machine, not only can a picture of the particle shape be displayed, but also a matrix relating to specific elements under investigation (see Chapter 10, Figs 10.3 and 10.4). In this case the technique is very definitely off-line and it requires preliminary preparation of the particle sample in order to avoid loss of the particles (ie. they need coating) and to prevent confusion from any oil which might remain in the sample. One example is given in Fig. 9.16; another one of a particle with some silicon content is shown in Fig. 11.11. The concentration of the silicon is displayed by the intensity of white dots in the lower photograph, the scale and positioning being identical in the two illustrations.

Spectrometric equipment is multi-value in the sense of providing, in many cases, a chart of a number of different elements which may be present in the sample. Laboratory analysis, such as provided by the kits mentioned in Appendix 1, provides more on the physical and chemical constituents of the oil rather than the particles; but again this is a valuable extra.

CHAPTER 12

Levels and Standards

The analysis of wear debris can take three forms:

Individual examination of particles
Trending the increase in debris concentration
Comparing a level of debris concentration with a Standard.

The first of these, the individual examination, has already been studied in previous chapters such as Chapter 9 where the elemental break down of the particles, and hence their origin, was discussed. Now we look at the other two forms of analysis. We shall examine first what trend analysis means, including multi-trends; then we look at the use of standards and how they have a place in both contamination control and also in condition monitoring.

TRENDING

Trend monitoring is a technique designed to enable the analyst to become aware of any significant change in the value of a parameter. Originally it was only concerned with the likelihood of exceeding an upper or lower limit. In a sense, all was considered well provided the value remained within the tolerance bands; but there was another key feature, where a steady value, which previously had remained almost level, suddenly took on a new slope on the graph. Warning signals would light up in the mind of the observer.

The occurrence of a change of rate, indicates a different process is taking place. Perhaps before, the wear rate was quite normal for a fully lubricated sliding surface. But now, maybe, oil starvation has occurred and the shedding of particles has increased dramatically; the wear debris concentration increasing accordingly. The value of a parameter (perhaps, 'total debris per 100 mL of oil') could still be well within the upper acceptance boundary, but one senses that that will not be the case for long. Disaster is imminent.

A word of warning. Whether an upswing (or a downswing) is significant or not, can only be judged from experience. We are all aware of those who, wishing to emphasise a point (usually political) greatly magnify the scale of a graph to show a 'major' effect; however, if the chart is re-plotted with a full axis, the effect becomes totally insignificant!

A trend needs to be consistent before it can be used as a reliable check. A change in trend also needs to be consistent. (The odd high point is often due to an instrument fault, and may need to be ignored.)

The levels chosen for the upper and lower limits are usually quite arbitrary at the start of an exercise. No experience exists, and an estimated guess has to be made. It may be right, it may be totally wrong, but most likely it will be meaningless until a failure actually occurs and the level of the value is known. The advantage of the trend examination is that it will vividly illustrate what the wear debris analyst is looking for — a significant change — and it will do that right from the start of a test.

Take, for instance, the following example:

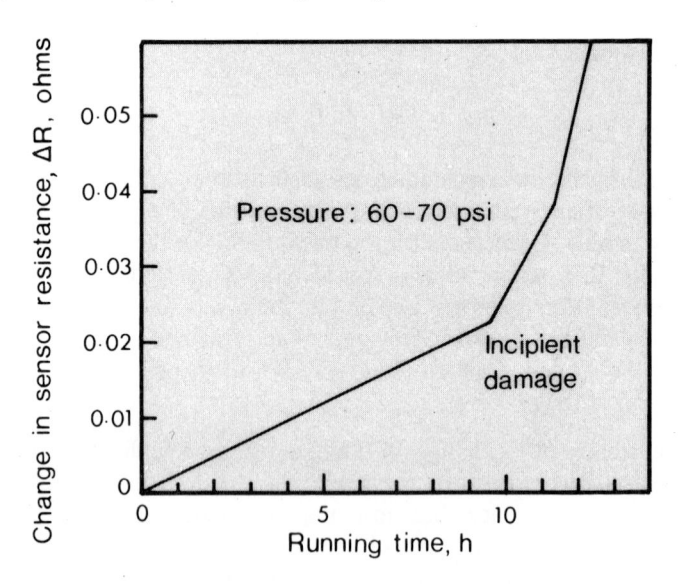

Fig. 12.1 *Rate of resistance change due to particle bombardment (Santilli 1989)*

Figure 12.1 shows an example of a successful trend monitoring relating to particle content in a fluid (Santilli 1989). This example involves the use of the Fulmer monitor (see Index for further detail).

Notice how consistent the initial wear rate is in this example. (In the actual test there would have been a slight unsteadiness in the plotted value, but, from my experience of the particular instrument, the overall impression would be as illustrated.) It gives a confidence in the acceptance of that rate as being reliable. Even if there had been a slight scatter in levels, the trace would still be valuable; the plot would have given a band within which no point would be considered a significant change, in a sense providing a 'thick' line which waited for an upturn. Then, as in the example, comes the upswing which again is consistent although probably the scatter will be greater because an area of trouble is being met.

Although it is not quite clear which came first — the noticing of the apparently significant change in electrical resistance across the surface, or, from other examination, the awareness of the incipient damage to the bearing — either way, a significant change has occurred. Whichever it was, it does not matter, the next time it happens, this level of upswing of the graph will be associated with the approach of failure and would be spotted by computer (or the same analyst) with no difficulty.

Trend monitoring has developed because so often the operator has been unable to make a decision from the occasional debris reading. It requires a data logger (human, mechanical or electronic), but that is no great disadvantage. Most engineers are happy with a certain amount of discomfort, and will have no qualms with such an extra. Provided the figures are ultimately contained on one page, preferably with some sort of graphical representation to highlight that 'vivid' change he seeks, the operator will be more than happy.

It is still possible that the operator will not be able to be sure that a change is really significant or not. This will improve with his experience. However, when a trend is monitored by computer, there is usually no problem at all in deciding the critical point — the danger signs of imminent failure. Someone, of reasonable intelligence and experience, has programmed the computer and the operator need only press the buttons after attaching the appropriate sensors.

Multi-Trends
The idea of trend monitoring could be considered as not just recording and noticing a significant upswing in the level of one parameter, but in observing a trend away from an acceptance level which consists of a complex signature, which has previously been identified as 'acceptable'.

In this case a 'significant' deviation from this signature would indicate a possible fault. It is a sort of trend, but is much more because it looks at many actual features at the same time. It is a 'Multi-Trend'.

Multi-Trend monitoring is worth considering. Its primary use has been in vibration and acoustic patterns where it looks at frequencies associated with different components. It is the trend monitoring of complex signals, able to detect perhaps only one variable when it occurs. It involves the significant trend away from some acceptable complex 'signature' indicating the point of no return when the machinery will break down. This is complex! It does require a computer with a sophisticated program, preferably one which is capable of learning or up-dating itself.

Some early work on this was done by Curtiss Wright on Aero engines in the 1960's, using sonic analysis. It involved one or more microphones listening to an aero engine running at a set speed, initially idle. The magnitude of signal levels against the individual frequencies of vibration, when new, were compared with later spectra after many hours service. A deviation in the height of the individual 'spikes' signalled a change which, depending on the calculated individual frequencies expected from gears and bearings, could be identified with an actual component. The idea was excellent and some successes were recorded. However, the aero engine is highly complex and although every effort was made to only test under identical conditions, changes in temperature and fit, became as significant as any change due to wear or malfunction.

The example of the aero engine vibration serves as a warning on multi-trend monitoring — beware of the complicated mechanisms! Be very careful to consider what other features may influence the different factors being measured.

It might be thought that such complex trend monitoring, the multi-trend, would have been only useful with vibration signals. But this is not true — it is also being applied to debris analysis. Century Oils CENT programme, for instance, involves the regular sample by sample trend analysis of a number of different parameters, such as water content, viscosity, but also the different elemental contents of the solid debris. A change in the pattern of this debris would be seen as not only simply significant but it would also lead towards identifying what is actually the problem, as shown by the following example (Fig. 12.2). It is apparent in this hypothetical example that something was going wrong

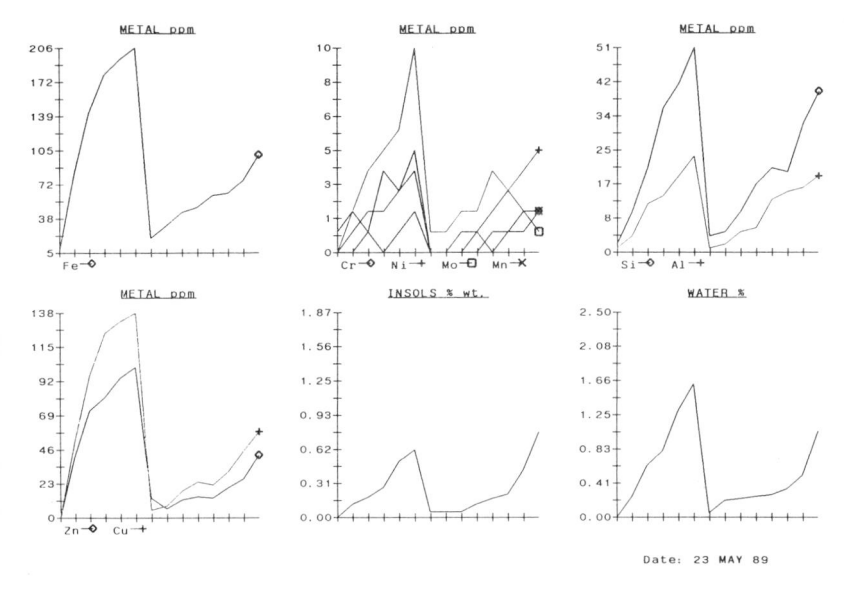

Fig. 12.2 CENT example of trend indicating fault

half way through the time scale shown; this was then fully corrected, but then continued to become a problem in all aspects! (If this were true, then there would probably be more than one fault, eg. a problem with air filtration, severe wear and leakage.)

Another oil analysis laboratory, SGS Redwood, have a similar approach. The following (Fig. 12.3) is an example relating to a long term bus engine. The four plots show a trend from

CAUTION — NORMAL — NORMAL — CAUTION — CRITICAL.

And the four features plotted are the Sooty Insolubles (% weight), Contaminants (considered silicon and sodium, ppm), Top End Wear (considered iron, aluminium and chromium, ppm) and the Bottom End Wear (considered copper, lead and tin, ppm).

The diagnosis SGS Redwood made was along the lines of suggesting that there was an increasing abrasive contaminant in the oil which would cause excessive wear, actual top-end (piston/liner/ring) and bearing wear was also increasing. The action proposed was three-fold

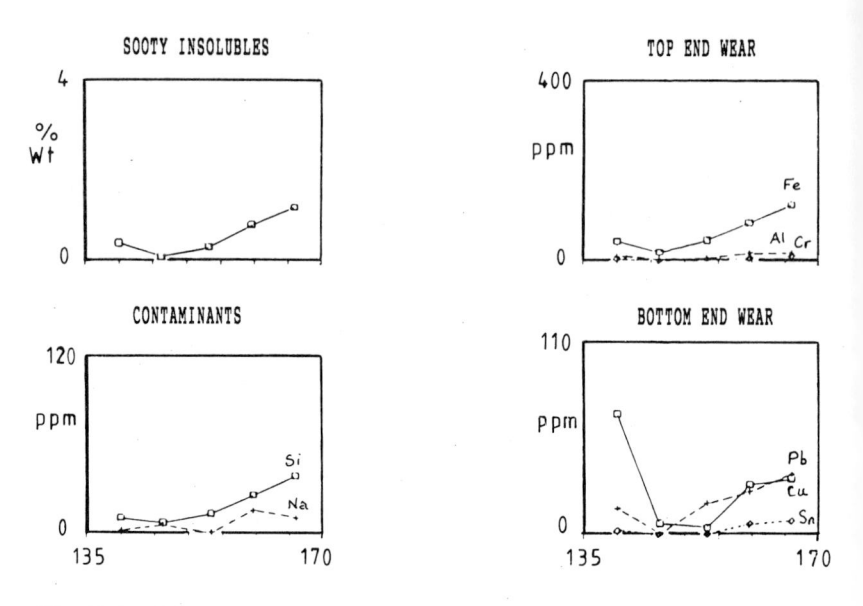

Fig. 12.3 Oil monitoring of a bus engine (thousands of miles) (SGS Redwood)

    Change lubricant
    Check for leaks in air intake system
    Submit a new sample at a reduced interval.

All these suggestions were only possible because several trends had been plotted for many thousands of miles. Samples had regularly been submitted and analysis was proving valuable.

CONTAMINATION CONTROL STANDARDS

Hydraulic Fluid Contamination Standards
Particles in liquids have long dogged the hydraulics engineer. In the early days they did not matter too much because the clearances between moving components were large. But once the fluid changed from water to oil and, with the change, a much higher pressure and closer tolerances, it did matter. Perhaps at first the operator knew nothing about them, although he was seeing their result in the premature breakdown of his equipment. Gradually education has spread to the fluid power industry through such institutions as the **BHR Group**, with its series of Condition Monitoring and Contamination Control conferences, and the

Bath Fluid Power Centre, with its courses on, amongst other subjects, contamination in hydraulic systems; the British Fluid Power Association has issued official Guidelines on the contamination levels which can be accepted in industry (BFPA P5/1991) and how those levels can be achieved. Chapter 4, of course, went into this in great detail.

Once it was known that particles affected the life of components then official Standards became necessary. They have been around a long time in the United States and in the aero-industry. Two which are now widely used around the world are NAS 1638 and ISO 4406 (Figures 12.4 and 12.5). NAS 1638 examines the range of sized particles up to 100 μm, ISO 4406 currently just looks at the 5 μm and 15 μm counts. (5 μm representing the silt level in the liquid and 15 μm the larger wear and blocking particles.) Three others which are quite extensively used are the SAE, DEF-STAN and the Conpar classes (Figures 12.6, 12.7 and 12.8). The graphs are scaled Log/Log^2 which tends to give a straight line with some naturally found distributions like ACFTD.

The numbers associated with these graphical charts are given above each figure. It should be pointed out that all the Standards, apart from the ISO 4406, quote numbers per 100 mL. The ISO committee chose 'per 1 mL'. However, in order to make comparison easier, the numbers in the tables are all given 'per 100 mL and on the graphs 'per 1 mL'. The class is chosen for counts 'up to and including' the number quoted; a higher number places the contamination level into a higher class.

The NAS 1638 standard has been extended by the Aerospace Standard AS4059, to include a lower class, 000, and counts down to 2 μm instead of the previous lowest of 5 μm. See Fig. 12.4 on page 278.

The ISO 4406 is quoted as two numbers (maybe three). An example of the two numbers would be 18/14 where the first number, 18, refers to the class of 5 μm particles and the second number, 14, to the class of 15 μm numbers. (If three numbers are quoted the first could be the class relating to the 2 μm particles.) The numbers connected to the classes are shown in Fig. 12.5 on page 279.

The defence standard DEF STAN uses the counts at 15 μm per 100 mL to identify the different classes. Thus Class 15000 has 15000 particles greater than 15 μm per 100 mL. See Fig. 12.7.

These are recognised standards for stating hydraulic fluid cleanliness. As stated in Chapter 4 the level which is chosen as 'acceptable' varies from industry to industry. The mechanisms which have very fine clearances,

Class μm	000	00	0	1	2	3	4	5	6	7	8	9	10	11	12
> 2	164	328	656	1310	2620	5250	10500	21000	42000	83900	168k	336k	671k	1·34M	2·69M
> 5	76	152	304	609	1220	2430	4860	9730	19500	38900	77900	156 k	311k	623k	1·25M
> 15	14	27	54	109	217	432	864	1730	3460	6920	13900	27700	55400	111k	222k
> 25	3	5	10	20	39	76	152	306	612	1220	2450	4900	9800	19600	39200
> 50	1	1	2	4	7	13	26	53	106	212	424	848	1700	3390	6780
> 100	0	0	0	1	1	2	4	8	16	32	64	128	256	512	1024

NAS 1638 Table modified by AS4059 – see Fig. 12.4 – per 100 mL.

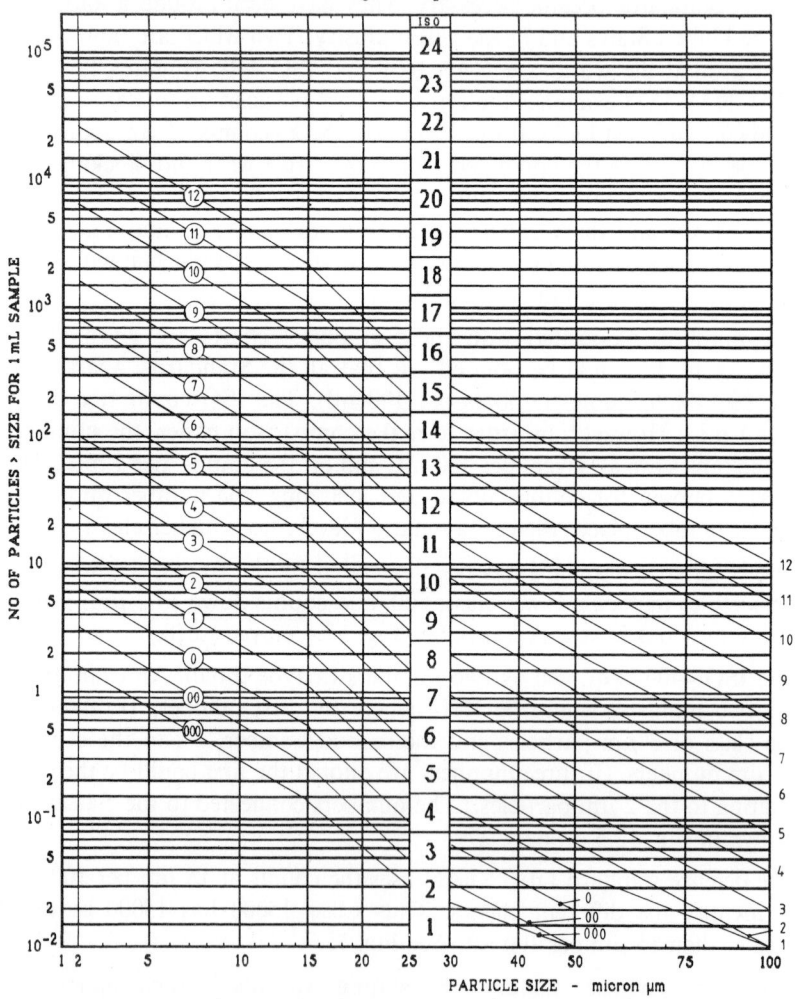

Fig. 12.4 The NAS 1638 Contamination Level Classses for solid particle modified by AS4059

1	2	3	4	5	6	7	8	9	10	11	12	13	14	15	16	17	18	19	20	21	22	23
2	4	8	16	32	64	130	250	500	1k	2k	4k	8k	16k	32k	64k	130k	250k	500k	1M	2M	4M	8M

ISO 4406 Table (modified to 'per 100 mL') – see Fig. 12.5.

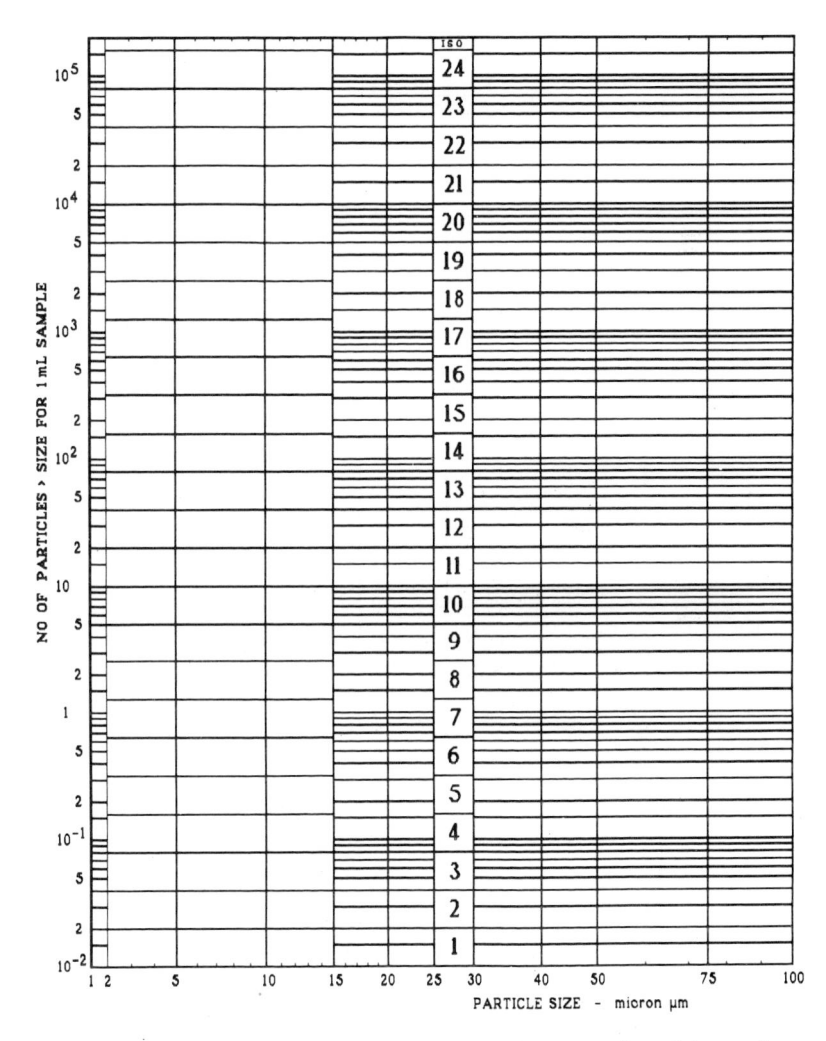

Fig. 12.5 *The ISO 4406 Contamination Level Classses for solid particles*

Class μm	0	1	2	3	4	5	6
> 5	3480	6180	12820	30300	44500	112000	177600
> 10	780	1581	3120	6260	12460	25000	49600
> 25	110	241	441	901	1756	3600	7590
> 50	17	31	61	121	246	471	1092
> 100	1	3	5	11	21	41	92

SAE 749D Table – see Fig. 12.6 – per 100 mL.

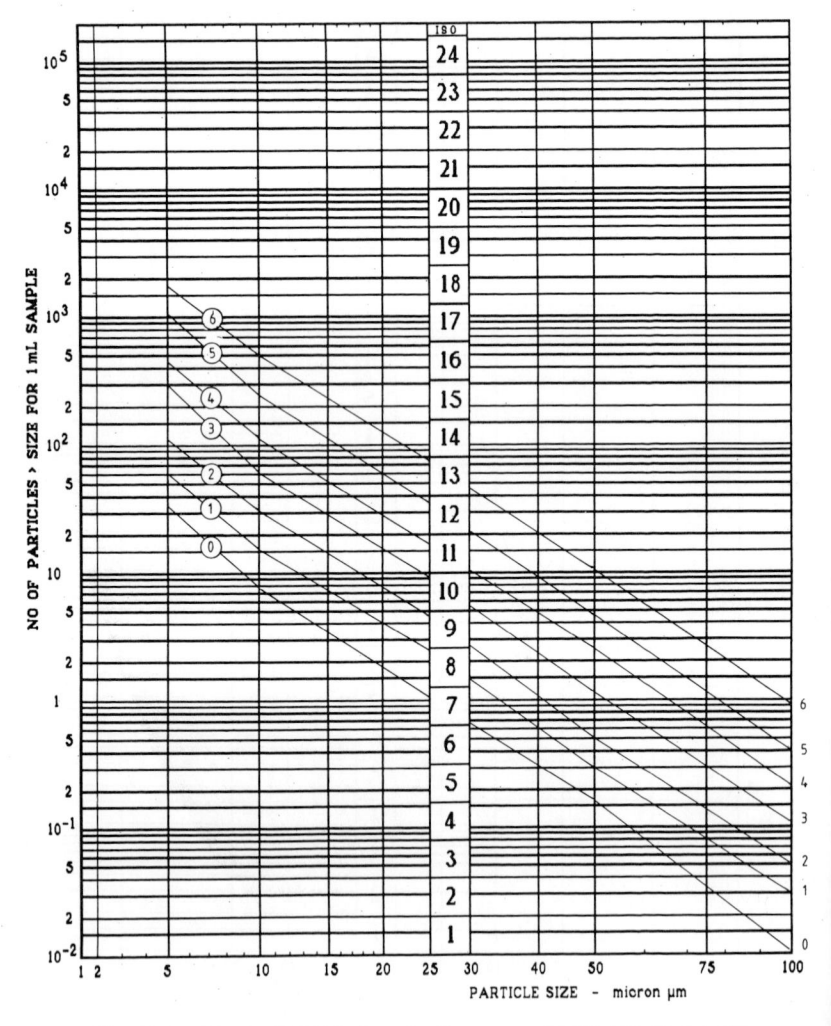

Fig. 12.6 The SAE 749D Standard (same as NAV.AIR 10-1A-17)

Class μm	400	800	1300	2000	4400	6300	15000	21000	10000
> 5	20000	39000	81000	190000	480000	741000			
> 10	1900	3800	6300	12000	33000	41000	105000	171000	
> 15	400	800	1300	2000	4400	6300	15000	21000	100000
> 25	100	180	300	530	950	1260	2600	4600	15000
> 50	20	30	50	80	120	160	300	600	1000
> 100	5	7	12	20	32	50	80	140	200

DEF STAN 05-42 Table – see Fig. 12.7 – per 100 mL.

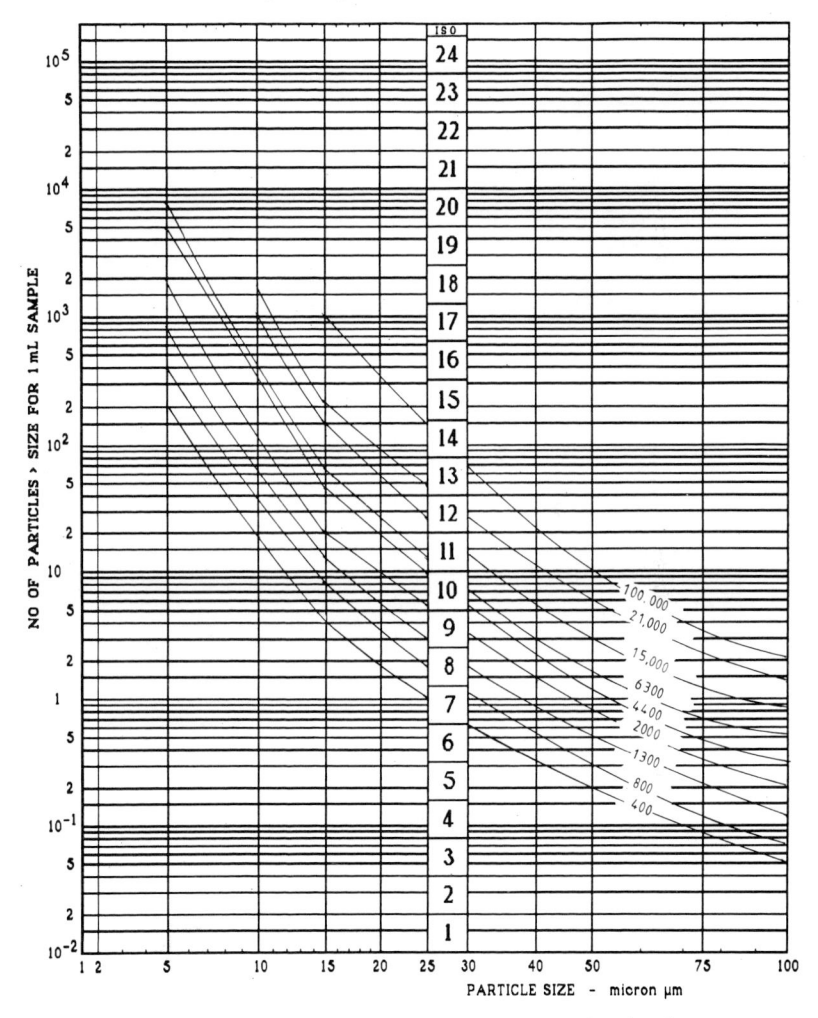

Fig. 12.7 The Defence Standard 05-42 Classes for cleanliness

Class μm	1-H	2-H	3-H	4-H	5-H	6-H	7-H	8-H	9-H
> 1	33600	445000	915000	3·02M	9·05M	15·04M	18·05M	24·10M	29·20M
> 5	2720	24600	54400	179600	484000	763000	915000	1·22M	1·525M
> 10	1107	2310	4360	14390	33600	48800	58100	77100	96100
> 15	406	496	684	2260	4430	6610	7700	9880	12060
> 25	116	126	155	512	950	1391	1609	2047	2488
> 50	22	22	22	73	121	171	194	242	292
> 100	7	7	7	22	32	43	47	57	68

CONPAR Table – see Fig. 12.8 – per 100 mL.

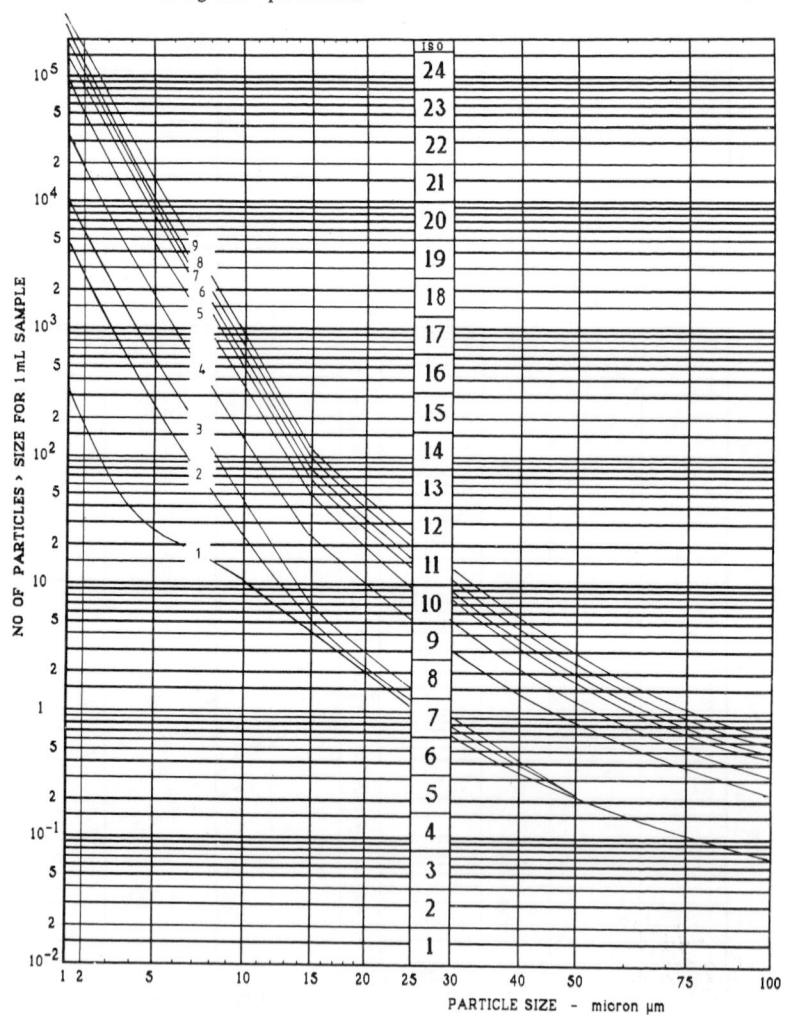

Fig. 12.8 The Howden Wade 'Conpar' Classes for cleanliness

require a lower class (less particles allowed) and those with heavy machinery with large dimensional tolerances, only need the higher classes which allow for more and bigger particles. As an example of the type of acceptable levels, the following may be taken:

ISO 4406	NAS 1638	ACCEPTABLE for
21/14	11 or 12	General Industrial
18/13	8 or 9	Transmissions
16/11	6 or 7	Servo Control
15/9	5 or 6	Critical Control

CONPAR	DEF STAN	NAS 1638	S.A.E. (NAV AIR)	I.S.O.
		00		8/5
		0		9/6
1H(G)		1		10/7
		2		11/8
1H(R)		3	0	12/9
2H(G)				13/7
		4	1	13/10
		5	2	14/11
3H(G)				15/8
2H(R)	400			15/9
		6	3	15/12
3H(R)	800			16/10
4H(G)	800			16/10
		7	4	16/13
	1300			17/11
		8	5	17/14
5H(G)				18/11
4H(R)	2000			18/12
		9	6	18/15
6H(G)				19/12
7H(G)				19/12
5H(R)	4400			19/13
		10		19/16
6H(R)	6300			20/13
7H(R)	6300			20/13
8H(G)	6300			20/13
9H(G)	6300			20/13
		11		20/17
8H(R)				21/14
9H(R)				21/14
		12		21/18
	15000			22/14
	21000			22/15
	100000			23/17

Fig. 12.9 *Correlation of hydraulic Standards (Baker 1991)*

Note that there is no absolute direct relationship between the different Standards, particularly because they are based on different calibration procedures (ie. latex spheres, ACFTD and direct microscope visual counts on real particles). However, Fig. 12.9 from Baker 1991 gives a reasonable overall comparative view.

These standards are trying to state an acceptable level of cleanliness of fluid to enable the fluid power equipment to have the reliability that is expected of it. (It has been found that the major cause of unreliability in such systems is the presence of damaging particles.)

Lubricating Oil Contamination Standards

The level of debris in lubricating fluids is not normally given a standard. Historically this is because the clearances are much larger than in refined hydraulic equipment and also because there is other more serious contamination in the liquid. Water, fuel dilution, oxides, acids are all present to some degree, and when they reach a certain level, instead of the oil giving protection, it provides a means of destruction!

However, there have been marked strides in rectifying the situation regarding contaminant levels. The following are examples of published literature:

> Holmes (1985) describes oil analysis where the ppm of various elements has indicated excessive wearing; but this was off-line and time consuming. Gravimetric levels (ie. the total quantity of the agglomeration of insolubles, rather than the particle count) have been measured for many years using off-line glassware like that provide by Millipore; it is more rapid, although less accurate, but does give results (Foulds 1985). Bearing manufacturers, in particular, have noted the disastrous effect of foreign solid debris (Harvey & Herraty 1989). They now suggest certain filtration for really long life using their bearings (they have even gone so far as to suggest an infinite life is possible if the filtration is good enough!)

But each of these still lacks that final seal of approval given by a Standard. Hopefully, it cannot be long before Standards become as common place as is rapidly becoming the case in fluid power.

CONDITION MONITORING STANDARDS

So far we have only looked at standards which are accepted in industry for contamination control. In a sense these standards are PREVENTION standards; they prevent faults developing provided there is a dedicated adherence to the standard. This is not the standard that is suggested for wear debris. For wear debris we need a DETECTION standard. We need to detect the presence of particles which indicate a fault has happened or a failure is soon to occur.

The point being made, though, is that standards are known to be of considerable value in the prevention of accidents, simply because they are absolute values. If a product fails the standard, it is rejected. If a product meets the standard it is accepted. No discussion, no debate, no mistake.

To return to the simpler single function trend, and repeat some thoughts made earlier. How large a change in the slope of the trend has to be shown before one can say that the level of debris being generated indicates a major fault, and the stop button should be pressed? Who makes the decision? What is the penalty if it is wrong? Is it better to do nothing and receive a cursory comment, or do something and maybe be severely reprimanded? This is where the fixed or absolute level is really valuable, because so many of these questions then no longer apply.

The proposal is that there should be a critical level, a 'go/no-go' level, a trigger point, a standard.

Undoubtedly, the detection of that significant upswing in the trend is the earliest sign of excessive wear developing, and, provided it is genuine, it is the best monitoring method. But is there not a case for a less esoteric method, which the unskilled engineer can use with confidence? Is not the 'absolute' the answer?

After all, are we not familiar with using absolutes in other engineering. Not just dimensional measurements (which may even have an National Physical Laboratory standard at a mean temperature and pressure), but such commonplace parameters as flow, temperature and pressure with their absolute gauges. Even the more exotic parameters like light, magnetism and radioactivity have their gauges. Why not wear debris? To be able to monitor to an absolute level is far superior in that the level is preset and the measured level is either acceptable or it is not — there is no last minute debate.

The absolute particle level is certainly worth considering. As we saw above when talking about contamination control, it is used in the fluid power industry and helpful in lubrication, too. Decisions are made in all sorts of engineering because an absolute level has been reached — the valve which releases at a set pressure, the fuse which blows at a fixed current, the solder which melts at the pre-agreed temperature, the tie which breaks at a known load — these are absolutes which we readily accept. Great comfort can be taken by the maintenance engineer, if he knows that all is well providing Debris Class X, or whatever, is not exceeded.

Trend Monitoring undoubtedly carries a bonus with its ability to pictorially display what is happening, but the absolute may be easier to use. The absolute may, in fact, be all that is necessary in some cases. It is simple. But the inventions of the past, which have been so successful, have so often been the most simple. The best condition monitoring will be the simplest, for two very good reasons —

It is less likely to go wrong itself
It is less likely to be misunderstood.

The absolute value is much less likely to be misunderstood, and, provided it is truly related to the system in hand, it will be much more reliable because it does not depend on human intervention.

Although I am not aware of any internationally recognised Wear Debris Standards, there is nothing to stop a company experimenting with their use. The actual values would again depend on factors similar to the contamination control criteria, ie. the fineness of the engineering process, and the seriousness of failure. For instance, if even a slight drop in efficiency is going to cause distress in the board room, then a much lower debris level will be the critical one. If however, it is only a major breakdown which will raise eyebrows, then there is no need to take action until the debris level is much higher.

Kohlhaas 1991, discussing hydraulic oil, deals with the debris in the oil as a measure of the condition of the machinery, ie. in aero-engines. He suggests that four features in the oil can be monitored with absolute standards, ie. viscosity, water, chlorinated solvent and 'dirt'. The 'dirt' is the wear debris. He states the combat aircraft as having an acceptable level of NAS 5 or 6, but used oil could actually be at NAS 8. The STANAG 3510 standard (2nd issue) quotes an acceptability level for used oil as NAS 5 between 15 μm and 100 μm, but allows a higher value above and below these sizes. The German aircraft requirement is NAS 6 throughout the size range and Panavia is NAS 7.

In a new machine monitoring programme, the following could be tried as a starting point:

ISO 4406	NAS 1638	ACCEPTABLE for
21/18	12	Heavy Industrial
19/16	10	Light/med. Transmissions
18/14	8 or 9	Critical Jacks
16/13	7	Control Mechanisms
15/12	6	Long life precision
14/11	5	Critical precision

But this is only a trial suggestion. We need to consider more deeply what actually happens during the lifetime of lubricated mechanical system.

The Bath-Tub criterion

Wear debris generation, over the life of a machine, is often illustrated by the 'bath tub' curve (named after the shape of the old fashioned bath tub that used to be found in every home). Basically there are three regions of wear in the life period portrayed by the 'tub' (Fig. 12.10) —

> There is the initial settling down, the running-in period, when debris is released as the asperities and high spots are knocked off and a smoother action takes place. This wear is usually beneficial and provides an improvement for later running. This is particularly true for gears and bearings.

> The second, hopefully much longer, period is the useful working life of the component. The debris release is quite small and insignificant in size and quantity. Lubrication is doing a good job and no disasters are impending.

> The final stage is the disaster stage, the death throws of the component as it endeavours to last a little longer! At this stage fatigue begins to show with the resulting large chunks of debris; the quantity of debris increasing exponentially.

Now if these three regions of life are considered, the standards to be associated with them will be quite different. It is as if it is necessary to have three separate Standards. But there is a problem — when does one

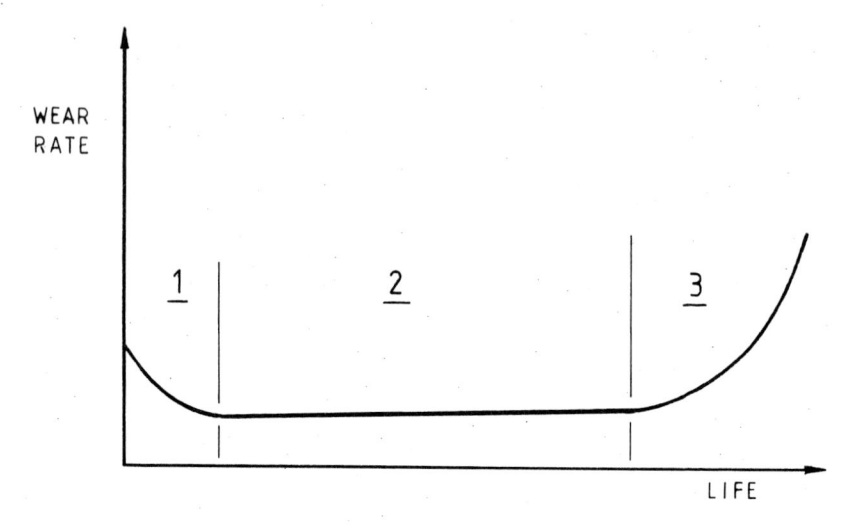

Fig. 12.10 The Bath Tub wear curve

know that a progression has been made into the next region? An experienced engineer is required here, and he may well suggest that only one standard is needed because he can sense when stage 1 moves into stage 2; all he wants to check is when the final stage is reached.

The opening running-in period will initially show some erratic behaviour, followed by considerable small size wear. This, in a sense, does not need monitoring except to ensure that the mechanism is allowed to run on in order to run in. But then, as stage 2 commences, the absolute setting needs to be chosen. It needs to check for Stage 3 because that is the trouble period. So although the checking is occurring in the low level OK region it is actually looking out for the start of stage 3.

If the above standards of acceptability, or something similar, can become vogue, then it is a short step be able to fit an absolute oil debris monitor.

How the monitor would work is by constantly checking the machinery to ensure that the debris level is below the Standard (by suitable filtration), but when that level is exceeded by an agreed amount (although the filtration is still present and operating) then the monitor must signal a problem has arisen. For optimum results the monitor must be fitted between the units requiring monitoring and the filtration, but, otherwise, there is little other complication.

In practice this is already happening, without the official approval of British Standards or ISO. Companies are finding the acceptable level of particle contamination, and then agreeing that the level may rise, say, two classes of ISO 4406, before shut down is actuated. It could conceivably be due to the filtration breaking down, rather than the machinery, but either way a disaster is impending. (Wilkes & Hunt 1990 give an example of this in Chapter 14).

CALIBRATION

Before listing some relevant standards, it is important to understand something of what is happening as regards calibration. After all, if we have an instrument which displays a certain standard level, how can we be sure it is correct? Is some kind of calibration necessary? Do I have a choice in which calibration I use?

National and international standards invariably discuss calibration in their documents. They may well state a 'preferred' calibration, followed by a secondary, and maybe a third possibility. In other words, although it is recognised by the learned bodies which produced the standards that one type of calibration is best, the same bodies are sufficiently aware that the user may not be able to manage the best. Take for instance the BS 3406: Part 7: 'Recommendations for single particle light interaction methods'. Here the 'primary calibration' is that of using "particles of the material under test suspended in the test fluid" (the particles having been independently sized in a way which is traceable to national standards of mass and length). This would be incredibly difficult. A 'preferred secondary calibration' uses latex spheres dispersed in water. An 'alternative secondary calibration' uses ACFTD powder mixed in some suspension fluid. Both of these latter two are feasible.

However, because there are different methods which are acceptable to the standard, it is necessary to actually quote which method of calibration has been used — BECAUSE THEY GIVE DIFFERENT ANSWERS! It is well known that a light obscuration instrument calibrated with latex spheres will give different results, to if it is calibrated with ACFTD (size change 1·54 : 1·00). There is also a difference depending on

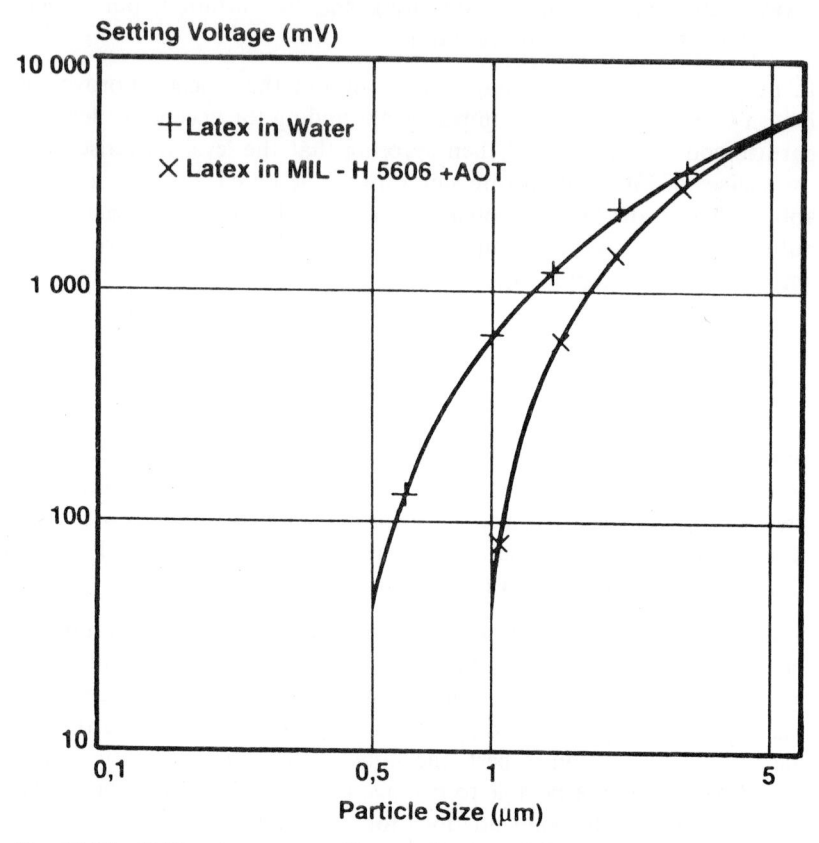

Fig. 12.11 *Calibration curves of latex spheres in different media (Day & Tumbrink 1991)*

the media in which latex spheres are suspended; Day & Tumbrink 1991 illustrate this for particles below 5 μm (see Fig. 12.11). It has been suggested that if a hybrid calibration fluid is used (ie. latex spheres in oil) a lot of the problems with differences might disappear. Be that as it may, for the moment it must be repeated that it is important to always say how the instrument is calibrated.

Some instruments have built-in calibration procedures. Be careful! These may only refer to electronic tests, which, although important and helpful, do not fully provide a total calibration check. There is no substitute for an actual fluid containing 'known' debris being passed through an instrument, and seeing what the instrument displays. There may be no means of changing the display, but at least there will be a confidence (or otherwise) that what is being displayed in the real tests is likely to be correct for that instrument, or else within a certain percentage of the truth. (A cautionary note must be added, that the way the 'known' debris was assessed must also be stated.)

Remember that how one instrument measures a particle is not the same as another. You cannot necessarily correlate one instrument with another of different type — it may correlate with one type of debris shape, but be totally different when the debris changes.

Calibration Particles
Reference has already been made to ACFTD and latex spheres as suitable particles for calibration of particle counters. Although spheres are carefully sized with their deviation from the mean quoted, the 'dust' is less precise. ACFTD is available from AC Spark Plug Division of General Motors, USA or AC Rochester, Southampton, UK, and the quoted percentage variations in size versus percentage weight (numbers) of the as-supplied ACFTD is as follows:

0–5 μm	39 ± 2%	ie.	37%–41%
5–10 μm	18 ± 3%		15%–21%
10–20 μm	16 ± 3%		13%–19%
20–40 μm	18 ± 3%		15%–21%
40–80 μm	9 ± 3%		6%–12%

Which means that in the larger sizes, for instance, we could note that the count can be up to ± 30% in error! And that is the standard! However, in terms of the logarithmic plotting used for particle counting, this need not cause too much worry. The actual particle count associated with this distribution of ACFTD is shown below (based on Fitch 1979). It shows the mean number of particles per 1 mL sample when the calibration liquid has been seeded with 1mg/L of ACFTD (numbers given to four significant figures):

SIZE (μm)	MEAN NUMBER	SIZE (μm)	MEAN NUMBER
1	1752	51	1.199
2	1397	52	1.115
3	991.8	53	1.039
4	708.1	54	0.9682
5	516.7	55	0.9035
6	385.7	56	0.8439
7	294.0	57	0.7890
8	228.2	58	0.7383
9	180.0	59	0.6914
10	143.9	60	0.6481
11	116.5	61	0.6079
12	95.37	62	0.5707
13	78.82	63	0.5362
14	65.72	64	0.5042
15	55.23	65	0.4744
16	46.78	66	0.4467
17	39.83	67	0.4209
18	34.14	68	0.3968
19	29.42	69	0.3744
20	25.48	70	0.3535
21	22.18	71	0.3339
22	19.39	72	0.3156
23	17.01	73	0.2985
24	14.99	74	0.2825
25	13.25	75	0.2675
26	12.76	76	0.2534
27	10.46	77	0.2403
28	9.340	78	0.2279
29	8.360	79	0.2162
30	7.503	80	0.2053
31	6.751	81	0.1950
32	6.089	82	0.1853
33	5.504	83	0.1762
34	4.986	84	0.1676
35	4.526	85	0.1595
36	4.117	86	0.1519
37	3.751	87	0.1447
38	3.425	88	0.1379
39	3.132	89	0.1315
40	2.869	90	0.1254
41	2.632	91	0.1196
42	2.418	92	0.1142
43	2.225	93	0.1090
44	2.051	94	0.1042
45	1.892	95	0.0995
46	1.748	96	0.0952
47	1.617	97	0.0910
48	1.498	98	0.0871
49	1.389	99	0.0833
50	1.290	100	0.0798

Using such dust requires very careful preparation. To weigh out the dust and just add it to the required quantity of liquid, just does not work. The dust remains together, and forms large agglomerations of particulate which will give extremely erroneous results. The correct process is that of adding a small quantity (few mL) of the liquid to the weighed ACFTD, and gently mixing it with a spatula on a non-shedding surface. A little more liquid is added, and the process repeated. Finally the whole concentrated mix is washed by the remaining liquid into a container. Additional mixing using shaking and ultrasonics then enables the calibration fluid to be fully mixed and available for test. (A word of warning about using commercial shakers; quite often they do virtually nothing if they only have a circular motion! Hand shaking is far superior because the complex routine in more than one plane provides thorough mixing.)

Latex spheres are available from a number of sources. Appendix 2 includes the following:

>Advanced Polymer Systems (USA)
>Brookhaven Instruments Limited (UK)
>Duke Scientific Corporation (USA)
>Dyno Particles A.S. (Norway)
>Metachem Diagnostics Limited (UK)
>Seradyn Inc (USA)

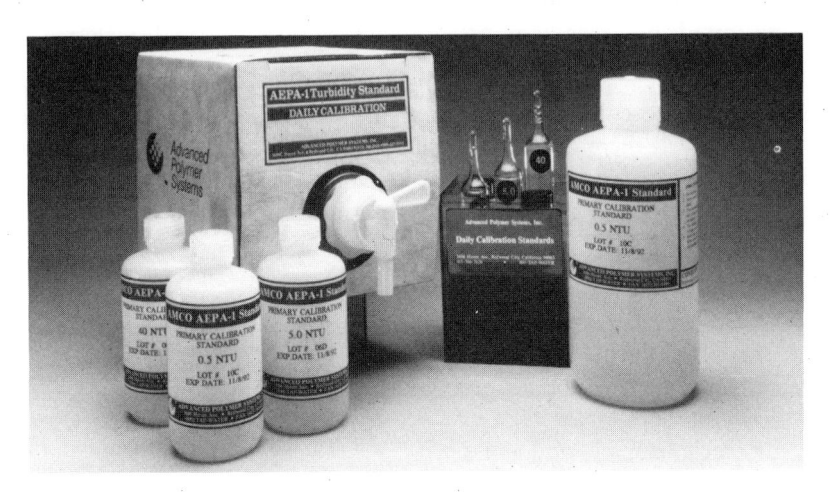

Fig. 12.12 Different supply methods for spheres (Advanced Polymer Systems)

Fig. 12.13 *Typical latex spheres (Dyno Particles)*

The spheres available for calibration are mainly a mixture of styrene and divinylbenzene monomers that are polymerized in an aqueous medium (ie. polymer microspheres in water) (see West 1990). They are called a number of different names, such as, PSL particles, beads, latex particles, latex spheres, polymer suspensions, polystyrene particles, polystyrene spheres. (Other types are glass spheres and pollens.) The latex spheres are hydrophobic and hence require a wetting agent before being mixed; however, they are normally supplied in small vials ready mixed in a liquid.

Spheres can be coloured or translucent. The liquid in which they are suspended will therefore need to be carefully controlled if problems relating to optical variations are to be avoided, as shown in Fig. 12.11 above. The types of instruments which are regularly calibrated by latex spheres are:

    Blood analysers
    Colorimeters
    Hazemeters
    Light scattering and absorbing instruments
    Microscopes
    Particle counters and sizers
    Spectrophotometers
    Turbidimeters

LIST OF SOME RELEVANT STANDARDS

ASTM E 11:87	Wire mesh sieve sizes.
ASTM E 20:85	Particle size analysis of particulate substances in the range 0·2 to 75 micrometres.
ASTM E 161:70	Precision electroformed sieves.
ASTM E 323:80	Perforated plate sieve sizes.
ASTM F 312	Particle shape index by microscopical examination on membrane filters.
ASTM F 321:72	Automatic particle counter size setting.
ASTM F 651:80	Standard method for particle counter single-point calibration by the median method.
ASTM F 658:87	Standard practice for defining size calibration, resolution and counting accuracy of a liquid-borne particle counter using near-monodisperse spherical particulate material. [In an aqueous solution.]
ASTM F 660:83	Standard practice for comparing particle size in the use of alternative types of particle counter.
BS 410	Standard sieve sizes.
BS 3406	Methods for the determination of particle size distribution (abbreviated titles): Part 1: Powder sampling. Part 2: Liquid sedimentation methods (gravitational). Part 3: Air elutriation methods. Part 4: Optical microscope method. Part 5: Electrical sensing zone. Part 6: Liquid sedimentation methods (centrifugal). Part 7: Single particle light interaction methods. (Details on all the above are given in Lines 1985; the author has kindly updated his paper and provided more recent information for inclusion in this section of the book.)

BS 5540	Evaluating particulate contamination of hydraulic fluids: Part 1: Qualifying and controlling of cleaning methods for sample containers. Part 2: Method of calibrating liquid automatic particle count machines (using ACFTD). [Same as ISO 4402.] Part 3: Methods of bottling fluid samples. Part 4: Method of defining levels of contamination (solid contamination code). Part 5: Method of reporting contamination analysis data. Part 6: Method of calibrating liquid automatic particle count instruments (using mono-sized latex spheres).
DEF-STAN 05-42	Particulate contamination classes for fluids in hydraulic systems.
DEF-STAN 05-43	Standard procedures for taking samples of hydraulic fluids for evaluation of particulate contamination.
DEF-STAN 05-46	Determination of particulate matter in hydraulic fluids using an automatic particle size analyser employing the light interruption principle.
DIN 66 165 1983	Sieve analysis Sieving procedure Photoelectric sedimentation classifier Photometric methods Sampling and sample preparations
ISO 565	International test sieve sizes.
ISO 3310	Sieve sizes: Part 1: Wire mesh series. Part 2: Perforated plate series.
ISO 3722	Hydraulic fluid power — Fluid sample containers — Qualifying and controlling cleaning methods.
ISO 3938	Hydraulic fluid power — Contamination analysis — Methods for reporting analysis data.

ISO 4021	Hydraulic fluid power — Particulate contamination analysis — Extraction of samples from an operating system.
ISO 4402	Hydraulic fluid power — Calibration of liquid automatic particle count instruments — Method using ACFTD contaminant.
ISO 4405	Hydraulic fluid power — Fluid contamination — Determination of particle contamination by the gravimetric method.
ISO 4406	Hydraulic fluid power — Method for coding level of contamination by solid particles.
ISO 4407	Hydraulic fluid power — Fluids — Determination of solid particle contamination — Counting method using a microscope.
ISO 4783	Industrial wire screens and woven wire cloth — Guide to the choice of aperture size and wire diameter combinations Part 1 — Generalities Part 2 — Preferred combinations for woven wire cloth.
ISO 5884	Aerospace — Fluid systems and components — Methods for measuring the solid particle contamination of hydraulic fluids.
ISO 9044	Industrial woven wire cloth — Technical requirements and testing.
NAS 1638	Cleanliness requirements of parts used in hydraulic systems.
SAE AS 4059	Aerospace-cleanliness classification for hydraulic fluids.
SAE ARP 1192	Aerospace Recommended Practice — Procedure for calibration and verification of a liquid-borne particle counter. [Latex spheres in an aqueous solution 1987.]
STANAG 3713	Determination of particulate matter in aerospace hydraulic fluids using a particle size analyser.

CHAPTER 13

Examples of Successful Experience

Financial success with debris monitoring, as with any condition monitoring, can normally only be measured in hind sight. If only I had .... But the failure has come, caught everyone unawares, and heads must roll.

This is a great shame, because so often a manager is not willing to include a monitor at the design stage, feeling that there is no evidence that it will achieve anything. After all, no designer is designing-in features to fail his system; if he were, then obviously a monitor could be seen as an essential corollary! The manager, therefore, sadly proceeds optimistically in the hope that no failure will occur and hence no monitor is necessary.

His head is in the sand!

There are no 100% reliable systems. The human being is not capable of ensuring total continuous 100% reliability in anything, let alone mechanical systems. This is because, to be gracious, he does not have full control on all the factors involved, he does not know what tomorrow will bring. Take, for example, illness, weather, blind spots, lapse of memory, freewill (for bad, as well as good) — these can all affect a machine at any stage — design, manufacture and final operation.

Machines will fail, but if that failure can be detected at an early stage, then at worst, financial loss will be minimized, and at best, nothing serious will happen. That is what real condition monitoring is all about. It is

- NOT Observing a failure has happened,
- NOT Noticing a problem which has gone so far that failure must occur.
- IT IS Detecting a change early enough, so that, instead of failure, corrective action can be taken.

The examples outlined in this chapter are of different categories of success, but they serve as an incentive to management to take that leap of faith — investing a small amount of money NOW (on monitoring) to save a great deal of money LATER (which otherwise would have been lost due to a failure).

It should also be added that no monitor is 100% reliable, either. The fact that sophisticated aircraft with multi-back-up computer facilities can still crash due to error, is a harrowing thought. But what must be emphasised is that the number of such crashes is nevertheless considerably reduced because of monitoring.

Although each example in this chapter relates to one particular application, it will be apparent that the success story could just as likely have been realised in a totally different field. It is pointless, therefore, to order these examples into the particular industries involved. Rather it is for the reader to sense the growing awareness of debris analysis and contamination control, and hence the examples are in chronological order.

There are also examples where different techniques have been used to see which method was the most effective. With these, one needs to think about the particular application very carefully. Is the kind of failure which occurred, the kind which might occur on my machine? If so then the findings could well be very appropriate.

However, before giving the detailed examples of success, two examples are given relating to why debris monitoring was not taken on. It highlights some of the built-in prejudices and lack of commitment and understanding that managers may still have because they have never really considered the real value of the subject, and hence do not examine the subject properly.

WHY DEBRIS MONITORING DID NOT WORK

Raubenheimer 1989 from Shell Research Limited reported that when lubricating oil monitoring was considered on one of their operations, it was completely rejected. However, this was not surprising, he said, because "the sampling of the oil for analysis was left to unskilled personnel without giving them precise instructions as to where the samples were to be drawn. They naturally took the samples from the most easily accessible points — which were test cocks on the bearing oil supply lines, downstream of pumps, coolers and high efficiency filters. Needless to say that the input data to the monitoring system was unrepresentative of the areas of interest, and therefore quite useless for a condition monitoring strategy." The management were convinced that there was no value in implementing such a scheme.

The story also goes of a fitter who was told to send a sample of his machine oil, for analysing every month. He was keen to show initiative and save money, so he drew off a large supply of oil at the beginning of the program, kept the oil in his room, and sent off a small sample from it every month as requested! (The absurdity of this was spotted by the personnel in the oil analysis laboratory; they could not understand the complete lack of any increase in debris.) I do not blame the fitter, I blame the manager!

EXAMPLES OF SUCCESSFUL DEBRIS MONITORING

Summerfield & Mathieson 1981
describe the use of the Debris Tester. They suggest its value for detection of flakes from ball-bearing failures and then illustrate the point with an example from a marine gas turbine (Fig. 13.1). It was estimated that the bearing would not have lasted more than another 50 hours before complete failure.

Yarrow 1982
from the Ministry of Defence at Gosport, reports on the value at that time of Magnetic Chip Detectors (MCD). He appreciated that various techniques overlap, especially in terms of their particle size response,

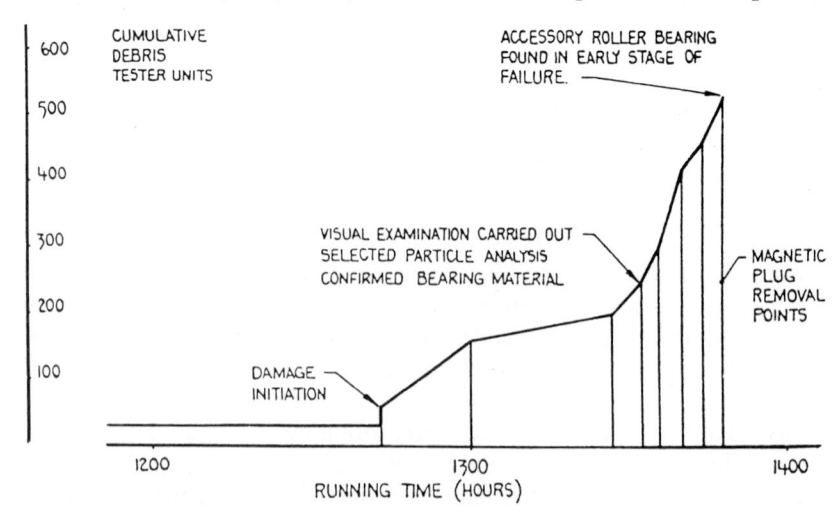

Fig. 13.1 Debris Tester used on a marine gas turbine test

and each has to be chosen for the machine in question. "Chip detectors are the basic monitoring method for military aircraft and have been used with considerable success." The MCD is placed in the scavenge line of the lubrication system, rather than in the sump, in order to only assess the current debris being generated. Although the electric chip detectors had been tested, they were not used due to the experience of a number of erroneous indications. This, however, was not a fault in the chip detector but rather because of confusion with the debris present — the Ministry of Defence only required certain metallic wear debris to be detected whereas the detector indicated the presence of built-in debris and other metallic non-critical debris, not surprisingly!

A comprehensive system of examination had been developed for the Ministry of Defence using the MCD to collect the debris, and an on-site Debris Tester to monitor the quantity of metallic wear debris present after each operational duty. (Strictly speaking, this is the Magnetic Chip COLLECTOR.) The debris was removed by means of adhesive tape after the MCD had been 'dried' by solvent. When the debris level was excessive, then the particular particles could be examined under a microscope for identification. Filter examination was also used. Many examples of significant debris being collected and identified are known, but there were some failure modes which were not being detected, such as fretting of keyways. (It should be noted that British Airways also operate a similar monitoring procedure on their civil aircraft.)

Morley 1982
illustrates the gradually improving situation in British Rail from the analysis of oils in 1982. At that time BR used more than 10,000,000 litres of oil per annum, not just from leakage, but primarily from oil changes. However, due to regular analysis of the oil the oil drain period had been extended from 30,000 miles to about 150,000 miles — a considerable saving. The analysis of the oil included viscosity, fuel dilution, water content and pH (and TBN if the pH was low), as well as elemental determination. British Rail sets limits on the chemical elements which are acceptable in new oil, eg. 10 ppm for Chromium, Aluminium and Copper, and then an absolute level above these is used as a warning, for instance 20 ppm for Aluminium would require a stop for examination.

The analysis programme had revealed a number of potential failures. However, coolant leakage (shown by an increase in Sodium in the oil) was much more common than component wear; liner sealing was the main problem. Detected mechanical faults had been a piston crown, main bearing and piston pin failures. British Rail had experienced only a small number of false indications or undetected faults as exampled by the following list for one period:

— 94 faults correctly detected
— 15 faults were missed (some were fractures)
— 6 indicated faults were not found on examination.

"The success rate for detecting wear is, therefore, good. If the spectrographic results are satisfactory the overhaul period for a specific engine may well be extended. the programme is also worthwhile in monetary terms giving a saving of £1,500,000 for a cost of £100,000."

Stafford-Allen 1982
from Cummins Engines at Daventry, describes a valuable analysis technique using X-ray Fluorescence. A monitoring programme is used in the production of the 4-stroke diesel engines (up to 1600 hp) using a method known as the 'mini-patch' system. Within the lubrication of the system, a large pore size fibre disc is inserted to retain the larger debris particles. These are removed at two different stages within the test cycle, and then examined by XRF to determine the levels of Fe, Ag, Cu, Al and Mn present. (A silver sub-surface layer is useful in bearings to indicate debris scoring of the surface.)

"A running average of the results is kept and deviations used to detect 'rogue' engines. In fact the limits are set below the acceptable level so that some good engines are stripped for examination. The strip-down rate is typically 5%. Faults that have been identified using this system include scuffed pistons, crank burrs and a cam follower seizure." The programme saved at that time £30,000 to £40,000 per year in warranty claims.

Lewis 1984
from the British Coal, describes a comparative test undertaken on a large British Coal gear box. The box was purposely run at excessive levels, both in load and in contamination of the oil, in order to cause failures. Eight different monitoring techniques were used throughout the six month test, over 2000 hours. The particular failure which would cause complete stoppage was the fatigue fracture of a reduction gear shaft. Six of the techniques showed a change, and these are reproduced graphically in Fig. 13.2, a to f; only the last 10 days readings are given.

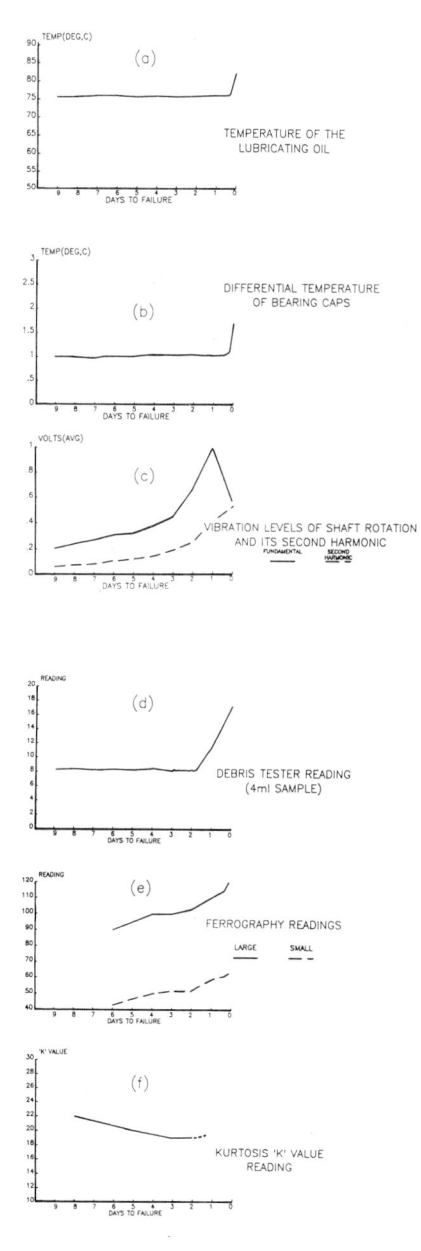

Fig. 13.2 *Comparative monitor readings for a gear box shaft failure (Lewis 1984, pp. 475, 476)*

Whilst the vibration monitor (a series of accelerometers) did show up a gradual change, it would require much experience and complex analysis equipment to identify at what point something important is happening. The ferrous wear techniques, however, were more discerning, although a saturation effect was caused on the ferrographic equipment at one stage which spoilt the sharpness of its rise.

It is interesting that Lewis was advocating a double check on debris monitoring, to see whether the high debris level was due to contamination in the oil or due to wear debris. He cites the example of a colliery where they got a rise in debris reading in two different oil systems. The oil was changed in both. However, whilst the debris reading then remained low in the one (a supply gate gearbox), the other one went back up again (haulage section). The haulage box was dismantled and the bearing on the bevel shaft assembly was found to be on the point of collapsing. In this case the debris picked up was bearing wear debris. The financial saving in otherwise lost production time, by this one prediction alone, paid for that mining region to finance the introduction of oil debris analysis in all its twelve pits.

Holmes 1985
from Leyland Vehicles Limited, was involved in a regular oil and engine condition checking programme during the manufacture of diesel engines. By using ferrography the company was able to change its running-in routine by being able to assess when sufficient controlled wear had occurred. In this way their annual overall cost of the programme was reduced by over £100,000.

Holmes also cites the value of oil analysis in fleet use. Two different operators overseas had sent samples on a regular basis from thousands of miles away. Without any need of visiting the users, it had been found that the servicing of fuel injection was faulty with one garage causing both extra wear metal and fuel dilution. Bringing the service up to standard enabled extended oil drain periods to become acceptable, with considerable financial saving.

Lewis 1986
of British Coal, published an extension to his two-level monitoring idea (first propounded in 1984). However, this time it involved different monitoring techniques rather than a repeat of the same one. Although the paper is primarily concerned with a variety of vibration monitoring

techniques, he does show that debris monitoring should also be used in two levels. He describes the British Coal use of taking oil samples regularly (weekly), of changing the oil if the debris content is found high (as in 1984), of rechecking by taking a sample and then analysing the oil more thoroughly if a high reading is recorded. This could take the form of sending the oil sample to a spectrometric laboratory for elemental analysis. Alternatively a combination with vibration or endoscopy (borescope) would be possible.

Lewis also lists nine separate gearbox faults which had been found on strip, the gearboxes having been brought out of service because of high debris readings.

Lloyd et al 1986

from the Central Electricity Generating Board, describe their experimental work on a 500 MW turbine-generator using, what was later called, the SDM 100. The sample oil was drawn from the 15,000 gallon oil tank, near the main oil scavenge return. XRF analysis of the patches produced indicated that it was possible to detect a change of as little as 1g in white metal content evenly dispersed in the whole system, between successive patches. The pressure curves were also able to detect these changes.

Astridge 1987

from Westland Helicopters Limited, gives a very comprehensive explanation of the monitoring of helicopter gear boxes, both in service and under test. The savings involved with such monitoring are considerable due to the human safety aspect as well as the aircraft costs. However, the addition of all but quite small monitors is not possible due to weight restrictions.

The type of monitoring being used in helicopters is still developing but the recommendations of Astridge relate to the QDM wear debris detector in conjunction with vibration monitoring. He stresses that no one monitor is able to detect all modes of possible failure; ie. a crack which does not release debris needs to be detected by enhanced vibration monitoring. However, the more common failure characteristics do release debris which may only successfully be detected by wear debris analysis.

Wyatt 1987

Oil debris condition monitoring in British Coal has been undertaken for many years, initially with ferrography and more lately with other techniques. In 1986 the Debris Tester provided some notable successes. Whilst a debris reading is logged, an additional microscope examination of the debris is always conducted when the levels are high. Over one period in North Yorkshire alone, 22 Debris Testers were used to monitor over 7000 oil samples. (Each 'sample' consisted of a 5 micron membrane filter through which 2 mL of the oil had been passed.) The results were —

> One gearbox was changed for consistently giving a high reading,
> 5 gearboxes had oil changes after showing a rise in reading, which resulted in an acceptable reading,
> 3 gearboxes had oil changes because of high coal debris levels,
> One gearbox gave a high reading which was associated with an unspecified external fault.

The savings could be described from the output and the cost of repair. Capital equipment, at the time, cost £3300 with consumables at £85 per week (ie. £4080 per working year). The 'unspecified external fault', mentioned above, related to a Dosco Mk.3 Roadheader gearbox. Had this broken down at an unscheduled time, it would have cost £22,000 for parts and a working loss of 325 man hours — vastly in excess of the monitoring costs. Even higher savings are reported on a shearer ranging arm unit where high debris reading had been obtained (some £61,753).

Schwabe & Asher 1987

Although the Thin Layer Activation (TLA) technique is relatively expensive to install per component, Schwabe of Cormon Limited has been able to demonstrate considerable savings in test time in a number of applications. In the case of the measurement of tool wear, layers of ^{56}Co were laid down at around 0·12 mm depth on the cutting edge of a variety of tools. A 50-fold reduction in time taken to determine the wear rate was achieved.

Another application was the measurement of wear of an inner cylinder wall of a Ruston engine. A 3 mm × 6 mm elliptical spot of ^{56}Co was formed to a depth of 75 μm. The rate of wear was accurately and rapidly measured without any engine dismantling. In addition the influence of an oil change was monitored and an unexpected engine fault from poor combustion was identified.

Johnson et al 1988

report the work undertaken at British Rail on the automatic spectrographic analysis of diesel engine oils. The purpose of the analysis is primarily to determine the wear metals in the oil in the most efficient manner. ARL in Switzerland, designed and produced a special ICP based oil analyser for this high-speed automated analytical system — the Fast Oil Analyser.

British Rail estimate that savings amounting to approximately £4,500,000 per year are achieved using this system. The cost to analyse 65,000 samples p.a is around £350,000.

Before deciding on the use of ICPOES (Inductively Coupled Plasma Optical Emission Spectrometry), the advantages and disadvantages in comparison with the Rotrode OES were assessed. As far as British Rail were concerned they saw these as

>   Advantages
>>   Sensitivity
>>   Repeatability
>>   A true solution technique
>>   Compatible with the automation of the whole analytical process
>>   Analytical speed
>   Disadvantages
>>   Relatively high running costs
>>   A more complex analytical process than the Rotrode

In practice they also found that some cross contamination occurred in the automatic sampling process. However, as this only amounted to an average of 2·7%, this was felt to be acceptable in view of the exceptional high speed of analysis achieved.

Lasentec 1988

outlines the use of the LAB-TEC® 100ME analyser to measure oil droplet size in an undiluted emulsion. The mechanical process involved is that of the sheet metal industry. An oil-in-water emulsion is used for both cooling and lubrication, and it is the control of this which enhances the finished product with consistent properties and improved surface quality. Contamination of the oil-in-water emulsion (both solid debris and chemical) can seriously degrade the lubricant performance. Fig. 13.3 shows the results of such testing where both droplet size (1) and solid debris contamination (2) were successfully detected and correction made before serious damage ensued.

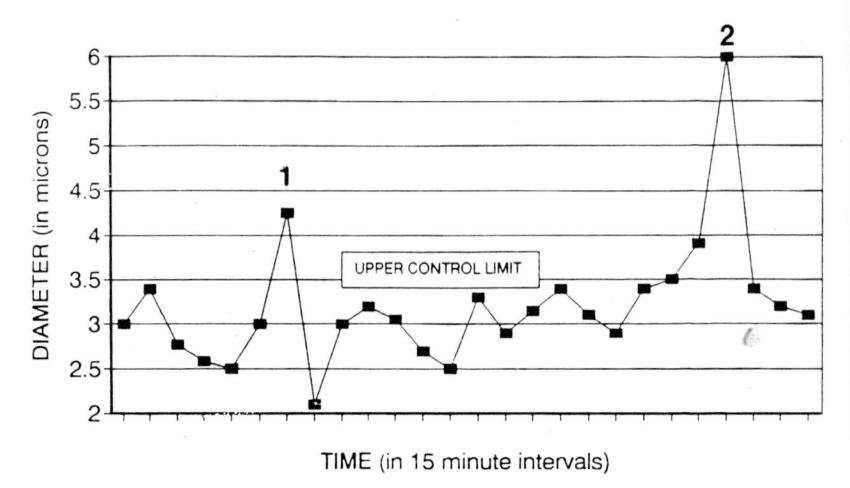

Fig. 13.3 Particle size (diameter) against time (Lab-Tec®)

Jones & Larkin 1988

describe work in a steel plant where there is a need to monitor the 60 or so engines relating to cranes, locomotives, large trucks, etc. Initially various chemical, spectrometric and microscopic techniques were investigated but it was felt that these could be misleading as the larger particles were not given a priority. Analysis since 1986 has been undertaken using the Model 56 Wear Particle Analyzer from Tribometrics. The steel company report this to provide substantial help in predictive maintenance, and, when the results are added to other information, alarms are being triggered at the right time for rebuild and overhaul.

Xu, Huo & Qiao 1989

describe what they call the Ferrogram Quantitative Analyser, mentioning that it had been "effectively used in the Huo Lin He coal mine" in China. Basically the idea is that of automatically determining the optical density of a Ferrogram slide over its full working length; ie. in essence DR Ferrography but with many reading points and computer analysed. Some 75 ton trucks were monitored every 250 hours at the usual service using this ferrographic method, but also at 50 hour intervals by using oil samples. A unique severity-of-wear index, which they call the 'J parameter' enabled them to identify excessive wear in one of the trucks when the J value exceeded the 'alarm' setting (Fig. 13.4).

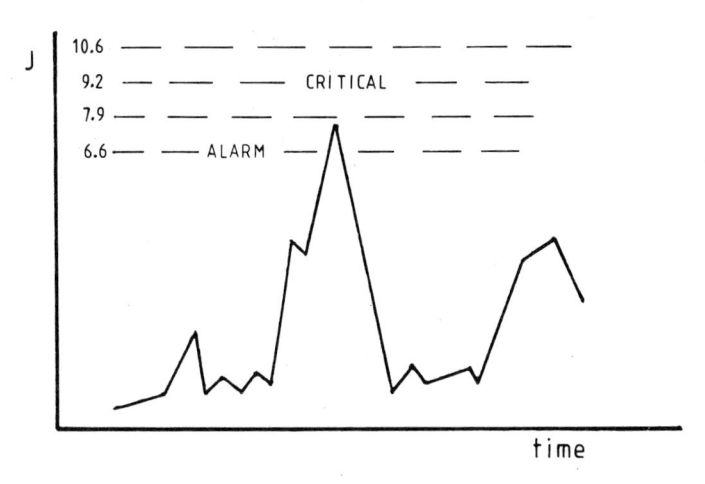

Fig. 13.4   Parameter J versus time for a HD-680 truck 1986-87

Nicholls 1989
from NIKAT Associates, reports the typical annual plant savings obtained in the USA using predictive maintenance techniques. Whilst this is not restricted to debris analysis, it will nevertheless constitute a fair proportion of it:

Paper	$US 500,000 per paper machine
Power	$US 500,000 per turbo generator
Chemical	$US 1,000,000 plus
Refinery	$US 1,000,000 plus.

Nicholls goes on to describe around twenty specific examples where savings have been incurred through condition monitoring. Although vibration will have been the method in many cases (the particular expertise of Nicholls), the examples indicate the financial importance of monitoring. He also cites Michael Neale and Associates who had found in a study that a reasonable monitoring cost for a typical industrial plant was about 1% of the total capital value of the equipment which is being monitored; the typical cost/benefit ratios of a well managed programme achieving approximately £5 saving per £1 expended.

Wilkes & Hunt 1990
Rover plc have been concerned with the cleanliness levels of their robots for several years. They use an absolute cleanliness acceptability level of NAS 1638 Class 7. By using a Lindley Flowtech Fluid Condition Monitor

to rapidly check their cleanliness/debris levels they were able to identify a major hazard when an exceptional debris level was demonstrated. By suitably flushing and correcting they were able to continue with the minimum of delay and no disastrous failure, as shown below:

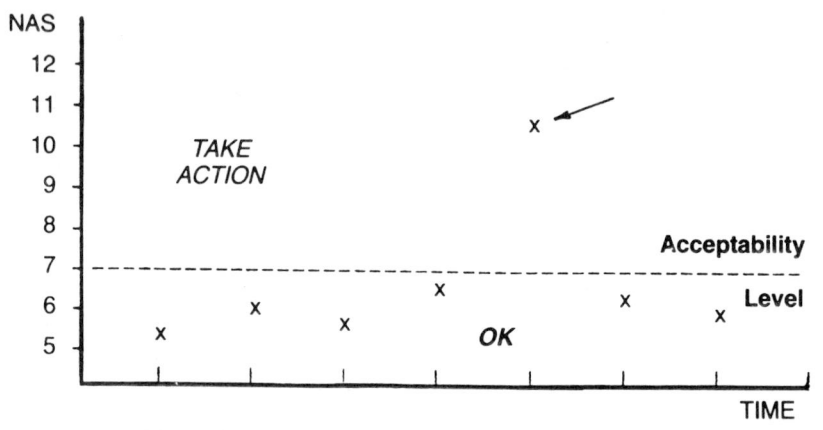

Fig. 13.5 Detection of exceptional debris level (FCM)

Ranco Control 1990
reports some highly successful trials of two of their DM10 Continuous Debris Monitors fitted to drum shearers at the British Coal Maltby Colliery. In this case no suitable oil supply line was available, so an adapted lube pump provided a flow from the reservoir into a brass manifold in which the DM10 was fitted. Over a period of six months the DM10's provided data as shown in Fig.13.6

The graph is simplified but does convey that up to week 20 the flushing cycle time, t_n, was around 444 minutes (giving a Flush Frequency Rate of 0·23 — a typically acceptable value). At week 22, t_n dropped to 253 minutes (FFR of 0.4) and worsening. At week 30 t_n was only 15 minutes (FFR of 6·67). Week 34 was the prescheduled change time (t_n then showing 16 minutes and an FFR of 6·3).

On strip the gear box was found to have a damaged roller bearing which had caused gear damage through shaft misalignment. The majority of the wear debris had come from these features, and it was considered that the gear box would have totally failed after just a few more hours. The new box was fitted with a DM10 and as the gears were run-

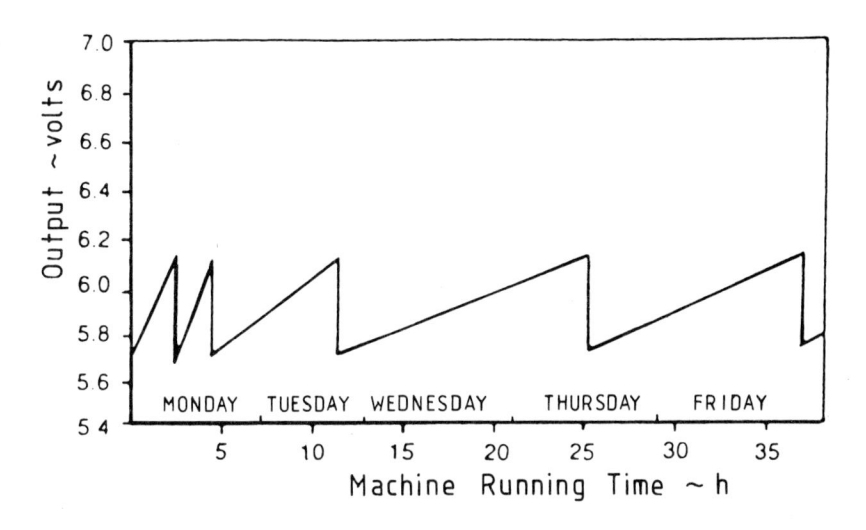

Fig. 13.6 Flush frequency rate of the CDM over gearbox life

in the t_n improved from 345 minutes to 450 minutes. It was estimated that had the shearer failed in service, losses could have amounted to some £200,000 in damage and downtime.

Cumming 1990
gives a review of monitoring in British Airways. The debris monitoring is a valued and respected technique. Oil samples are regularly analysed spectrometrically, but the type of spectrometry depends on the engine because of the different normal rate of wear debris. For instance the JT9D (Boeing 747) normally produces a ferrous count of 0·2 ppm to 0·4 ppm, and an increase of 0·6 ppm would be significant — for this AA analysis is used. Conversely, the JT8D (Boeing 737) produces 10 ppm regularly and a 10 ppm change would be significant — hence the use of AE analysis.

British Airways have used magnetic chip detectors ('collectors') for many years on a routine basis. This has been effective because most of the rotating parts in the engines are made from magnetic sensitive parts. Cumming reports that "a large number of bearing failures" have been saved through this technique, which would otherwise have incurred considerable expense. For example, the cost of an engine removal and repair will be in the region of £20,000, whereas a missed bearing failure could be in excess of £500,000, and, if uncontained, would cause a much higher financial penalty.

Faulkner & MacIsaac 1991

in their description of the GasTOPS FerroSCAN®, give an example of an unexpected failure on a pump bearing. The pump was part of a system being monitored by the FerroSCAN®. It was apparent to the operators, from superimposed transients on the transmitted signal, that bursts of debris were being released into the system. When the pump was later stripped and examined, 5% damage was visibly detectable. This shows an early detection ability, which is so critical for effective monitoring.

Day & Tumbrink 1991

outline the findings of two much earlier industrial investigations where the presence of dirt in the fluids caused a considerable drop in life of the components. A comparison is made between one fork lift truck operator who cared little about the high debris content in his hydraulic oil, only to have to experience a very high level of unreliability (a mean time between failures of 100 hours), against a pump in high pressure use but with a good oil cleanliness level, where the manufacturer could find no wear after 50,000 hours!

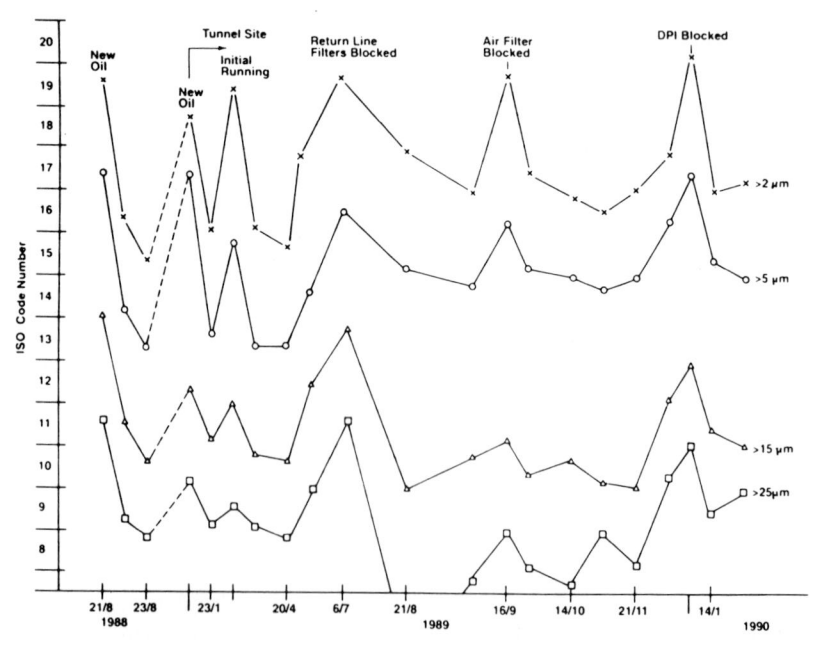

Fig. 13.7 Particle counts on a Tunnel Boring Machine

A later example of a Tunnel Boring Machine which needed to be built and run at a suitable cleanliness level, shows how valuable particle analysis can be for achieving optimum life for machinery. Fig. 13.7 shows the particle counts over an eighteen month period.

The initial period (21/8–23/8) represents the build of the TBM at the manufacturers. The fresh oil had to be filtered down to achieve the required cleanliness of ISO 4406 13/10 (at 5 μm/15 μm). The same process was repeated at the build in the tunnel. Thereafter, various features caused a rise and fall in the counts — there was running-in, gradual increases due to filter blockage, excessive ingestion from the debris laden atmosphere causing the air filter to block, damage to the filter with a differential pressure indicator (DPI). In each case corrective action was taken promptly, and the periods of increased wear have been kept to a minimum. The enormous financial penalties of such a project are only too apparent, and they have been avoided by suitable monitoring of the debris in the oil.

SGS Redwood

By nature of their work, oil laboratories cannot normally divulge customer information. However, SGS Redwood do publicise some of the results they are aware of, either of their Redwood Oilscan or of other laboratories, where regular oil checks have been made. The following are some of their findings:

> In early 1977, oil analysis was introduced to monitor the powershift transmission on a fleet of 26 45-ton dump trucks in an open cast mine in South Africa. This resulted almost immediately in an estimated saving of $US 325,000 per annum. By 1982, this lube oil monitoring programme had expanded to cover some 1,430 units of miscellaneous mining machinery at an estimated annual saving of $US 8.75 million!

> Early in 1982, an oil refinery in the USA established a lube oil analysis pilot programme to cover some 30 major machines at the refinery. By the completion of the one year project, an estimated $US 300,000 was saved in machine repair costs.

> In the 6 months prior to the introduction of lube oil monitoring in 1983 at a UK colliery, 17 failures were recorded, causing production

losses of some 100 hours. In the 6-month period following the introduction of lube oil analysis, production losses due to gear box failures had been reduced to zero.

And so the list goes on. Perhaps, when the second edition comes out, the reader, or his or her company, will be among the new stories of success!

CHAPTER 14

Putting Wear Debris Analysis into Practice

The previous chapters have outlined techniques and means for debris analysis. They have described specific instruments and methods which will do the job. Chapter 13 gave many examples of successful debris monitoring, but now it needs to be put into practice in the application appropriate to the reader. But how does one start?

THE CME APPROACH

The CME Approach (specially formulated for Chartered Mechanical Engineers!) looks at three separate steps which help in the choosing of a monitor for a system:

COMPONENT CASUALTIES	Step 1
MEASUREMENT METHODS	Step 2
ECONOMIC END-PRODUCT	Step 3

In other words one needs to look initially at what failures have occurred or are likely to occur (or may occur) with various components. These must be identified first. There is no point in monitoring a fluid which has only one route, and that passes a highly reliable component, which has never failed and, even if it did, would be of no great loss.

Secondly, there is the need to consider the possible ways of measuring any signs of failure? Because an engineer must first look at the modes of failure of the components, he or she will be able to sense what sensor will best detect the fault. Will debris monitoring detect the fault or miss it? Would another type of monitor be preferable? Ultimately the most valuable monitor is the one that can detect the earliest signs of impending disaster.

The third aspect relates to money being wisely spent. Is the monitor cost-effective?

Neale 1987 puts the three aspects slightly differently, but certainly along the same lines. He divides the 'component casualties' and the other two categories he joins together, as shown below. ('Time to repair it' must of necessity include detection and cost.)

Component casualties — { The likelihood of failure
{ The effect of failure

Measurement methods —
Economic End-product — } The time to repair it

In a little more detail, the CME approach can be seen as:

Step 1
Examine the system
Note the likely failures (or even possible ones)
See whether the failures would shed debris into the oil or hydraulic fluid (if not an alternative type of monitoring will be necessary).

Step 2
Identify places where the debris could be monitored, either because there is already a sampling point present, or because one could be fitted (if not, perhaps the sump or reservoir could be used).
Decide the type of monitor which best suits the debris or contaminant which is to be monitored.

Step 3
Work out the overall costs, including monitor purchase, fitting, running, maintenance, disposal (ie. the terotechnological cost).
Work out the cost of failure, should one occur.
Work out the economy of having a debris monitor.
Get approval from management (using the evidence contained in this book).

Let us now elaborate on the listing made above.

Step 1:

In examining the system, there are certain components which need to take one's attention. They are the ones which could lead to failure —

Components which MOVE on their own
Microscopically, as with creep
or Macroscopically, with obvious motion.

The motion may be linear, rotary, vibratory
It may be an expansion or contraction

Components which MOVE AGAINST something else (Relative Motion)

    It could be a sliding action, or rubbing
    It could be an indenting or pulling

Components which experience LOAD

    By pressure (including vacuum) or force
    By temperature expansion or contraction

Components requiring LUBRICATION or COOLING or HEATING

    which if removed would cause a failure

Components requiring a SKILLED operator

    who may be off-colour or be replaced for the day

Obviously, the greater the magnitude of the feature, the more likely is a failure. Low speed, low load, heavily constructed gearing is unlikely to ever fail. Harvey & Herraty 1989 suggest in their provocative paper, that there is no need for even high speed modern bearings to ever fail, if they are correctly used (but they are not, so they do fail!). A combination of a number of the above features would carry higher failure rating.

Take, for example, the following system (Fig. 14.1):

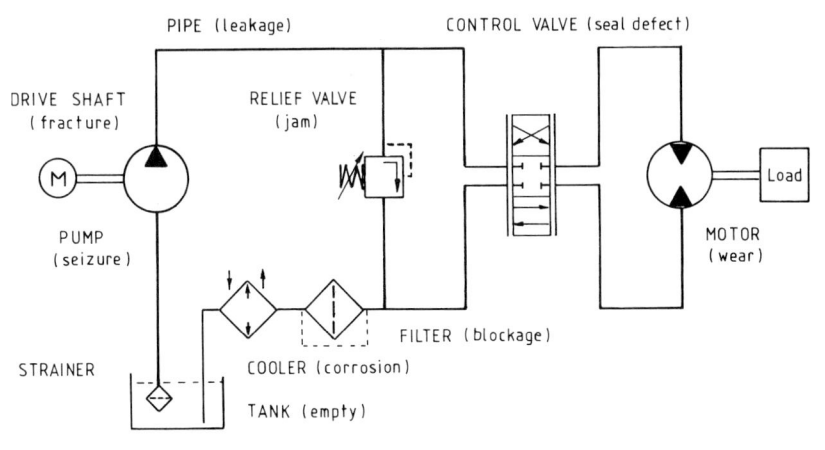

Fig. 14.1 Some possible failure regions in an hydraulic system

Note how each one would fall into one or more of the above categories. But it must be emphasised, that unless previous experience has indicated specific failures, the most likely are where the magnitude is exceptional. Another key problem is where the type of skill required by the operator is complex; there may not be a fool proof method to stop a replacement operator misusing the equipment.

The actual failures can usually only be determined from past experience. This experience need not necessarily be with precisely the same system; the value of a good engineer is that he or she can relate one experience to another. One gets a 'feeling' that something is likely to happen, or, at least, could happen. By 'failure' is meant a 'loss of function', be it by complete fracture, a developing fault or by an inadequate fit.

Failures can be a total disaster or, perhaps, a dwindling reduction in efficiency. Either way, a financial loss will be incurred. Even worse there could be some injury or health hazard.

Consider the following as possible 'failures' (Fig. 14.2):

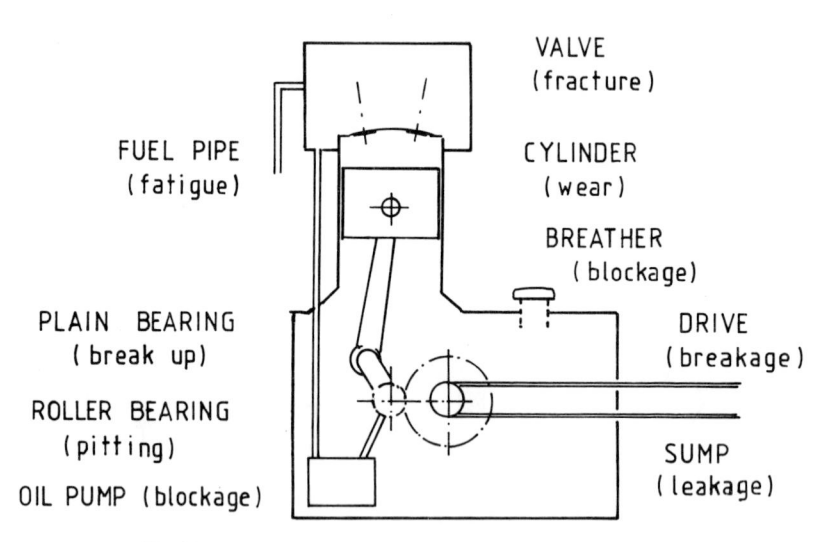

Fig. 14.2 Some typical 'failures' in mechanical systems

Now the crunch question — In the approach to the failure (or the failure, itself) will debris be generated into an oil or grease or other liquid? If not, then another type of monitoring, apart from this one, must be examined. If the debris is there, then we can go ahead with some form of wear debris monitoring.

The debris needs to be transportable in the liquid. Some large particles get no further than the immediate vicinity of the break-up region. On the other hand, some smaller particles may not mix with a liquid, and may be skimmed off or attracted away from the main flow. However, the majority of those shed into a flowing liquid will be carried along around the system, until stopped by the filtration or by a sampler. Grease lubricated bearings can also be monitored using syringe withdrawal (see page 104).

Step 2:

Figures 14.3, 14.4 and 14.5 show examples of where one or more sample points could be fitted into a system. The relevance of the particular points will differ slightly as to whether the debris needs mixing by an artificial flow mixer or whether a sharp bend may be sufficient. It will also differ depending on the level of identification required. If it is essential to check on one very special component, then a sampling point immediately after that component will be necessary. This has already been discussed in Chapter 5.

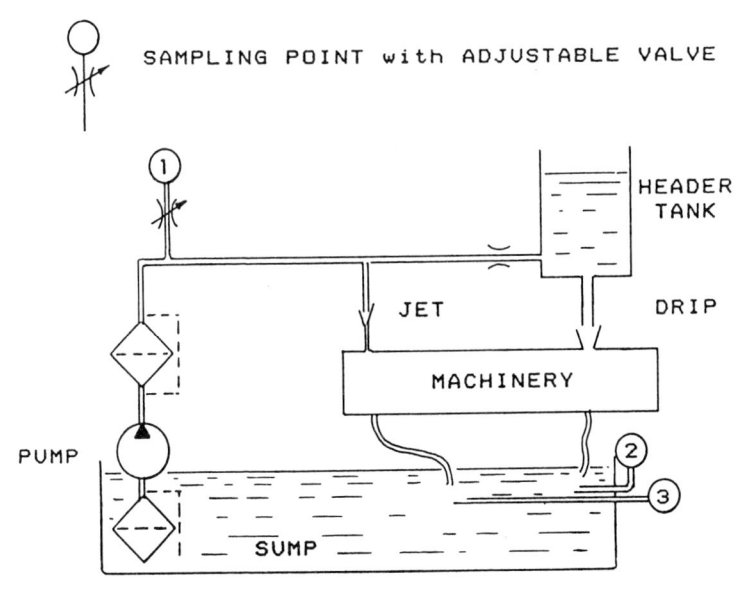

Fig. 14.3 Sampling points in a lubrication system

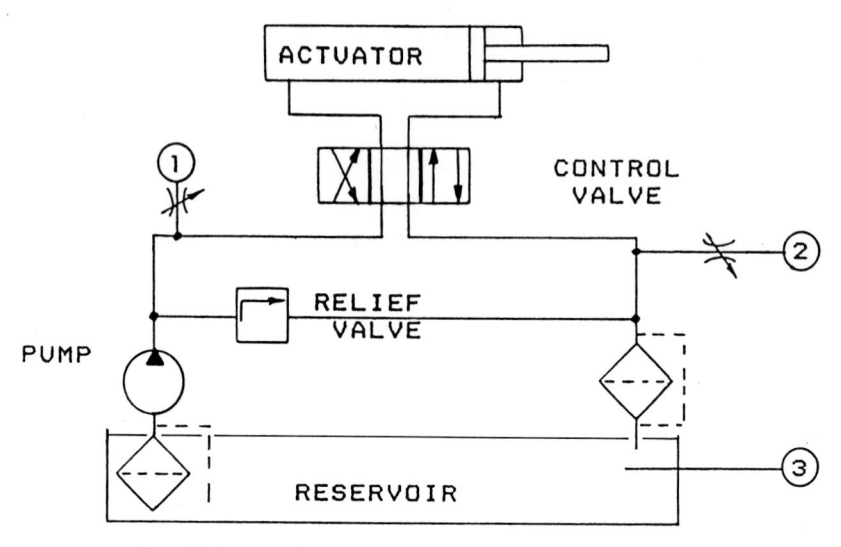

Fig. 14.4 Sampling points in an hydraulic actuator circuit

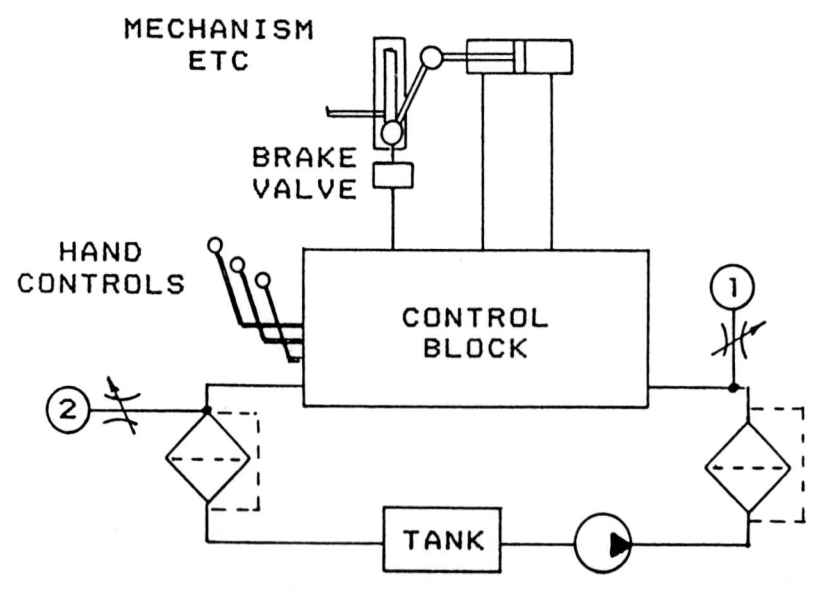

Fig. 14.5 Sampling points in a multi-purpose mobile machine

The second part of Step 2 is the choice of a monitor. The best monitor for the job may be the hardest decision, or it would be, without the information contained in this book! There are some 15 key factors to take into consideration; and the weighting of them will significantly vary between one industry and another, between one application and another. Below are listed these factors:

1. Able to monitor all faults
2. Easy to fit
3. Easy to use
4. More reliable than the system being tested
5. Able to be checked and calibrated
6. Able to give an immediate answer
7. Able to compensate for variables
8. Adaptable to different systems
9. Able to store evidence for later checking
10. Inexpensive
11. Small and light
12. Non-iatrogenic
13. Safe
14. Economically viable
15. Easy to obtain and maintain.

The order of these is arbitrary, here, but should be changed to suit the situation in hand. For instance, a kilometre or two underground will mean that accessibility is critical, and hence numbers 2 and 15, 'Easy to fit' and 'Easy to .. maintain', are paramount. Or again, if the monitoring is required in an aircraft, available space and permitted weight is limited, and number 11, 'Small and light', comes into its own.

A word of explanation on each of the above, will help in making this choice:

1. Able to monitor all faults.

   There are few monitors/systems to which this would apply. But if there is one, then the monitor is a winner.

2. Easy to fit.

   Quite often monitors are not used because, either the machinery cannot be stopped long enough to fit the unit, or the instructions are so complicated the fitter would rather not attempt it!

3. Easy to use.

One hand held monitor that used to be around was fitted very easily, but in use it was not only difficult to understand, but it was positively hard work to actuate! So it wasn't.

4. More reliable than the system being tested.

This is how electrical chip plugs got a bad name originally (they are OK, now!). They failed almost as much as the machinery they were monitoring. There can be no confidence in a monitor which behaves in that way.

5. Able to be checked and calibrated.

Some monitors do not need calibrating. But there must be some sort of means to check that the output stated, does bear a direct relationship to the situation.

6. Able to give an immediate answer.

This may, or may not, be necessary. For any system there is a time limit beyond which a knowledge of the system condition is meaningless. The monitor must react well within that acceptable time.

7. Able to compensate for variables.

Variables could be anything from the people who operate the system or monitor, to the possibility of a typhoon, not forgetting the machine itself. The likely variables (or possible variables) need to be assessed for each system.

8. Adaptable to different systems.

This is not the same as number 1. We are now looking at the use of the monitor as a portable. Using the same monitor will bring the cost down (fewer units would have to be purchased), it would be more convenient (the maintenance engineer will get used to it), and its results would be better understood (ie. a similar output to present to management).

9. Able to store evidence for later checking.

It is all very well for a monitor to display some 'go/no-go' result at the time of test, but is it really true? Or, later, when other results are obtained, can the original results be confirmed? In a sense, it could be important to be able to perform an autopsy, well after the event has happened. In the case of wear debris analysis, it is the debris which needs storing.

10. Inexpensive.

In English, 'cheap' is not the same as 'inexpensive'. 'Cheap' implies it not being made very well. 'Inexpensive' conveys a device which is not over priced by the manufacturer. It is value for money.

11. Small and light.
    We have already looked at the aircraft requirement. There may be other situations, as with portables, where this must apply.
12. Non-iatrogenic.
    It is the medical field which probably coined this expression. It refers to the unfortunate individual, who is put into a worse condition by being given an exploratory operation. It can happen with a monitor on a machine; it is possible to generate a more serious failure by the fitting of the monitor, than the monitor will ever be able to detect!
13. Safe.
    This is always a requirement. But in places where people's lives are put at risk, it becomes even more important. Electrical BEAB approval may be one aspect, but the very mechanical construction could be another.
14. Economically viable.
    This is really our Step. 3. The cost of the monitor could be in the thousands of pounds sterling, but if it saves millions, then it is viable. On the other hand, it might cost less to replace a simple machine tool, should it fail, than to fit and run a monitor.
15. Easy to obtain and maintain.

The supplier is another factor. If he cannot provide the consumables or spares on time, then the monitoring will be a disaster.

Step 2 is all about actually getting a monitor. It must be remembered, though, that that monitor may not be a wear debris monitor. There are other techniques which could be more appropriate, as discussed in Chapter 1. This book only looks in detail at the wear debris aspect.

Step 3

The calculated costs of monitoring are quite frequently in error. It is very easy to forget certain components in the cost cycle. Obviously first cost and operator time cost are there, but what about the maintenance cost of the monitor? Have we considered the cost of disposables involved in the monitor? Is the cost of the final disposal also included? This is what 'terotechnology' is all about — the science of total cost from first thought to final disposal, with all the hardware and manpower in between.

In order to see whether it is worth buying and installing the chosen monitoring system, an estimate must also be made of the likely savings to be made. In other words, what would an unwanted and unexpected failure cost? Suppose the mechanical system were to suddenly fail, what would it cost to replace including, hardware, fitting, man-hours and, very importantly, lost production?

Nicholls 1989 reports the two (or is it three) commonly used questions raised by management as being

"1. What savings can be achieved?

2. What will it cost, and what will it involve?"

Some of the examples given in Chapter 13 may well help to answer these questions, particularly if there has been no first hand experience of failures within one's own company. Fig. 14.6 gives an idea of a typical breakdown with and without monitoring.

Then comes the presentation to management, in order to obtain approval for the condition monitoring system. Remember, it is primarily

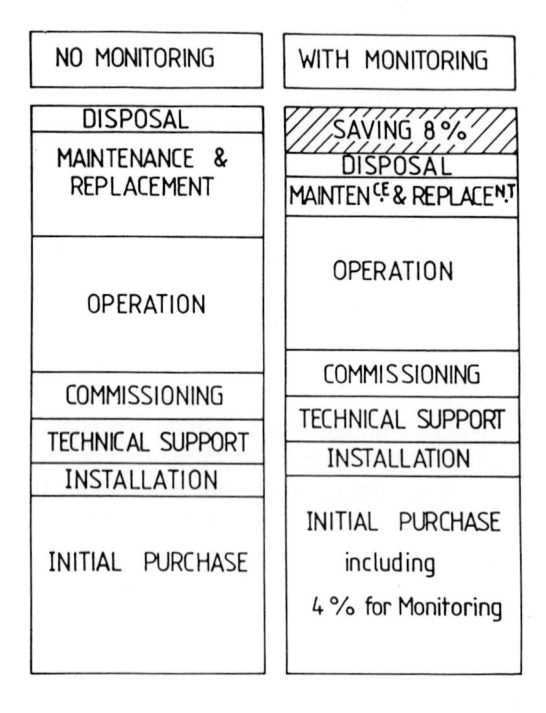

NO MONITORING	WITH MONITORING
DISPOSAL	SAVING 8%
MAINTENANCE & REPLACEMENT	DISPOSAL
	MAINTEN^{CE} & REPLACE$^{N.T}$
OPERATION	OPERATION
COMMISSIONING	COMMISSIONING
TECHNICAL SUPPORT	TECHNICAL SUPPORT
INSTALLATION	INSTALLATION
INITIAL PURCHASE	INITIAL PURCHASE including 4% for Monitoring

Fig. 14.6 Typical overall costs with and without monitoring

the financial saving which is of importance, although, of course, if there are possible serious safety hazards involved, the expense may not be quite so important.

USING WEAR DEBRIS ANALYSIS

Assuming management has given the go-ahead for a condition monitoring system, and it is to be (or include) wear debris monitoring, what happens next? There are several items to be addressed.

Money and manpower
The money allocation needs deciding. It is not sufficient to agree just the purchase of the monitor; manpower must be considered in order to action the monitoring. The maintenance team (not a bad name for those responsible for the monitoring) must be of sufficient calibre to understand what is happening and be accepted by higher management. They must feel their job is essential by a full back-up from top management. Maintenance must be seen as synonymous with reliability and cost saving. The team must be clear about what maintenance role they actually have. Before going into action, it is helpful to review those maintenance roles, and to get approval from senior management.

Maintenance and reliability
Reliability may have been an objective of the 70 and 80's, but it became a practical reality only when condition monitoring was accepted — accepted as the major feature of maintenance. True, there were certain financial short term benefits by using time based maintenance with its regular check on the gauges; maybe even the unplanned maintenance was not too much of a disaster providing sufficient spares were in the stores and the fitters enjoyed getting up in the middle of the night. But preventive maintenance involving condition monitoring was a bonus which lifted the whole concept of machine reliability and acceptability to a new plane. (Note 'preventive' can also be spelt 'preventative' — they both mean the same!)

Of course, machine based maintenance need not include condition monitoring. Provided the specialist tools are to hand, corrective maintenance, which 'tweaks' the settings into alignment, may be used each time the machine breaks down. This is, however, of limited success

```
                              MAINTENANCE
                    ┌──────────────┴──────────────┐
              PLANNED MAINTENANCE          UNPLANNED MAINTENANCE
         ┌──────────┴──────────┐                   │
    TIME BASED            MACHINE BASED            │
    MAINTENANCE           MAINTENANCE              │
      ┌───┴───┐            ┌─────┴─────┐           │
  MACHINE  CALENDAR    CORRECTIVE   PREVENTIVE  EMERGENCY
   HOURS    HOURS     (BREAKDOWN)  (CONDITION) (BREAKDOWN)
```

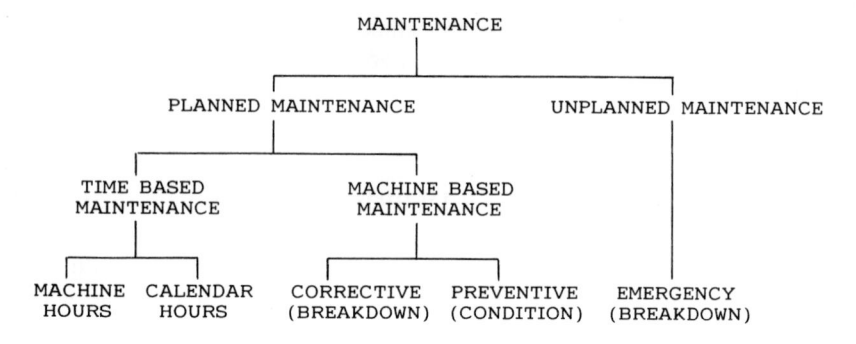

Fig. 14.7   The different types of maintenance

because it does not address the whole condition of the machine but only those parts which are readily adjustable.

It is sometimes said that there are three areas of maintenance operation (eg. Walsh 1991) — breakdown maintenance, planned maintenance and predictive maintenance. I do not use the separate idea of 'predictive maintenance' because I believe complete maintenance is predictive, ie. it is all there to make some advance assessment of likely future condition and to do something now to alleviate the problem. I do recognise, though, that what is actually meant by the idea of predictive maintenance is what I call machine based maintenance — the machine being assessed for the necessity of corrective action, whilst it is still running. That is important. That is, for many systems, the most advantageous type of maintenance.

Machine based maintenance requires on-line monitoring to maximise its effectiveness. Fig.14.8 from Walsh 1991 gives a graphic representation of the improvement in effectiveness and the raising of reliability.

Proactive Maintenance, is the terminology given to looking at what causes failure rather than the failure mode itself. Basically this is the contamination level assessment in hydraulic and lube oil systems (see Chapter 4), because it is that very debris which is the major cause of failure.

Training
This book provides an awareness of what systems are available and how to use them effectively. It suggests techniques and approaches which can

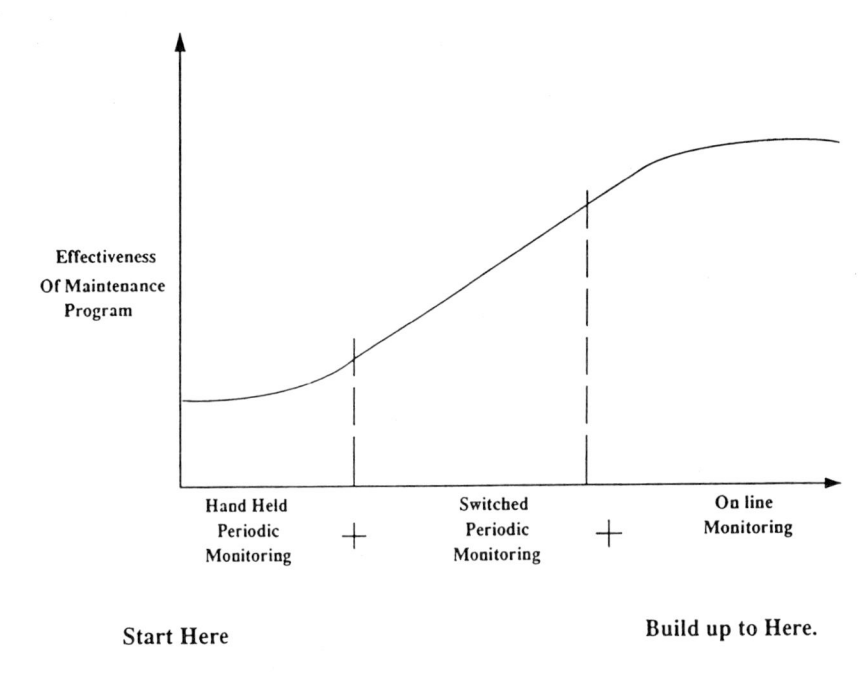

Fig. 14.8 The effectiveness of a maintenance programme (Walsh 1991)

be taken. This may well be adequate to commence a programme and avoid the pitfalls that are met by the uninitiated. However, each company which sets out on a major investment for the future by, for one, incorporating condition monitoring such as wear debris analysis, needs to go a little further. Training of its personnel becomes essential.

There are several university, college or training company based courses available in condition monitoring which can run for a few weeks to a year or more. There are also short course of a few days or a week to whet the appetite and highlight the possibilities available. However, a word of warning, here. All courses, of whatever length, are going to mislead attendees unless the organisers are prepared to show the whole spectrum of condition monitoring. Condition monitoring is not 'vibration' or 'performance' or 'thermography' or ... 'wear debris analysis'. The maintenance team must consider all viable possibilities before choosing. It could be, and quite often is, that the optimum is a combination of several types, each complementing the other.

As examples of short and medium length courses on Condition Monitoring, there are those at Brunel University, Sheffield University and Swansea University. No doubt there are many others, too.

Year courses (and shorter ones) are mentioned by Clifton 1990 (Longlands College of Further Education) and Rao 1991 (Southampton Institute of Higher Education).

Although all these course cover monitoring in various aspects, as suggested earlier, the particular part of the syllabus relating to wear debris would cover many aspects included in this book. Southampton Institute has included five modules (covering a total of some two hours actual teaching time for each module):

1. Introduction to wear debris monitoring
    Part 1 — Machine health monitoring and wear debris analysis
    Part 2 — Fluid condition

2. Collecting and describing wear debris
    Part 1 — Methods of collecting debris
    Part 2 — Classification of debris
    Part 3 — Identifying debris

3. Quantifying wear debris
    Part 1 — Methods of detecting debris
    Part 2 — Quantifying debris
    Part 3 — Counting debris by size

4. Wear debris control and contamination control
    Part 1 — Levels and standards
    Part 2 — Wear debris control and contamination control

5. Wear debris in practice
    Part 1 — Examples of experience
    Part 2 — Putting wear debris analysis into practice

The total time will be much longer than the 10 hours teaching. Longlands College suggest 60 hours study time for 'Tribology Monitoring Methods' with even more time on a project.

The Fluid Power Centre at the University of Bath, has a regular series of two to four day courses on fluid power aspects, including contamination control as one of the major subjects.

Into action

The supplier of the monitor, or monitoring system, should provide a full back-up with necessary instruction on use and abuse. If he does not, because of the relative low cost of the unit, say, then additional help can be provided by consultants or colleges specializing in condition monitoring (some examples are given above). Most consultants would be quite able to provide a minimum of time, which would save many hours of wasted internal labour.

A monitoring technique has been chosen. Perhaps a monitor has been purchased. Now launch out! Begin monitoring the results and acquire experience. Develop confidence. Discuss the results with the supplier or consultant involved — there may be odd features which are not right and which could spoil the whole programme; they need to be corrected as soon as possible.

Remember the purpose of the monitoring — improved reliability of the machinery. If that purpose is to be achieved, positive action must be taken when the monitor indicates a problem. Once a confidence in the monitor is there, then the actions must be taken. If a significant deviation from the acceptability level, or from the normal, is measured, then the person responsible must be called in.

That is what the monitor is for. That is how the wear debris analysis can be the success in your system, that it has been in so many others.

CHAPTER 15

Bibliography

AHEM — see BFPA.

Albidewi, A., Luxmoore, A.R., Roylance, B.J., Wang, G. & Price, A.L. (1991). Determination of particle shape by image analysis — the basis for developing an expert system. *Proceedings of the Condition Monitoring '91 international conference*, Erding, Germany, 14–16 May 1991. (Ed. Jones, Guttenberger & Brenneke) Pineridge Press, pp. 411–422.

This is an introductory paper on the technique developed at the University College of Swansea and Tribology Centre, to evaluate the shape of a particle automatically and to thus determine its origin.

Allen, T. (1975). *Particle Size Measurement*, 2nd edition, Chapman and Hall, London, 1975. (Note: 3rd Edition 1981).

An early classic on particle size measurement. Discussion on a wide range of particle sizing techniques, primarily concerned with powders.

Alsop, P. (1992). Condition monitoring — downtime reduction guaranteed. *Plant Engineering & Manufacturing Management*, February 1992, Trinity Publishing, Ruislip.

The author laments the ostrich-like attitude to condition monitoring, and goes on to show how the various ways have provided financial benefit to many and how a start can be made. Oil analysis is described due to the estimated 70% of all bearing failures being due to lubrication problems.

Anderson, D.P. (1982/91) — see 'Wear Particle Atlas'.

Anderson, M. & Stecki, J. (1990). A diagnostic key for wear debris obtained from oil and grease samples from operating machinery. *Proceedings of the 3rd International Conference on Condition Monitoring*, London, 15–16 October 1990. BHR Group.

A detailed step by step diagnostic approach to particle identification from observations made by visual inspection through a microscope. Work developed at Monash University in Australia.

Asher, J., Conlon, T.W., Tofield, B.C. & Wilkins, N.J.M. (1983). Thin-layer activation — a new plant corrosion-monitoring technique. *On-line Monitoring of Continuous Process Plants*, the proceedings of the 1st International conference on on-line surveillance and monitoring, London, September 1983, ed. D.W. Butcher. Ellis Horwood Limited, Chichester, pp. 95–105.

Thin layer activation (TLA) is defined with a clear distinction made between its use of trace quantities of radio-isotope generated in a thin surface layer, and ion and neutron implantation which normally would have a much higher activity level. TLA, thus has a higher sensitivity to material loss. Practicalities are discussed including its use for wear monitoring.

Astridge, D. G. (1987). HUM — health and usage monitoring of helicopter mechanical systems. *Condition Monitoring '87*, proceedings of an international conference, University College of Swansea, 31 March–3 April 1987. (Ed. M.H. Jones) Pineridge Press, pp. 143–162.

A range of monitoring techniques used on helicopters are described including the QDM and 'fuzz-burner' chip detectors for transmission health assessment.

B

Baker, R. (1991). *Correlation of hydraulic cleanliness standards and solid particle sizing and counting equipment.* Howden Wade Ltd, Thermal Control Division note issued September 1991, 4 pages.

The difficulties in correlating cleanliness/contamination standards are discussed with particular reference to the Conpar standards. The ranges covered by the different standards are listed, and a correlation chart is presented.

Bayvel, L.P. & Jones, A.R. (1981). *Electromagnetic scattering and its applications.* Applied Science Publishers, London & New Jersey, 289 pages.

Whilst briefly indicating the basic features of light scattering and how it can be used in the determination of particle physical properties, the book is mainly an in-depth advanced description and discussion of the theories involved. Mie theory, relating to spheres (13 pages), and Fraunhofer diffraction theory are both covered, as well as many other ideas not mentioned in this current book.

Beerbower, A. (1975). *Mechanical failure prognosis through oil debris monitoring.* Exxon Research and Engineering Company report prepared for the Eustis Directorate, US Army Air Mobility Research & Development Laboratory USAAMRDL-TR-74-100 1975, 134 pages.

A substantial report on the methods then available for wear debris analysis. It covers, to a varying degree, chip detectors, spectrometric oil analysis (AAS and AES), light scattering, particle counters, Ferrography, microscopy, wear particle characterisation, all with wear in helicopter gearboxes in mind.

Belman, H.M. (1991). Use of ultra-high particle capture versus flow-through sensors in modern aircraft lube systems for real time early failure detection. *Condition Monitoring '91*, proceedings of an international conference, Erding, Germany, 14–16 May 1991. (Ed. Jones, Guttenberger & Brenneke) Pineridge Press, pp. 297–305.

A thorough discussion from 40 years' company experience of magnetic plugs and sensors, indicates the need for both flow-through sensors (which do not keep the debris) and actual debris collectors. Comparisons are made between large and small particles, and between shape of debris and type of failure.

Berg, R.H. (1991). Precision in particle size analysis. International *Labmate*, Vol. XVI Issue 1.

A detailed comparative analysis is given by the chairman of Particle Data Companies on three types of particle detection. These are Electrical Sensing Zone, Laser Diffraction and Sedimentation. Features discussed are fidelity (shape, roughness, optical properties, particle and fluid density and conductivity, instrument malfunction), calibratability, resolution and repeatability. The precision of the ESZ is emphasised.

BFPA P48 (1988). *Guidelines to the cleanliness of hydraulic fluid power components*, British Fluid Power Association, London.

A particular series of directions to assist in the preparation and build of fluid power components without the serious effects of contamination.

BFPA P5 (1991). *Guidelines to contamination control in hydraulic fluid power systems*, British Fluid Power Association, London.

A practical working document based on the findings of a combined contamination control programme in the early 1980s, undertaken jointly by the National Engineering Laboratory, BHR Group and the Universities of Aston and Bath.

BFPA P9 (1992). *Guidelines for the flushing of hydraulic systems*, British Fluid Power Association, London. (Completely new edition. First issue May 1978).

The system features which need thorough cleaning and how that can be undertaken, are outlined in a logical sequence. Flushing fluids and displacement oils are discussed.

Blatchley, C. & Sioshansi, P. (1990). Monitoring wear with gamma rays. *Machine Design*, 25 October 1990, Penton Publishing Inc., 4 pages.

After describing the concept of Thin Layer Activation (Surface Layer Activation), suggestions for its application are made. Particular emphasis is made on the unique advantage of the idea for in-line detection of wear. An actual example cited is that of steam flow valves in a turbine, where different treatments were successfully compared in-line.

Bogue, R.W. (1984). An Improved magnetic plug for the continuous monitoring of wear debris, *Condition Monitoring '84*, proceedings of an international conference, University College of Swansea, 10–13 April, 1984, (Ed. M.H. Jones), Pineridge Press, pp. 628–636.

An early description of the original Continuous Debris Monitor. Much of the data is related to the characteristics of the initial device which used magnetoresistors in the sensor head. A viable instrument is indicated.

Bott, S. & Hart, H. (1990). Sizing up to problems with particle analysis. *Laboratory Equipment Digest*, Vol. 28, 1990, pp. 25–29.

An extensive review and description of the COULTER® LS Series equipment, including a sketch of the optical system used.

Bowen, E.R., Bowen, J.P. & Anderson, D.P. (1978). Application of Ferrography to grease lubricated systems. *Proceedings of the 46th Annual Meeting of NLGI*, 1978.

Early examples are given of the use of Ferrography to determine the presence of ferromagnetic wear particles in a grease.

Bowen, E.R. & Westcott, V.C. (1976). *Wear Particle Atlas*. Naval Air Engineering Center, Lakehurst, NJ.

The original wear debris Wear Particle Atlas prepared for the Center to help in the use of the Ferrographic equipment. It contains not only pictures but a thorough list of suggestions relating to the sourcing of particles. (See Wear Particle Atlas for later version).

British Coal (1984). *Wear Particle Atlas*. British Coal Headquarters Technical Department, Burton-on-Trent. 16 pages. (£30 in 1990).

A small pamphlet with 17 colour photographs, showing the type of wear debris and ingested debris experienced in machines working underground. Primarily related to the Debris Tester, it includes typical readings.

Bruno, C.A. (1982). Monitoring cleaning liquids — benefits of on-line detection. *Circuits Manufacturing*, February 1982, Morgan-Grampian Publishing Company, 2 pages.

Examples are given of the advantages of monitoring on-line in the sense of rapid detection of a fault such as a filter bed failure. Micro-Pure equipment was used.

C

Campbell, P. (1991). On-line monitoring of ferromagnetic debris concentration. *Condition Monitoring '91*, proceedings of an international conference, Erding, Germany, 14–16 May 1991. (Ed. Jones, Guttenberger & Brenneke) Pineridge Press, pp. 324–334.

The idea and operation of the Sensys/GasTOPS® FerroSCAN® is described from its initial conception. Calibration, trapping efficiency, temperature effects, flow velocity and viscosity, particle sizes and oil/air mixtures are discussed.

Castrol (UK) (1992). Oil analysis solves blockage mystery. *Plant + Works Engineering*, January 1992, Industrial Trade Journals Limited, Westerham.

A short article showing how oil debris analysis paid off when used to check a refuse compactor whose hydraulic machinery filtration became blocked too frequently. Not only was the problem solved (excessive wear with oil additive depletion) but regular checking now provides a much more reliable system.

Childs, E.B. & Krulish, J.A.C. (1969). *Automatic fuel filter monitor.* United States Patent 3,452,586 July 1st 1969 (assigned to Mobil Oil Corporation, New York).

By examining the pressure drop across a fuel filter, the presence of unwanted debris in the fuel could be detected. At the end of each run the filter was back flushed using separate precleaned fuel.

Clayton, Z.A. (1980?). *Particle-size analysis of silts — an optical method of size analysis.* Document prepared by Spectrex Corp., 7 pages.

The near-forward light reflection principle is described with examples of its use on sediment from the continental shelf after sieving through a 63 μm screen. Further information is given on the use of the Spectrex Laser Particle Counter.

Clifton, R. (1990). *An educational response to condition monitoring and diagnostic techniques,* a paper presented at the *COMADEM 90* international congress, Brunel University, July 1990, 44 pages (not included in the proceedings).

The whole concept of teaching condition monitoring and diagnostic control is discussed. Particular emphasis is placed on industrial requirements and a number of courses are cited, from a few days to a whole year. Various syllabuses are presented.

COMRAD (1984). Monitoring oil lubricated machinery using magnetic plugs. The Condition Monitoring Research and Development Group notes on condition monitoring. *The British Journal of NDT,* May 1984, pp. 287. 229.

One of a series of helpful notes on different techniques. This one illustrates the simple chip plug by means of typical debris and rise in debris quantity with growing deterioration of the machine.

Coulter, W.H. (1949/1953). United States Patent 2,656,508 October 20th 1953.

A patent specifications referring to the Electrical Sensing Zone technique.

Cowan, M.P. & Wenman, R.A. (1990). The technique of mesh obscuration for debris monitoring in the Coulter LCM II. *COMADEM 90,* proceedings of the 2nd international congress on condition monitoring and diagnostic engineering management, Brunel University, 16–18 July 1990. (Ed. Raj B.K.N. Rao, Joe Au & Brian Griffiths) Chapman & Hall, pp. 86–91.

The LCM II's use of the filter blockage principle (mesh obscuration) is explained and examples of its accuracy given over a range of ACFTD seedings (10–150 mg/L) at 5 μm, 15 μm and 25 μm. Reproducibility is also reported over the same range for both particle counts and viscosity assessment.

Cumming, A.C.D. (1990). Condition monitoring today and tomorrow — an airline perspective. *COMADEM 90,* proceedings of the 2nd international congress on condition monitoring and diagnostic engineering management, Brunel University, 16–18 July 1990. (Ed. Raj B.K.N. Rao, Joe Au & Brian Griffiths) Chapman & Hall, pp. 1–7.

A review of condition monitoring used in British Airways indicates the potential savings which can be achieved. A major part of that monitoring involves magnetic chip detectors (MCD) and oil analysis. The use of both atomic absorption and atomic emission spectrometry enables a good coverage of fine wear debris. X-ray dispersive equipment is also used for later analysis of suspect debris obtained from the MCDs and from filter element examination.

Czarnecki, J.V., Seinsche, K., Loipführer, C. & Frank, H-J. (1991). Automated condition monitoring by SEM/EDX analysis of wear particles and pattern recognition techniques. *Condition Monitoring '91*, proceedings of an international conference, Erding, Germany, 14–16 May 1991. (Ed. Jones, Guttenberger & Brenneke) Pineridge Press, pp. 399–410.

Whilst enumerating the elemental contents of a range of aero engine alloys, the paper ultimately indicates that the elemental debris pattern change can be the precursor of an engine condition change. About 5000 jet engines and gear boxes are under test with 25% having magnetic plugs whose debris is analysed.

D

Dadd, A.T. (1989). Automatic environment monitoring. *COMADEM 89*, proceedings of the 1st international congress on condition monitoring and diagnostic engineering management, Birmingham Polytechnic, 4–6 September 1989. (Eds. Raj B.K.N. Rao & A.D. Hope), Kogan Page, pp. 139–143.

Mass spectrometry is described, and its use for the analysis of a number of volatile organic compounds is discussed.

Day, M.J. & Tumbrink, M. (1991). Options for contaminant monitoring of hydraulic systems. *Condition Monitoring '91*, proceedings of an international conference, Erding, Germany, 14–16 May 1991. (Ed. Jones, Guttenberger & Brenneke) Pineridge Press, pp. 246–268.

Results of industrial studies undertaken in 1971 and 1983, showed that over 55% of failures were attributable to debris in the hydraulic fluid. A variety of options for its measurement are briefly described. Calibration of optical automatic particle counters is seen as not quite so straightforward as had previously been thought.

Dickson, J. (1991). Monitoring and early failure detection using the Debris Tester II. *Condition Monitoring '91*, proceedings of an international conference, Erding, Germany, 14–16 May 1991. (Ed. Jones, Guttenberger & Brenneke) Pineridge Press, pp. 449–465.

Typical bath-tub curves are illustrated and described for debris generation detectable by the Debris Tester. The technique used is described, as well as the device's ability and achievements.

Dunn, A.R. (1980). Selection of wire cloth for filtration/separation. *Filtration & Separation*, September/October 1980, Uplands Press Ltd. 8 pages.

A detailed description is given of the various types of wire cloth/mesh available. This covers square mesh, plain Dutch single and double weaves, reverse plain Dutch weave, twilled Dutch double weave and Betamesh. The various parameters and characteristics of each are discussed.

Dwyer, J.L. & Connelly, R.F. (1966). *Apparatus for determining the silting properties of liquids containing minute suspended particles.* United States Patent 3,271,999, September 13th 1966 (assigned to Millipore Filter Corporation, California).

The patent describes the technique commonly called the 'Silting Index'. Details of the glassware and associated piston and markings, as well as the necessary equation, are given.

E

Endecotts (1977/1989). *Test Sieving Manual.* Available from Endecotts Limited, latest update and reprint August 1989.

A comprehensive 32 page manual covering; Scope of test sieving, Terminology, Standard sieves, Sampling methods, Techniques of test sieving, Data analysis, Particle 'size', Industrial screening, Specifications, Accessories and Useful data.

Evans, N., Wicks, K. & Smith, G.J. (1963). *Methods of detecting suspended solid material in liquids.* United States Patent 3,111,839 November 26th 1963 (assigned to the British Petroleum Company, London).

An optical absorption technique is described which is able to cope with the presence of suspended water in an aviation fuel, when looking for solid particulate. The idea is that of separating the two liquids by suitable heating prior to the optical analysis taking place.

F

Faulkner, D. & MacIsaac, B. (1991). On-line simultaneous detection of failure mechanisms producing large and fine ferromagnetic debris. *Condition Monitoring '91*, presented to an international conference, Erding, Germany, 14–16 May 1991 (see Campbell 1991 for the basic data in printed form).

A full description is given of the principle of operation of the GasTOPS® FerroSCAN®. A particular example on a pump is cited where initial stages of bearing failure were detected.

Faure, L. (1991). Different types of wear — how to classify? *Europower Transmission*, April 1991. Techniche Nuove, Milan.

A comprehensive survey of the types of wear associated with gear teeth meshing. The article from CETIM is also fully illustrated with worn, or partly worn, gear wheels showing the extent of the various wear modes. No attempt is made to discuss the debris generated.

Fitch, E.C. (1979). *An Encyclopedia of Fluid Contamination Control for Hydraulic Systems.* Fluid Power Research Center, Oklahoma State University. Publishers — Hemisphere Publishing Corp.

A comprehensive survey of the background and usage of contamination control in fluid power. Particles which constitute a hazard in hydraulic systems are described.

Fitch, J.C. (1991). *Proactive maintenance can yield more than a 10-fold saving over conventional predictive/preventative programs.* Technical Application Note, Diagnetics Inc., 16 pages.

The value is stressed of spending time and money on tests which look at what might cause failure, rather than on the failure itself (even if the early stages of it). In other words dangerous contamination in the oils should be monitored, by such as the Contam-Alert®.

Fuell, F.J. (1968). *Apparatus for assessing the level of solid contaminants in hydraulic systems.* British Patent 1,098,186 January 10th 1968 (assigned to Fairey Engineering Limited, Heston, UK).

The technique explained is a combination of Filter Blockage and Silting. Fluid from a working pressure line is forced through two equal fluid resistances, one being a fine gap and the other a long spiral orifice of larger bore than the shorter gap. When the gap became silted the flow would predominantly occur from the larger bore, and a signal would be generated.

G

Gaucher, D.E. (1984). Real-time particle monitoring in an ultrapure water system. *Ultrapure Water*, September/October 1984, Tall Oaks Publishing Inc., pp. 26–28.

The method used by the Micro-pure ultrasonic monitor is described. Because the monitor is able to detect up to 6 points in any desired sequence and because of its sensitivity, it is shown to be very suitable for in-line measurements of particles in water.

Glacier Metal Company Ltd (1985). *Centrifugal Filters, Winning the Fight for Fitter Engines.* Designers Handbook No. 3 Glacier Metal Company Limited.

The effect of debris in bearing tests indicated that the major size to cause wear was of the order of 3 μm to 15 μm, hence the need for suitable filtration.

Goldsmith, A. (1984). Fluid power particle identification by ferro-kinetics. *Contamination Control in Hydraulic Systems*, proceedings of a conference at the University of Bath, September 1984. I.Mech.E, pp. 51–55.

The various usual features of microscope analysis of particles are described, and also the special rotating magnet stage developed by the author to detect ferro-magnetic particles in a freshly prepared membrane sample.

Goldsmith, A. (1986). Catching the engineer's eye. *Royal Naval Hydraulic Engineering Conference*, Manadon, Plymouth, April 1986.

A description is given of microscope analysis of debris held on a membrane. Particle levels, back-projection and the SEM are discussed with a survey of the use of Conpar.

Graham, A.L. & Hanna, T.H. (1962). The micro-particle classifier. *Ceramic Age*, September 1962, 2 pages.

A detailed description of the AB Bahco (Sweden) air-centrifuge-elutriator is given along with its operation and calibration.

Gregory, J. & Nelson, D.W. (1984). A new optical method for flocculation monitoring. *Solid–Liquid Separation*, 1984, Ellis Horwood, Chichester, pp. 172–182.

This is an introduction to the Photometric Dispersion idea for measuring the concentration of particles in a suspension. The suspension is passed through a transparent tube and the way light is transmitted (with a fluctuating ac component) determines the particulate present.

Guide to Wear Particle Recognition — see Swansea Tribology Centre.

Gulla, A. R. & McCadden, J.F. (1987). A large particle detection system. *Condition Monitoring '87*, proceedings of an international conference, University College of Swansea, Pineridge Press. (Ed. M.H. Jones), pp. 259–271.

A technique is described for extending the useful particle size range of an emission spectrometer using a technique of integrating the response signal above a preset threshold. The actual threshold is varied for the system under test but up to 6 separate elements were successfully identified in the 'large' region.

Guy, N.F. (Ed.) (1990). *Hydraulic System Contamination Bibliography*, Elsevier Science Publishers.

A range of chapters covering the practice, effects, measurement and analysis of contamination control. Fluid and system cleanliness is another heading.

H

Hanseler, J. & Preikschat, E. (1987). Particle size measurement with focused laser beam- and back-scattering geometry: on-line readings versus laboratory sampling. *Proceedings of the 18th annual meeting of the Fine Powder Society, 1987*, Particulate/Powder Technology, Boston, August 1987, 9 pages.

Although representing the LASENTEC® equipment, the authors provide a much more general insight into scattered light analysis, sampling techniques, statistical accuracy and possible adverse effects on particle analyses. Considerable agreement was found in one comparison test between the LABTEC 100®, pipette sedimentation, the Sedigraph" and the Microtrac®.

Harvey, S.N. & Herraty, A.G. (1989). Why put up with bearing failures? *COMADEM 89*, proceedings of the 1st international congress on condition monitoring and diagnostic engineering management, Birmingham Polytechnic, 4–6 September 1989. (Ed. Raj B.K.N. Rao & A.D. Hope), Kogan Page, pp. 358–63.

It is suggested that if conditions are as they should be, then there is no need for severe wear or failure of bearings at all! Reasons that there are failures could be lack of suitable lubrication, contaminant ingress, local vibration, electrical current leakage, wrong loads and speeds. Oil debris analysis is seen as one way to monitor these faults developing, along with acoustic emission, orifice testing, performance, power consumption, local temperature and vibration.

Henry, T. (1985). Advances in monitoring bearings. *Proceedings of the Bearings: Search for Longer Life Conference*, M.E.P. conference 1985.

Various methods of monitoring bearings are discussed including lubricant debris monitoring. The total size range of debris is seen as being monitored by spectroscopy, Ferrography and chip detectors. A typical growth pattern of the sizes is illustrated.

Heron, R.A. & Hughes, M.L. (1986). A contaminant monitor for fluid power applications. *Proceeding of the International Conference on Condition Monitoring, Brighton, May 1986*, BHRA, pp. 57–71.

An extensive description is given of the silting valve contaminant monitor developed by BHRA (now B.H.R. Group). The problems and achievements are explained. In practice it was found suitable for particles between 3 μm and 20 μm.

Hipkin, E.L. & Hermann, A. (1978). Safeguarding the health of machinery: plant condition monitoring in a chemical works. Written on behalf of Albright & Wilson Ltd. in *Noise Control and Vibration Isolation*, April 1978.

Although primarily concerned with vibration monitoring, the brief article also discusses the place of spectrometric oil analysis as regards which elements relate to which engineering components.

Holmes, K. (1985). Oil analysis techniques used in the development of automotive diesel engines, and for their condition monitoring in service. *Proceedings of the conference on Vehicle Condition Monitoring and Fault Diagnosis*, March 1985, London. I.Mech.E., pp. 95–100.

The experience of a regular routine oil analysis on Leyland vehicles is described. Details of metals detected, as well as oil properties, are given. Instruments used were ferrography, the Fulmer Wear Debris Monitor, a dielectric monitor and spectrometry.

Horowitz, A.J. & Elrick, K.A. (1986). Evaluation of air elutriation for sediment particle size separation and subsequent chemical analysis. *Environmental Technology Letters*, Vol. 7, Science & Technology Letters 1986, pp. 17–26.

A very thorough evaluation is given of the Bahco air elutriator. There is both a brief description of its working and also various examples of the results achieved, both in size and chemical contents. A full bibliography is included.

Huller, D. & Weichart, R. (1981). Quantitative Shape Analysis of Particle Contours. *Proceedings of Particle Size Analysis 81*, Loughborough Univ. of Tech. September 1981, Heyden, London.

An examination of the means of describing sphericity and roundness of a shape in association with roughness and lumpiness. Fourier coefficients and the Parseval theorem are mentioned and compared.

Hunt, T.M. (1982). Condition monitoring of hydraulic pumps. *Proceedings of the 1st Royal Naval Engineering Conference*, Manadon, February 1982.

A detailed summary is given of the early stages of the three-year investigation into the best methods for monitoring a high pressure high efficiency fluid power pump.

Hunt, T.M. (1985). Contamination and viscosity monitoring of automobile and motor cycle oils using a portable contamination meter. *Proceedings of the conference of Vehicle Condition Monitoring and Fault Diagnosis*, March 1985, London. I.Mech.E., pp. 41–8.

The results of an unusual series of tests on car and motor cycle oils using an on-line monitor. Both viscosity changes and debris changes were detected with the filter blockage device.

Hunt, T.M. & Kibble, J.D. (1986). Condition monitoring of hydraulic systems. *Proceedings of 2nd Royal Naval Hydraulic Engineering Conference*, Manadon, April 1986.

The final report on the work initially summarised in Hunt 1982. The findings relate to the comparison between the different test techniques. Although wear debris is mentioned as a possibility, the best method involved pressures.

Hunt, T.M. (1987). *Particles in Fluids — Instruments and Techniques.* Book produced and published by the author, April 1987. 72 pages.

Information supplied by manufacturers and suppliers on 34 different devices for assessing the particle content in liquids, all set out in comparative form including particle size covered and costs.

Hunt, T.M. (1988). Condition monitoring of British Coal high pressure water pumps. *Proceedings of the 1st Bath International Fluid Power Workshop*, University of Bath, September 1988. (Editors: C.R. Burrows & N.D. Vaughan), pp. 99–112.

A brief description of various techniques examined which might have been suitable for monitoring the type of failure experienced. An in-depth discussion of the fluid monitoring which was finally identified as the best method.

Hunt, T.M. & Bowns, D.E. (1983/1987). *Contamination level indicator.* European Patent Specification, 0116580, European Patent Office, Filing 4th August 1983, granted 9th December 1987. 11 pages.

The full patent specification of the filter blockage technique used or could be used in contamination monitors. Work developed at the University of Bath Fluid Power Centre under a grant from the Department of Trade and Industry.

Hunt, T.M., Rice, C.G. & Tilley, D.G. (1983). *A review of instruments and measurement techniques for the characterisation of mineral oil contaminant.* Fluid Power Center, University of Bath. Private document for the Department of Trade & Industry.

A comprehensive manual covers particulate sizing, level assessment, identification, water content and fluid condition and degradation, primarily concerned with the fluid power industry.

I

Ionnides, E. & Jacobson, B. (1989). Dirty Lubricants — Reduced Bearing Life. *SKF Ball Bearing Journal Special Issue — 89*, pp. 22–27.

The authors provide an extensive review on the wear effect on bearings produced by debris in oils. They also cite the significant deterioration of life when there is even 0·01% of water present in the oil.

J

Jantzen, E. & Buck, V. (1983). Influence of particle size on wear assessment by spectrometric oil analysis, I Atomic Absorption Spectrometry. *Wear*, 87 (1983), pp. 331–338.

The upper limit of particle size is discussed as regards spectrometric analyses. In this brief article it is demonstrated that 1 μm is the maximum size for AAS if the optimum reliability is required.

Jarvis, E.E. & Lewis, D.C (1984). Routine condition monitoring. *The Mining Engineer*, August 1984.

A description is given of the current monitoring recommended for underground and surface use with coal mining. Gearbox debris is monitored for quantity to determine trends. Savings have been experienced through this method. Ferrography and in-line ferrous wear monitoring is also discussed.

Jin, Y. & Wang, C. (1989). Spherical Particles Generated during the Running-in Period of a Diesel Engine. *Wear* 131, pp. 315–328.

Although primarily using debris from a diesel engine test, other processes are also examined. The detailed examination of the spheres, using Ferrography, optical microscope and SEM, enables the authors to identify melting and rapid cooling as a key reason for the formation of spheres.

Joffe, S.H.D. & Allen, C. (1990). The wear of pump valves in fine particle quartzite slurries. *Wear* 135, pp. 279–291.

Quartzite slurry, with particles of nominally 200 μm size, were used to investigate the wear of a disc poppet valve. As the wear was associated with impact wear, a reduction could be achieved by control of the operating procedures.

John (85). An account of the life of Jesus Christ. *Proceedings of the Holy Bible*, 34 pages.

Numerous examples, albeit limited, are given of the many instructions given by the subject on the health of, and ultimately the life of, human beings.

Johnson, T.J., Vogel, W., Feran, N. & Burton, M.K. (1988). The routine application of an automated system for the determination of wear metals in oil. *Proceedings of the 2nd International conference on Condition Monitoring*, London, 24–25 May, 1988. BHRA, paper 18, 14 pages.

British Rail use of spectrometric equipment started in the early 1960's and has had considerable success. A further development of this is described in the form of the ARL Fast Oil Analyser (using ICP) which is in routine use. Although high capital cost is involved, the running costs are significantly reduced.

Jones, M.H. (1990). Grease — a suitable case for condition monitoring. *COMADEM 90*, proceedings of the 2nd international congress on condition monitoring and diagnostic engineering management, Brunel University, 16–18 July 1990. (Ed. Raj B.K.N. Rao, Joe Au & Brian Griffiths) Chapman & Hall, pp. 98–107.

The description is given of the method used to obtain a representative grease sample from a bearing and the successful results achieved.

Jones, W. & Larkin, M.C. (1988). Machine wear shown in oil by magnetics. *Plant Services*, March 1988, 2 pages.

An example is given of the use of the Tribometrics Model 56 Wear Particle Analyzer in a steel plant on some 60 engines a month. It was seen as a very effective way to improve maintenance by catching just the wear debris. Alarms would be triggered when there was a need for engine rebuild or overhaul.

K

Karasikov, N. & Krauss, M. (1989). Examining the influence of index of refraction on particle size measurements — using a time-of-transition optical particle sizer. *Filtration & Separation.* March/April, 1989, pp. 121–124.

The Galai CIS-1 is used to demonstrate a consistent reading even when the particle index of refraction changed. It is suggested that this is a great improvement over the light scattering or diffraction analysers.

Kasten, W. (1967). *Contamination indicator.* United States Patent 3,357,236, December 12th 1967 (assigned to the Bendix Corporation, Madison Heights).

Two filter units are use to separately remove and measure the presence of water and solids in a fuel. The pressure drops across the individual filters give a measure of how much of each contaminant has been absorbed.

Kohlhaas, G. (1991). Field experience concerning oil condition monitoring of aircraft hydraulic systems. *Condition Monitoring '91*, proceedings of an international conference, Erding, Germany, 14–16 May 1991. (Ed. Jones, Guttenberger & Brenneke) Pineridge Press, pp. 269–286.

An emphasis is placed on the monitoring of the oil viscosity, dirt, water and chlorinated contamination. 14 different reasons for changing oil because of possible adverse influence are discussed and some examples given. Various company oil cleanliness requirements are also given.

Kreikebaum, G. (1990). Sub 0·1 micron particle detection and analysis using high sensitivity nephelometry. *Proceedings of the International Conference on Particle Detection, Metrology and Control*, Arlington, USA, 5–7 February 1990, Institute of Environmental Sciences Parenteral Drug Association Inc., pp. 601–625.

A description is given of the modulating laser diode method allowing a sensitivity of 0·00001 NTU in turbidity measurements. Light scattering theory is also given. Examples of use include the real time examination of filter characteristics. Calibration is discussed. Discussion related to Climet instruments.

Kwon, O.K., Kind, H.S., Kim, C.H. & Oh, P.K. (1987). Condition monitoring techniques of an internal combustion engine. *Condition Monitoring '87*, proceeding of an international conference, University College of Swansea, 31 March – 3 April, 1987. (Ed. M.H. Jones) Pineridge Press, pp. 461–475.

After discussing the relative merits of a number of different techniques examined, the suggestion is made that the combination of the RPD and PQ is the most suitable for detecting abnormal wear in an internal combustion engine.

L

Lantos, F.E. & Lantos, J. (1971). Method for determining 'free carbon' and 'oxidised matter' in used lubricating oils. Lubrication Engineering, Vol. 27 No 6 June 1971, *Journal of the American Society of Lubrication Engineers*, pp. 184–189.

The technique involves depositing oil on filter paper strips in a certain manner, and observing the three specific zones of evidence. Distinct differences are apparent for obtaining information on combustion, engine cleanliness, piston ring sticking, hermeticity of combustion chambers, filtration efficiency, contamination and lube oil life.

Lasentec (1988). Control of oil droplet size in lubricants for sheet mills using the LAB-TEC 100ME analyser system. *Application Bulletin*, Red. #B-99-0011, Laser Sensor Technology Inc., November 1988, 4 pages.

After a description of the importance of maintaining an optimum oil droplet size in sheet mill applications, the LAB-TEC® is shown as ideal for the monitoring. Not only is oil droplet size measured (around 2 μm to 4 μm) but also the presence of solid debris (at 6 μm).

Lewis, D.C. (1984). Machine health monitoring at MRDE. *Condition Monitoring '84*, proceedings of an international conference, University College of Swansea, 10–13 April, 1984. (Ed. M.H. Jones) Pineridge Press, pp. 470–484.

The monitoring of a gearbox up to failure, at the Mining Research & Development Establishment, enabled a comparative assessment of several monitors to be made. Temperature, vibration, kurtosis, ferrography and ferrous debris testing were all examined. Both vibration and oil debris analysis were found to be successful. A further illustration of successful oil debris analysis in practice is also given.

Lewis, D.C. (1986). Condition monitoring on two levels. *Proceedings of the 1st International Conference on Condition Monitoring*, Brighton, 21–23 May 1986. BHRA, pp. 143–158.

The idea of using two stages in monitoring is developed in both wear debris and vibration. There is the initial immediate indication and then the need for closer more detailed examination. As regards wear debris, whilst something like the Debris Tester could given the first indication, the more advanced test would be spectrometric or borescope, or possibly verification from vibration.

Lewis, R.T. (1991). Particle size considerations in wear particle analysis. *Condition Monitoring '91*, proceedings of an international conference, Erding, Germany, 14–16 May 1991. (Ed. Jones, Guttenberger & Brenneke) Pineridge Press, pp. 350–363.

In order to cover the larger particles missed by spectrometric analysis, the model 56 wear Particle Analyzer from Tribometrics Inc. has a twin feature giving figures for both small and large debris (even over 100 μm). It was found that this analysis was independent of automobile running miles where normally the fine debris would mask the larger particles.

Lines, R.W. (1985). British Standards for particle and surface characterisation. *Particle Size Analysis 1985*, proceedings of the 5th PSA conference, University of Bradford, 16–19 September 1985, pp. 171–179.

A full listing is given, with some amplification, of all the then current British Standards. This covered both powders and particles in liquids. Sizing methods, terminology and sampling is included.

Lloyd, O. & Cox, A.F. (1981). Monitoring debris in turbine generator oil. *Wear*, 71 (1981), pp. 79–91.

A main constituent of the debris generated from an approaching failure of generator bearings and hydrogen seals appears to be tin (or white metal). This is not detectable by the magnetic sensors. Both spectrometric (XRF) and visual methods are possible but they provide insufficient time for remedies. A capacitive transducer was swamped by carbonaceous debris. The development of an entirely new filter blockage device was proposed. [See next reference.]

Lloyd, O., Cox, A.F. & Hammond, W.A. (1986). An automatic on-line debris in oil monitor and sampler. *Proceedings of the 1st International Conference on Condition Monitoring*, Brighton, 21–23 May 1986. HRA, pp. 73–84.

Detailed information is given on the development of the debris-in-oil monitor principle and practice. Some field experience on large rotating plant helped quantify the reasons for teething troubles on the plant. [The monitor was taken up by Muirhead Vactric as the SDM 100 Debris Monitor.]

Lloyd, P. J. (1988). Trends in particle size analysis. *The Complete Works of Coulter*, proceedings of the Coulter scientific symposium 'Particle Technology', Stratford upon Avon, 13–14 June 1988, pp. 47–51.

A general survey of past and developing techniques in particle size measurement. An emphasis on volume measurement and some discussion on current deficiencies in other instruments.

Lukas, M. & Anderson, D.P. (1991). Techniques to improve the ability of spectroscopy to detect large wear particles in lubricating oils. *Condition Monitoring '91*, proceedings of an international conference, Erding, Germany, 14–16 May 1991. (Ed. Jones, Guttenberger & Brenneke) Pineridge Press, pp. 372–398.

The authors, from Spectro Incorporated, describe eight factors which affect large particle analysis in conventional spectroscopy. A description is then given of three methods which go some way in improving the situation — single spark, ashing rotrode and acid digestion differential.

M

Maier, K., Jantzen, E. & Schröder, H. (1991). A contribution towards obtaining and evaluating wear particles from engine filters. *Condition Monitoring '91*, proceedings of an international conference, Erding, Germany, 14–16 May 1991. (Ed. Jones, Guttenberger & Brenneke) Pineridge Press, pp. 514–526.

Debris retained by filters in aircraft engine oil systems contains valuable evidence of parts wearing. Although it is possible to release the debris by backflushing a removed element, it is shown that whilst mere shaking (50 Hz) did little to help, ultrasonics (50 kHz) provides a significant improvement. A basic, but effective, rig for regular assessments is described.

McFadyen, P. (1989). Laser particle sizing goes on-line. *Process Industry Journal*, May 1989, pp. 27–29.

Laser diffraction (Fraunhofer) particle detection is explained, and its high particle size range emphasized in comparison with other methods. Whilst such a technique was previously only for laboratory use, some new models are able to go on-line (eg. the Malvern MasterSizer IP) with automatic control of the process.

Miller, B. (1988). Development of a commercial instrument using the mesh obscuration Technique. *The complete Works of Coulter*, proceedings of the Coulter scientific symposium 'Particle Technology', Stratford upon Avon, June 1988, pp. 127–129.

A discussion on the early findings of the Coulter Electronics development of the LCM II, in particular the comparison of the LCM II with the Coulter Counter® and the HIAC.

Miller, C.C. & Rumberger, W.E. (1972). *Chip detecting and monitoring device.* United States Patent 3,686,926 August 29th 1972 (assigned to the Boeing Company, Seattle).

A cylindrical grid is described, where the warp and weft differ by one being of a conducting material and the other a non-conductor. When a conducting metal particle is trapped by the 'mesh', a change in resistance is detected.

Mills, G.H. (1985). Development of a continuous debris monitor. *Condition Monitoring in Hostile Environments*, proceedings of the seminar, 26th June 1985, ERA Technology Ltd, Leatherhead, section 2.3, 12 pages.

A later description of the Continuous Debris Monitor (see Bogue 1984). A full discussion relating to the use of Hall Effect sensors and the results achieved after the initial trials, is enhanced by the printing of the informal discussion which followed the presentation.

Minns, H. & Stewart, R.M. (1972). An introduction of condition monitoring with special reference to rotating machinery. *Workshop on On-Condition Maintenance*, Southampton University 1972.

A very thorough review of techniques of the day, including wear debris monitoring as seen by chip detectors and spectrometric analysis.

Morley, G. (1982). British Rail experience of engine monitoring *Wear Debris Analysis Group Newsletter*, November 1982.

A summary is given on the many years' experience of British Rail in monitoring the oils used in the main line diesel engines. This involved both looking at the oil itself as regards its continuing effectiveness, and also at the mechanical component wear generating debris.

Mucklow, P. (1984). An overview of techniques for monitoring wear debris in oil. *PEMEC 84*, paper from the Factory Efficiency conference, Birmingham NEC, 25–28 September 1984, session 3E paper 1, 25 pages.

The experience shared is from the use of debris detectors and collectors for Rolls-Royce engines. A complete review of the need for wear debris monitoring is given, followed by a listing of the techniques then available; most of these being under development.

N

Neale, M.J. (1987). Trends in maintenance and condition monitoring. *Condition Monitoring '87*, proceedings of an international conference, University College of Swansea, 31 March–3 April 1987. (Ed. M.H. Jones) Pineridge Press, pp. 2–12.

Two major reasons for including condition monitoring are identified as where a system gradually deteriorates, and where statutory regulations require regular overhaul, the monitoring indicating priorities in the overhaul. Trends in wear debris monitoring are seen as a progression to on-line monitoring.

Newell, G.E. (1991). Computer aided oil analysis as a predictive maintenance tool. *Condition Monitoring '91*, proceedings of an international conference, Erding, Germany, 14–16 May 1991. (Ed. Jones, Guttenberger & Brenneke) Pineridge Press, pp. 225–234.

A portable kit, such as the Oilab kit, can enable a number of oil properties to be assessed rapidly on site. These properties include physical conditions such as viscosity, insolubles, TAN and TBN as well as the change caused by foreign additives such as water, fuels and abrasives. An example of a typical lube oil analysis is given where 50% is devoted to wear debris analysis.

Nicholls, C. (1989). Cost effective condition monitoring. *COMADEM 89*, proceedings of the 1st international congress on condition monitoring and diagnostic engineering management, Birmingham Polytechnic, 4–6 September 1989. (Ed. Raj B.K.N. Rao & A.D. Hope. Kogan Page, pp. 335–347.

Numerous examples are given of savings achieved by companies due to the introduction of a condition monitoring programme. Although mainly related to vibration monitoring, lubrication analysis, including wear debris measurement and oil particle analysis, is also mentioned in connection with gear, bearing and hydraulic system faults.

P

Perkins, W.D. (1987). Fourier transform infrared spectroscopy. *Journal of Chemical Education*, Vol. 63, January 1986, pp. A5–A10 and Vol. 64, November 1987, pp. A269–A271.

These two articles complete a thorough description of the FT–IR instrumentation and advantages, Extensive theory and relevant illustrations explain why the FT–IR is superior to the conventional dispersive instrument.

Peterson, M.B. & Winer, W.O. (Editors) (1980). *Wear Control Handbook.* ASME, New York.

A large book mainly concerned with reducing wear. However, there are useful passages on wear mechanism and measurement, including the classification of wear.

Pipe, K. (1987). Application of advanced pattern recognition techniques in machinery failure prognosis for turbomachinery. *Condition Monitoring '87*, proceedings of an international conference, University College of Swansea. 31 March–3 April 1987. (Ed. M.H. Jones) Pineridge Press, pp. 73–89.

Included in the general modelling argument are ideas which can be used for vibration analysis on turbine engine blades and gears. It particularly emphasises the importance of getting at the real problem, eg. fatigue, rather than just knowing that a vibration spectrum has changed.

Prakash, A. & Gandhi, O. P. (1991). Identification of hydraulic system malfunctions using ferrography. *Condition Monitoring and Diagnostic Technology*, Vol. 2, No. 2, October 1991, B. Inst.NDT., pp. 58–63.

A full survey is made of types of debris as regards their presence in a system. Two in particular are signalled out — the 'easy to dislodge' (ETD) and the 'difficult to dislodge' (DTD). Ferrography is shown to be able to identify these two, and help in determining the source of debris.

Price, A.L. & Roylance, B.J. (1984). The Rotary Particle Depositor — a response to problems experienced with wear particle deposition. *Condition Monitoring '84*, proceedings of an international conference, University College of Swansea, 10–13 April 1984. (Ed. M.H. Jones) Pineridge Press, pp. 596–607.

After a discussion on some of the features of conventional ferrography, the Rotary Particle Depositor, developed in Swansea Tribology Centre, is introduced. Results from the RPD are compared with the Analytical Ferrography. Particular emphasis is given to time of analysis, cost and ease of use.

Price, A.L., Roylance, B.J. & Zie, L.X. (1987). The PQ — a method for the rapid quantification of wear debris. *Condition Monitoring '87*, proceedings of an international conference, University College of Swansea, 31 March–3 April 1987. (Ed. M.H. Jones) Pineridge Press, pp. 391–405.

The technique of the Particle Quantifier is introduced and numerous examples given of its ability to assess quantitatively the quantity of ferrous debris in a small oil sample or from a filter membrane. Comparative tests were undertaken with the four-ball machine and I.A.E. gear machine.

Provost, M. J. (1989). COMPASS: A generalized ground-based monitoring system. *COMADEM 89*, proceedings of the 1st international congress on condition monitoring and diagnostic engineering management, Birmingham Polytechnic, 4–6 September 1989. (Ed. Raj, B.K.N. Rao & A.D. Hope), Kogan Page, pp. 74–87.

The means of integrating a number of different monitoring devices is described. As regards wear debris, SOAP and the magnetic chip detector are included.

Q

Quantachrome (1991). *Introduction to powder and bulk solid characterisation.* Booklet produced by the Quantachrome Corporation, 1991, 12 pages.

The following particle features are explained and discussed — bulk, tap and true densities, surface area (including multipoint and single point BET — Brunauer, Emmett and Teller), pore size distribution and measurement, and contact angle.

R

Ramos, J.G.D. (1990). Determination of particle size distribution of submicron particles by capillary hydrodynamic fractionation (CHDF). *CHDF Application Note 501)*, Matec Applied Sciences, USA, 15 pages.

An extensive historical survey is given with some comparison to other methods such as sedimentation field flow fractionation (SFFF) and hydrodynamic chromatography (HDC). The theory and experimental verification of CHDF is given.

Ranco Controls (1990). On-line lube monitoring is even better. *World Mining Equipment*, July/August 1990, pp. 46–47.

The manufacturer of the Continuous Debris Monitor explains its value for on-line gear box monitoring and then illustrates its success from a Coal Board example. [See Chapter 13].

Rao, R.B.K.N. (1991). New and novel postgraduate diploma in system monitoring and diagnosis. *COMADEM 91*, proceedings of the 3rd international congress on condition monitoring and diagnostic engineering management, Southampton Institute, 2–4 July 1991. (Ed. Raj, B.K.N. Rao & A.D. Hope), Adam Hilger, pp. 424–428.

A full outline is given of the Southampton Institute's one year postgraduate course in condition monitoring. The six modules cover balancing & systems, principles & practice of condition monitoring, diagnostic techniques, maintenance management, total quality management and industrially placed projects. Wear debris analysis is an integral part of the course with other techniques such as vibration, noise, corrosion, etc.

Raubenheimer, D.S.T. (1989). Condition-based Maintenance — Where to next? *COMADEM 89*, proceedings of the 1st international congress on condition monitoring and diagnostic engineering management, Birmingham Polytechnic, 4–6 September 1989. (Ed. Raj B.K.N. Rao & A.D. Hope), Kogan Page, pp. 20–24.

The five page article was a Key Note lecture. It includes a description of the problems associated with positioning sampling points in a system in order to achieve realistic results.

Raw, I. & Hunt, R.M. (1987). A particle size analyser based on Filter Blockage. *Condition Monitoring '87*, proceedings of an international conference, University College of Swansea, 31 March–3 April 1987. (Ed. M.H. Jones) Pineridge Press, pp. 875–894.

A comprehensive survey of a large number of devices and patents relating to the Filter Blockage technique covering over two decades. Although most had very limited applications, or were not produced, some items have been highly successful in the debris monitoring field.

Rideal, G. (1985). Today's techniques of particle size analysis. *Laboratory Equipment Digest*, February 1985, pp. 91–97.

Difficulties in deciding the 'size' of a particle and in its detection are illustrated by discussion on seven different types of particle analyser — Sieves, Microscopy (optical and SEM), Light obscuration, Electrical Sensing Zone, Sedimentation (gravity and Centrifugal), Photon correlation spectroscopy and light diffraction.

Riley, N.H. & Mann, M.D. (1990). *The role of the lubricant as a condition monitoring tool.* Century Oils Limited report, 23 pages.

The importance of obtaining an accurate sample, and how this can be achieved is discussed. From such a sample, good monitoring of machine health is possible. Examples of spectrometric ppm and particle count trends are described.

S

Saba, C.S., Rhone, W.E. & Eisentraut, K.J. (1981), discussed in Jantzen & Buck 1983.

Santilli, R. (1989). The Fulmer method of monitoring fluid abrasivity as an indication of fluid condition and machine health. *COMADEM 89*, proceedings of the 1st international congress on condition monitoring and diagnostic engineering management, Birmingham Polytechnic, 4–6 September 1989. (Ed. Raj, B.K.N. Rao & A.D. Hope.) Kogan Page, pp. 55–67.

A number of different uses of the Fulmer monitor are explained and the results described. The realistic range of debris/particle is discussed.

Sasaki, A. & Dunthorne, O.M. (1984). Electrostatic liquid cleaning of hydraulic fluid. *Contamination Control in Hydraulic Systems*, proceedings of the conference at the University of Bath, September 1984. I.Mech.E, pp. 111–118.

A full description of the Kleentek electrostatic cleaner as regards theory, design and application results. Its comparative efficiency with other types of filter is described particularly as regard particles close to 1 μm.

Sayles, R. S. & Macpherson, P. (1982). Influence of wear debris on rolling contact fatigue. *Rolling Contact Fatigue Testing of Bearing Steels*, ASTM STP 771, J.J.C. Hoo (Ed.), pp. 255–274

Considerable evidence is given on the effect that quite small debris can have on the life of a bearing.

Schwabe, P.H. & Asher, J. (1987). The use of Thin Layer Activation in condition monitoring. *Condition Monitoring '87* proceedings of an international conference, University College of Swansea, 31 March–3 April 1987. (Ed. M.H. Jones), Pineridge Press, pp. 852–862.

Some of the exceptional features of the TLA monitoring method are outlined, as well as different techniques and some examples of applications where considerable time saving has been achieved.

Seaman, A. (1987). An introduction to the application of 'EDXRF' methods to condition monitoring. *Condition Monitoring '87*, proceedings of an international conference, University College of Swansea, 31 March–3 April 1987. (Ed. M.H. Jones), Pineridge Press, pp. 831–841.

A comparison is made between debris collected and analysed in the approved manner using SOA, Ferrography, Magnetic Plugs and Filters, with that possible with EDXRF. It is concluded that the XRF technique holds many advantages.

Seow, S.T. & Kuhnell, B.T. (1990). Computer aided diagnosis of wear debris. *Proceedings of the 3rd Bath International Fluid Power Workshop on Computers in Fluid Power*, University of Bath, September 1990.

A comprehensive survey of the state of the art is given relating to automated image analysis of wear particles. Many real examples are given and the various relevant problems are discussed. The 31 references cover the field from 1975.

Shapespeare Corp. (1991). *Classical image analysis — the long march to nowhere.* A technical report from the Shapespeare Corporation, March 1991, 16 pages.

The report describes some of the classical means of describing both size and shape of a two-dimensional projection. Protected Area, Feret Diameters, Martin's Diameter, Perimeter and shape factors are shown to be not quite as helpful for real particles as they are for spheres. Brief comments are also added on 'physical behaviour' means of describing shape, as with Stokes' Law and sedimentation.

Sommer, H.T., Harrison, C.F. & Montague, C.E. (1991). Simplifying wear particle monitoring with laser diode light extinction sensors. *Condition Monitoring '91*, proceedings of an international conference, Erding, Germany, 14–16 May 1991. (Ed. Jones, Guttenberger & Brenneke) Pineridge Press, pp. 306–323.

The use of laser diodes for light obscuration sensors has doubled the possible concentration limits. However, even so, for the high levels experienced with debris monitoring, it had been found necessary to design a new volumetric sampler and a counter with adapted signal processing. In this way debris only counted above a certain size (e.g. 15 μm) was not confused by over concentrations that would have occurred at smaller sizes. Written from HIAC/ROYCO point of view.

Spair, J. A. (1990). *The A-B-C's of NTU's — a primer on Turbidity.* A booklet from Advance Polymer Systems, 1990, 42 pages.

A comprehensive description of turbidimetric methods and how the different methods compare. The means of calibration are also discussed in relation to the standards (primary and secondary).

Stafford-Allen, R. (1982). Oil analysis in engine production. *Wear Debris Analysis Group Newsletter*, November 1982.

A brief account is given if the XRF analysis programme on the debris in oils used during the production of diesel engines. The use is mentioned of a 'mini-patch' filter within the lubrication system to obtain the debris.

Stewart, R.M. (1975). Assessing the general condition of rotating machinery. *Proceedings of the Society of Environmental Engineers Conference SEECO 75.*

The paper deals with the investigation of both a rolling bearing rig and a gear rig. It gives a set of results for the bearing rig comparing, amongst others, vibration and wear debris.

Summerfield, A.H. & Mathieson, D. (1981). Monitoring lube oil debris in marine gas turbines. *Diesel & Gas Turbine Worldwide*, March 1981 USA.

The authors describe the value of magnetic chip detectors for assessing the failure of bearings. However, it is by a repeatable trend monitoring that the failure is most likely to be detected; this is illustrated by the successful use of the Debris Tester (MkI) on a marine gas turbine.

Swank, R.L. (1966). *Constant flow pressure filter apparatus.* United States Patent 3,266,299 August 16th 1966 (unassigned).

A third attempt by the inventor to develop a device for detecting unacceptable solids in blood. He uses a syringe full of blood and a constant movement of the piston forcing the blood through a filter membrane. The pressure drop recorded, indicated the presence of solids.

Swansea Tribology Centre (1990). *A guide to wear particle recognition — for use with the Rotary Particle Depositor*, University College of Swansea, April 1990, 40 pages.

Although written primarily for use with the RPD, the fully illustrated manual describes debris trapped by ferrographic means. Both SEM and optical microscope pictures are reproduced, in B/W and in colour.

T

Tauber, T. (1983). AIR 1828 — A guide to gas turbine engine oil system monitoring. *SAE Technical Paper*, 831477 1983.

The extensive use of chip plugs and sampling on board an aircraft is vividly illustrated by an all-inclusive diagram. However, the author points out that each engine has to have its own tailor-made system to work effectively.

Thomas, G. (1990?). Overcoming the limitations of particle size analysis. *International Labmate*, 1990?, pp. 9–10.

The COULTER® LS series particle size analyser is introduced showing how its range has been extended (over most other laser diffraction analysers) by using an additional lens to capture light scattered at greater angles. In addition, in order to improve resolution below 0·5 μm a polarisation intensifier was added.

Thomas, J.C. & Fairhurst, D. (1990). High resolution particle size analysis of colloidal materials using a disc centrifuge technique. *International Labmate*, 1990, pp. 25–29.

The authors from Brookhaven Instruments Corp. describe the basic theory of the disc centrifuge and illustrate how it is able to accurately determine size even for narrow distributions. It is suggested that the instrument is very applicable for the coatings industry.

Thompson & Walsh (1989). *Handbook of inductively coupled plasma spectrometry*, 2nd Edition, Blackie, Glasgow, 1989.

A full review of the ICP techniques covering the various types and comparing them as regards result and cost of running. The progress of DCP is mentioned.

Tilley, D.G. (1984). An investigation into the movements of contamination in hydraulic systems. *Contamination Control in Hydraulic Systems*, proceedings of the conference at the University of Bath, September 1984. I.Mech.E. pp. 33–40.

The results are given of a practical investigation into particle movement in pipes and reservoirs. A completely new design of reservoir is described enabling particles to be kept in suspension.

Tweedale, P. (1991). Clues to gear failure. *Professional Engineering*, Vol. 4 No. 9, October 1991, pp. 26–27.

There are several quite different modes of failure of gears. These are explained and charted on a Load v. Speed diagram. The type of surface changes and cracks which develop are described. Although not mentioning the debris generated, the mechanism of failure outlined provides a good understanding of the type of particle likely to be in the oil.

U

Uedelhoven, W., Franzl, M. & Guttenberger, J. (1991). The use of automated image analysis for the study of wear particles in oil-lubricated tribological systems. *Wear* 142, pp. 107–13.

Starting with Ferrography as a means of precipitating the debris, the use of automatic image analysis is seen as a success in investigating transmission lubricating oils. A significant change in particle size and shape is recorded as failure progresses.

Underwood, D.R., Newhouse, D.P. & Fisher, C.B. (1986). On-site completion/ workover fluid analysis using a portable laser particle size counter. *Society of Petroleum Engineers*, 61st annual technical conference, New Orleans, 5–8 October 1986, 8 pages.

Providing the particle concentration was below 999 counts per mL, and the opacity below or equal to 30%, then the Spectrex instrument was found to be a valuable portable device. It was found that if excessively large particles are present, they will mask the smaller particles until they are filtered out of the system.

V

von Bernuth, G. (1988a). Laser-Particle Sizer "analysette 22": A measuring device for laboratory and factory for measuring particle size distributions in mineral and synthetic raw materials. Published in *TIZ*, May 1988. 6 pages.

The theory behind the Fraunhofer diffraction technique using a convergent beam is fully explained and compared with the more conventional parallel beam.

von Bernuth, G. (1988b). *Particle size analysis in the laboratory technology and equipment.* Lecture prepared for Fritsch GmbH, July 1988, 8 pages.

A description of the full range of Fritsch equipment indicates an ability to provide a well run laboratory. Sieves, shakers, sedimentographs, centrifuges, etc. are described.

W

Walsh, C.T. (1991). Condition monitoring of machine systems for the 1990s and beyond. *Condition Monitoring '91*, proceedings of an international conference, Erding, Germany, 14–16 May 1991. (Ed. Jones, Guttenberger & Brenneke), Pineridge Press, pp.1–24.

A brief review is given of a variety of condition monitoring techniques including debris monitoring, temperature, flow, performance, acoustic emission, electric motor and vibration (which is given a somewhat supreme position!). On-line testing, rather than off-line, is suggested as best.

Wear Particle Atlas (1982/1991). Prepared by Daniel P. Anderson in 1982 with fourth, Revised, edition published in September 1991 by Spectro Incorporated, 160 Ayer Road, Littleton, MA 01460, USA. (price $100 in 1991).

This 192-page description of wear particles observed under microscopy is a classic reference book for all forms of ferrographic analysis. (It is a full revision of the original Bowen & Westcott Wear Particle Atlas in 1976). It includes over 200 individual photographs of debris, most of which are in colour. Operational procedures are described (including heating) and actual report examples given.

Wear Particle Atlas — see British Coal.

West, J. (1990). Practical considerations in evaluating a liquid-borne automatic particle counter. *Proceedings of the International Conference on Particle Detection, Metrology and Control*, Arlington, USA, 5–7 February 1990, Institute of Environmental Sciences Parenteral Drug Association Inc., pp. 372–94.

Light obscuration sensors are described and the means of determining their resolution discussed. Limitations on flow rate, sample volume accuracy, coincidence and calibration are outlined with features mentioned for their improvement, from the standpoint of Climet instruments.

Whittington, H.W., Flynn, B.W. & Mills, G.H. (1991). A novel on-line wear debris monitor based on an inductive transducer. *Condition Monitoring '91*, proceedings of an international conference, Erding, Germany, 14–16 May 1991. (Ed. Jones, Guttenberger & Brenneke), Pineridge Press, pp. 335–338.

Research undertaken at the University of Edinburgh has indicated the possibility of an inductive monitor able to differentiate between ferrous and non-ferrous metal on-line. The principle of operation being the use of the transducer coil as the resonating inductance in an oscillation whose frequency is compared with a fixed crystal controlled oscillation in a phase locked loop. Change in inductance as small as 0·01% can be acceptably detected.

Wilkes, P. & Hunt, T.M. (1990). Filter blockage technique protects robots at Rover. *Fluid Power*, Number 1, 1990. Fluid Power Intech Promotions Ltd., pp. 60, 61.

An example is given of the use of the Fluid Condition Monitor to successfully detect a fault developing in an industrial robot.

Williams, D.J. (1975). *Estimating the concentration of solid matter suspended in a fluid stream.* United States Patent 3,893,334 July 8th 1975 (assigned to the Australian Paper Manufacturers Limited, South Melbourne).

An ingenious invention for examining the concentration of fibres in a liquid stream using the blockage of a slowly rotating filter disc as a constant pressure upstream is applied.

Williams, P.M. (1988). The characteristics of particle behaviour related to transportation and analysis of wear debris. *M.Sc. thesis, University College of Swansea*, 1988.

A full description of the RPD is given, with the theory as well as practical results. Comparison is made with the Ferrograph and the idea of on-line analysis is discussed.

Wilmott, P. (1990). Presented at the 3rd International Conference on Condition Monitoring, Windsor, October 1990, reported by the *Condition Monitor* No. 47, November 1990, STI, p. 4.

A diagram is presented indicating a range of techniques, including vibration and wear debris, which may be used in a variety of engineering applications. It suggests that oil debris analysis is by far the most comprehensive.

Wyatt, S.J.F. (1987). Practical condition monitoring. *Condition Monitoring '87*, proceedings of an international conference, University College of Swansea, 31 March–3 April 1987. (Ed. M.H. Jones) Pineridge Press, pp. 30–57.

A full survey of British Coal monitoring techniques (including Vibration, SPM, Orifice, Tension, Performance, Thermography and Wear Debris) is given with some examples of the use of the Debris Tester for successfully monitoring a gearbox and a shearer.

X

Xu, X., Qiao, Y. & Huo, Y. (1989). Research on quantitative Ferrography Technology. *COMADEM 89*, proceedings of the 1st international congress on condition monitoring and diagnostic engineering management, Birmingham Polytechnic, 4–6 September 1989. (Ed. RAJ B.K.N. Rao & A.D. Hope), Kogan Page, pp. 450–452, 508–512.

The DR Ferrograph is attached to a microprocessor in order to be able to rapidly determine the quantitative nature of the debris in an on-going trend process.

Y

Yarrow, A. (1982). Magnetic chip detectors in use in the MoD. *Wear Debris Analysis Group Newsletter*, June 1982.

A brief account of Alan Yarrow's report given at the second meeting of this Group. Most of the content is given within the Chapter 13 in this book.

Yurko, R.J., Gilbert, T.R. & Sandoval, J.E. (1987). Direct analysis of wear metal particulates by the ashing rotrode technique. *Conditions Monitoring '87*, proceedings of an international conference, University College of Swansea, 31 March–3 April, 1987. (Ed. M.H. Jones) Pineridge Press, pp. 241–258.

This extensive report from Spectro Incorporated not only outlines the technique of ashing rotrode, but also indicates its distinct advantages for wear debris analysis. It is shown to be better than basic rotrode and direct dilution ICP but not as good as the acid dilution ICP. It was thought that particles up to 40 μm could be accurately detected.

APPENDIX 1

Listing of Instruments and Services

The appendix is divided up into three sections:

Section 1 — INSTRUMENTS FOR PARTICLE DETECTION AND ANALYSIS

Section 2 — IMAGE ANALYSERS

Section 3 — SERVICES AND LABORATORIES

(Names and addresses are given in Appendix 2)

A contents list is given at the start of each section. It is regretted that some manufacturers and suppliers were unable to confirm details of their instruments and services in time for inclusion in this book, and, in their case, they have had to be omitted from this appendix. The inclusion or omission of details does not imply that any one device or service is superior to any other. In addition it should be mentioned that whilst the manufacturers/suppliers have approved or even changed some of the descriptions of the equipment, the author has not necessarily been able to personally check the data.

Readers are advised that most instrument names are registered and the ideas are copyright or patented. Full details of these legal restrictions can be obtained from the manufacturers and suppliers concerned.

In some case the manufacturer has chosen to use a word to describe the principle of operation in a different way to the one chosen in this book. In this case his description is given in [square brackets].

SECTION 1
INSTRUMENTS FOR PARTICLE DETECTION AND ANALYSIS

(excluding spectrometric devices which are described in Chapter 9)

The instrument or kit is identified by its usual name, or, where that is not clear, by the manufacturer's name. (A full list of the manufacturers and their appropriate products is given in Appendix 2.)

(Note   PA = Particle Analyser
        PSA = Particle Size Analyser
        PSDA = Particle Size Distribution Analyser

INSTRUMENT NAME	MANUFACTURER
1. Actiprobe®	Cormon Ltd / A.E.A. Technology (Harwell)
2. Aerometrics Phase Doppler PA	Aerometrics Inc
3. Analite Turbidimeter	McVan Instruments Pty
4. Applied Imaging Disc Centrifuge	Applied Imaging International Ltd
5. Bahco Micro-Particle Classifier	Dietert Division of George Fischer Foundry Systems Inc
6. BHR Group On-line Contaminant Monitor	BHR Group
(Brinkmann PSA [N.America] — see Galai CIS)	
7. Bromley PSA	Bromley Instruments
8. Brookhaven BI-90 Particle Sizer	Brookhaven Instruments Corp
9. Brookhaven BI-DCP Particle Sizer	Brookhaven Instruments Corp
10. CCK 4 Kit	Hydrotechnik UK Ltd
11. Checker-Kit	United Air Specialists (UK) Ltd
12. CLIMET® CI-1000 Particle Counter	Climet Instruments Co
13. CLIMET® CI-1500 Liquid Quality Transducer	Climet Instruments Co
14. Climet® CI-2000 Particle Counter	Climet Instruments Co
15. CM 20	UCC International Ltd
16. Conpar® Particle Comparison Kit	Howden Wade Ltd
17. Contam-Alert® II	Diagnetics
18 Continuous Debris Monitor - Tecalert (CDM)	Ranco Controls Ltd
19. COULTER® LS Series	Coulter Electronics Ltd
20. COULTER® Multisizer Series	Coulter Electronics Ltd
21. COULTER® N4 Series	Coulter Electronics Ltd
22. Dantec Particle Dynamics Analyser	Dantec Elektronik
23. Dawn®	Wyatt Technology Corp.
24. Debris Tester (Digital Contam-Alert — see Contam-Alert)	Staveley NDT Technologies Inspection Instruments (NDT) Ltd
25. ELZONE®	Particle Data Inc.
26. Endecotts Test Sieves	Endecotts Ltd
27. FAS-CC100 Kit	Fairey Arlon Ltd
28. Ferrograph REO 1	VD AMOS
29. FerroSCAN®	GasTOPS Ltd
30. Fluid Condition Monitor (FCM)	Lindley Flowtech Ltd
31. Fritsch Analysette 3/18 Sieves	Fritsch GmbH
32. Fritsch Analysette 20 Photo-sedimentograph	Fritsch GmbH
33. Fritsch Analysette 22 Laser Particle Sizer	Fritsch GmbH

34.	Fulmer Wear Debris Monitor	Fulmer Systems Ltd
35.	Galai CIS-1, CIS-100, CIS-1000, DSA-10	Galai Production Ltd
36.	Hach Turbidimeter	Hach Europe S.A./B.V.
37.	HIAC/ROYCO® 8000 with Light Extinction sensors	HIAC/ROYCO Division of Pacific Scientific
38.	HIAC/ROYCO® 8000 with Submicron Liquid sensors	HIAC/ROYCO Division of Pacific Scientific
39.	Horiba Laser Diffraction PSDA	Horiba Ltd
40.	Horiba Photosedimentation PSDA	Horiba Ltd
41.	Horiba Liquid Particle Counter	Horiba Ltd
42.	LASENTEC® LAB-TEC®, PAR-TEC®	Laser Sensor Technology Inc
43.	Liquid Contamination Monitor (LCM II)	Coulter Electronics Ltd
44.	Magnetic Chip Collectors	Muirhead Vactric Components Limited
45.	Malvern Autocounter (ALPS)	Malvern Instruments Ltd
46.	Malvern Mastersizer	Malvern Instruments Ltd
47.	Matec CHDF 1100 PSA	Matec Applied Sciences Inc
48.	Met One Liquid Particle Counters	Met One Inc
49.	Micro Pure® MPS-D and MPS-3000 Particle Contamination Monitors	Monitek Technologies Inc
50.	Microscan II PSA	Quantachrome Corp
51.	Microtrac PSA 9200 SRA & FRA	Leeds & Northrup International Ltd
52.	Microtrac PSA 9200 UPA	
53.	Millipore® Fluid Contamination Analysis Kit (Gravimetric)	Millipore Corp
54.	Millipore® Patch Test Kit	Millipore Corp
55.	Monitek® Turbidimeter and CLAM Monitor	Monitek Technologies Inc
56.	Oilab Portable Lubricant & Fuel Analysis Kit	Oilab Lubrication Ltd
57.	Oilcheck	UCC International Ltd
58.	Otsuka Laser Particle Analyser	Otsuka Electronics Co. Ltd
59.	Parker Patch Test Kit	Parker Hannifin (UK) Ltd
60.	Particle Quantifier (PQ 90)	Swansea Tribology Centre
61.	PDA 2000 Photometric Dispersion Analyser	Rank Brothers Ltd
62.	PMS LBS-100	Particle Measuring Systems Inc
63.	PMS Micro Laser Particle Spectrometer	Particle Measuring Systems Inc
64.	PMT-3120, PMT-2120, PMT-3550, SKD6	Partikel-Messtechnik GmbH
65.	Polytec Particle Sizer PSE-1500	Polytec GmbH
66.	Purity Controller RC 1000	Hydac S.à.r.l.
67.	Quantitative Debris Monitor (QDM®)	Vickers Inc.
68.	Rion KL-01	Rion Co. Ltd

69.	Rion KL-20, KL-22	Rion Co. Ltd
70.	Rotary Particle Depositor (RPD)	Swansea Tribology Centre
71.	SediGraph 5100	Micromeritics® Instrument Corp
72.	Spectrex® SPC-510 Laser Particle Counter System	Spectrex Corp
73.	Spi-Wear®	Spire Corp.
74.	Sympatec HELOS PSA	Sympatec GmbH

(Tecalert — see Continuous Debris Monitor)
(Wear Debris Monitor — see Fulmer Wear Debris Monitor)

75. Wear Particle Analyser — Model 56   Tribometrics Inc

ACTIPROBE® Manufacturer & UK Supplier:
Cormon Ltd/A.E.A Technology
(Harwell)

———— 1 ————

Technique: RADIO ACTIVATION [THIN LAYER ACTIVATION (TLA)]

Description:

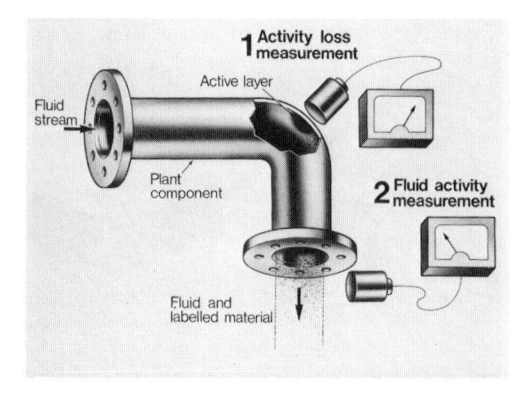

A thin layer of atoms in the surface of a component is made radioactive by bombardment with a beam of charged particles. This is necessary before build. The Actiprobe® system consists of a gamma ray detector which monitors the radioactivity of either the component surface or the fluid.

Range and sensitivity:
The depth of component wear can be monitored between 25 μm and 300 μm. The sensitivity is 1% of the active layer depth, eg. a 25 μm layer can be detected with a sensitivity of 0·25 μm. Debris loss or generation is detected.

Application:
It is an absolute wear debris monitor in the sense of being able to detect precisely where the debris originated. However, the radiation treatment must be applied before a component is attached to the machine system, and normally this would require shipment to an accelerator facility. It is thus primarily applicable to high cost test programmes or applications of considerable importance due to safety or future cost.

Cost:
The most expensive cost is that of the activation. This will depend on the component size and complication, and if a multiple order is involved. Single radiations are likely to cost well over £1000, but this reduces significantly with quantity.

Further details: [WARNING — for latest models check supplier]
See Pages 132, 252 and 306. Refs: Asher 1983, Schwabe & Asher 1987

AEROMETRICS PHASE
DOPPLER PARTICLE
ANALYSER (PDPA)

Manufacturer:
Aerometrics Inc.
UK Supplier:
Laser Lines Limited

Technique:

OPTICAL-PHASE/DOPPLER
SCATTER[LIGHT SCATTERING
INTERFEROMETRY]

Description:

A single component of a particle's velocity and its size may be measured using the He-Ne or Argon Ion laser. 2 or 3 orthogonal components of a velocity field may be measured simultaneously using the appropriate number of lines from an Argon laser.

Sample volume and time of arrival, along with velocity and size, are measured. Measurements may be made in forward or backward scatter, utilising refraction or reflection modes from spherical particles in gas or liquid. Three detectors pre-aligned behind a slit measure the Doppler shift and phase differences to give velocity and size respectively. A sophisticated frequency-domain digital processor avoids drift.

Range and sensitivity:
The size range is much larger than most other methods, covering 0·5 μm to 10,000 μm (10 mm) with an accuracy typically 4%. Maximum particle number density is 10^6/mL; maximum data rate is 100,000 samples per second.

Application:
Spray nozzle developments for fuels, chemicals and paints. Measurements in gas turbines and engines. Particle field characterisation of spherical particles or droplets in gas, and bubbles or droplets in water. Aircraft de-icing studies.

Cost:
Around £50,000 (1-D basic), £120,000 (2-D top of range)

Further details:
[WARNING — for latest models check supplier]
See page 128.

ANALITE® TURBIDIMETER

Manufacturer: McVan Instruments Pty Ltd
UK Supplier: Camlab Ltd.

Technique: OPTICAL -NEPHELOMETRY/ TURBIDITY

Description:

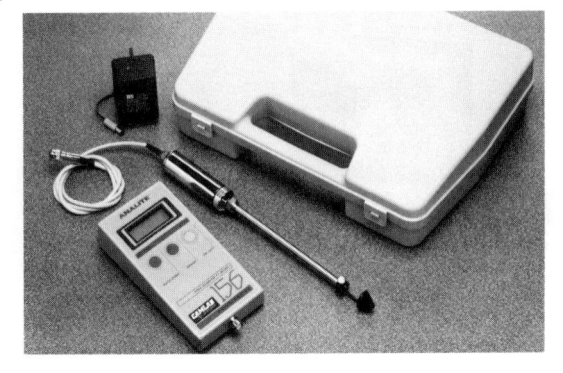

The Analite turbidimeter is a portable hand-held device using fibre optics to provide an instant digital reading of liquid turbidity. Measurements are obtained by simply inserting the stainless steel probe into the liquid sample. Light is transmitted through the optical fibres in the near-infra-red spectrum and back scatter (at 180°) carried back to be focussed onto a photo-diode. Interference from ambient light is avoided by using a modulated transmitting signal. Display is in Nephelometric Turbidity Units.

Range and sensitivity:
Two models cover 1-2000 NTU and 10-30000 NTU with resolution varying from 0.1 NTU (at 1 NTU) up to 100 NTU (at 10000 NTU). Repeatability is generally 2% ± 1 digit.

Application:
Wear particles in hydraulic and lubricating oils, transformer oil degradation, filter backwash monitoring, as well as many applications in water based fluids. The weight of only 520 g for the instrument and 210 g for the probe make the device extremely convenient for portable monitoring. A recorder output enables a more permanent record to be taken.

Cost:
Around £1000 to £1500.

Further details:
[WARNING — for latest models check supplier]
See pages 127 and 155.

APPLIED IMAGING DISC	Manufacturer & UK Supplier:
CENTRIFUGE DCF4	Applied Imaging International
	Limited (formerly Joyce-Loebl)

Technique: SEDIMENTATION
[DISC CENTRIFUGE]

Description:

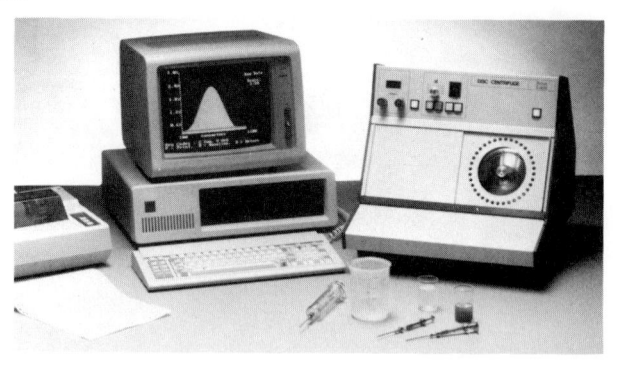

A hollow centrifuge disc is rotated at a pre-selected speed, and the particles to be separated start from a certain disc radius. They then settle through a spin sedimentation. The Stokes' Law is only true if the system is hydrodynamically stable, ie. no "streaming" occurs; in the Applied Imaging Disc Centrifuge this can be ensured by using the "Buffered Line Start" technique. This technique requires a thin buffer layer to be specially dispersed first (usually water), enabling the then added particle dispersion, to travel uniformly right from the start. The distribution is output in graphical form as well as numerically.

Range and sensitivity:
Size range 0·01 μm to 60 μm using a sample volume between 0·5–1mL. Acceptable particle concentration between 0·05–0·5% by weight. Speeds between 1000–10,000 rev/min with an accuracy ± 0·05%. An optional software package includes a calibration facility which compensates for difficult optical distortions, and makes the results comparable with the SEM and other standards.

Application:
Accurate determination of particle size and distribution. Particularly suitable for small size polymers, pigments and chemical dispersions.

Cost:
Around £15,000

Further details:
[WARNING — for latest models check supplier]
See pages 134 and 168.

BAHCO MICRO-PARTICLE CLASSIFIER	Manufacturer: Dietert Division of George Fischer Foundry Systems Inc.
	UK Supplier:
Technique:	SIEVING [AIR CENTRIFUGE ELUTRIATOR]
Description:	

At each pass through the device, a dry sample is separated into a fine fraction and a coarse residue in a rotor assembly driven by a 3500 rev/min 1hp motor. The rotation of the assembly forms a spiral of air moving towards the centre. After a primary dispersal in the feed mechanism, the sample is introduced symmetrically into this spiral. The size, form and weight of the particles determines whether they move outwardly, by centrifugal force, or with the air flow into the centre. By adjusting the size of the air inlet, the sample can be divided into perhaps eight fractions with limited grain size ranges.

Range and sensitivity:
Each fraction is completed in 15 minutes (ie. 2 hours for eight fractions). Sample size entered is between 1 g and 20 g. Particles, typically below 60 μm in size, can be calibrated by microscope, and then for future tests, the classifications will be known. Typical maximum deviation is of the order of 0·3%.

Application:
In the engineering field, the classification of fine metal powders, combustion dusts, abrasives.

Cost:
Four models of different voltage and frequency range between around $US 24,000 and $US 27,000.

Further details:
[WARNING — for latest models check supplier]
See pages 135 and 168. Refs: Graham & Hanna 1962, Horowitz & Elrich 1986.

BHR GROUP ON-LINE CONTAMINANT MONITOR	Manufacturer & UK Supplier: BHR Group

6

Technique:	SILTING

Description:

System fluid is passed through a carefully controlled cylindrical clearance between a specially designed spool and sleeve. Particles become trapped and eventually block the clearance. The volume of fluid passing through such a clearance before blockage occurs, serves as a reliable measure of fluid contamination level, largely unaffected by fluid pressure, temperature or viscosity variations. The device works cyclically by clearing the blockage at the end of each cycle, and resetting.

Range and sensitivity:

Sensitivity varies with clearance dimensions, the clearance normally being set to be slightly less than the most critical clearance in the system being monitored. Readings are given between 1 and 10 with a repeatability error (of the prototype) of less than 5% for the cleaner levels and better than 2% for the dirtier levels.

Application:
Primarily hydraulic systems where sensitive valves need to be protected from an excessive contamination level. It is considered that it could be designed into a portable, or used on lubrication systems once it has been determined which clearances are most suitable.

Cost:
[Not yet in production, March 1992]

Further details: [WARNING — for latest models check supplier]
See pages 136 and 158. Ref: Heron & Hughes 1986.

BROMLEY PSA   Manufacturer: Bromley Instruments
                               UK Supplier: Belstock Controls

Technique: SEDIMENTATION
Description:

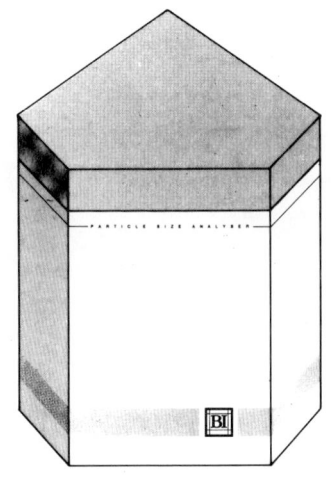

The liquid containing the particulate is thoroughly mixed in a cuvette by manual agitation. The cuvette is then placed into a thermally insulated measuring enclosure for analysis of size distribution. Gravity sedimentation is used with the particles sensed optically at two separated positions. The computer analysis takes account of density, viscosity, temperature and scattering theory to produce the size distribution.

Range and sensitivity:
Particle size range is 1–500 μm with a maximum gravimetric level up to 0.1 g in a sample of 150 mL. Wide angle silicon diode detectors are used to improve sensitivity.

Application:
As the analysis is simple to commence — by the press of a button — and requires no attention until the determination is complete, the instrument is ideal where labour intensive techniques are not acceptable. Process control of minerals, pigments, food, chemicals and pharmaceuticals are some possible applications, as well as engineering uses in such as mining.

Cost:
Complete with software, under £3000.

Further details:
[WARNING — for latest models check supplier]
See pages 134 and 168.

BROOKHAVEN BI-90 PARTICLE SIZER	Manufacturer: Brookhaven Instruments Corp.
	UK Supplier: Brookhaven Instruments Ltd.

———————————— 8 ————————————

Technique: OPTICAL – PHOTON CORRELATION SPECTROSCOPY

Description:

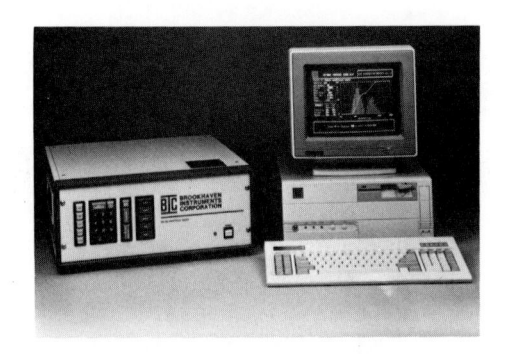

The BI-90 is a fast user friendly instrument. Once the sample ID has been presented and the start actuated, optimum run conditions are automatically set and retained. No calibration is required. Size distributions are calculated within one and three minutes without further intervention. Mean size and polydispersity are both updated on the screen every few seconds allowing trends in data to be rapidly identified. Fully model independent size distributions are also calculated allowing multimodal and skewed distributions to be analysed. A 5 mW He-Ne laser is used, focussed into a temperature controlled sample cell. 90° scatter is detected by a photomultiplier. The rugged construction allows the unit to be moved without the need to realign the optics. The BI-90AT includes a separate 80386, 40 MHz data processor (PC) and printer.

Range and sensitivity:
Size ranges 0·002 μm to 2 μm and 0·01 μm to 1 μm, with a reproducibility typically 1%. Sample volume between 1 and 3 mL.

Application:
Industrial processes and products involving dispersions, eg. polymer latexes, pharmaceutical preparations, paints, inks, etc. Size distribution produced off-line from sample.

Cost:
Around £20,000 (BI-90), £23,000 (BI-90AT)

Further details: [WARNING — for latest models check supplier]
See pages 130 and 200.

BROOKHAVEN BI-DCP PARTICLE SIZER	Manufacturer: Brookhaven Instruments Corp. UK Supplier: Brookhaven Instruments Ltd.
Technique:	SEDIMENTATION [DISC CENTRIFUGE MASS SEPARATION]

Description:

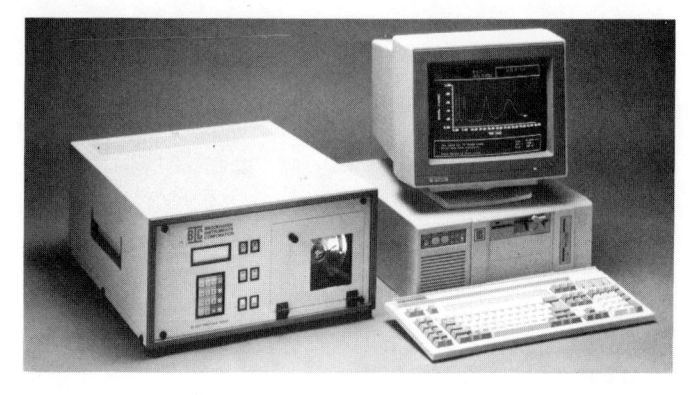

A variable speed centrifuge (500–15,000 rev/min) and a patented external gradient method, ensure that the BI-DCP yields a high resolution over the extensive size range. Weight distributions may also be determined. Typical analysis times range from 10 to 30 minutes. Size and weight are now reduced by using digital control, and the instrument is more stable and much quieter than an earlier version. No calibration is necessary. New homogeneous analysis mode and scanning detector further improve speed and simplicity of analysis and enable particles less dense than the medium to be measured.

Range and sensitivity:
Size range 0·01 μm to 60 μm maximum, with 0·1 μm to 10 μm being the most typical. Spin fluid volume is between 10 mL and 40 mL. The BI-DCP is able to resolve peaks only 13% apart.

Application:
All particles which are, or can be, dispersed in liquids, including polymer latices, printing ink particles, automotive sprays, magnetic oxides and emulsions.

Cost:
Around £27,000 with complete data system (including PC and software)

Further details: [WARNING — for latest models check supplier]
See pages 134 and 168. Ref: Thomas & Fairhurst 1990.

CCK 4   Manufacturer & UK Supplier:
                                        Hydrotechnik UK Limited

———————————————— 10 ————————————————

Technique: VISUAL APPEARANCE

Description:

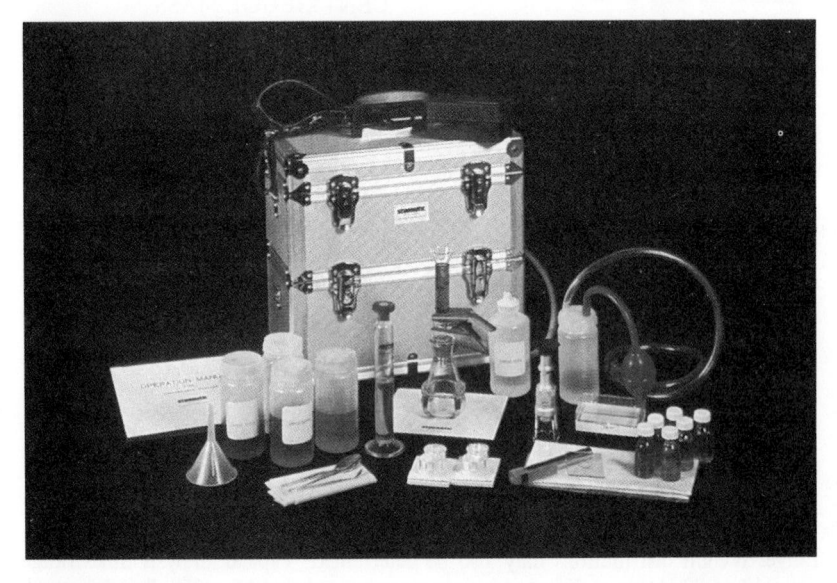

The CCK 4 checker kit is dedicated to contamination control of hydraulic systems and solid particle debris monitoring. The kit consists of a small hand-held microscope, a hand operated vacuum pump (electric pump optional), membrane filter holder, membrane filters, necessary glassware, etc. A set of photographic reproductions of debris enables direct comparisons to be made.

Range and sensitivity:
Generally 1 μm and upwards (ie. visible by optical microscope).

Application:
The kit may be used on site where oil or hydraulic fluid samples can be taken. Portable.

Cost:
Around £650 complete with hand operated pump or around £750 with the electrically operated pump.

Further details: [WARNING — for latest models check supplier]
See pages 138 and 148.

CHECKER-KIT　　　　　　　　Manufacturer & UK Supplier:
　　　　　　　　　　　　　　United Air Specialists UK Ltd

―――――――――――――――― 11 ――――――――――――――――

Technique:　　　　　　　　　　VISUAL APPEARANCE
Description:

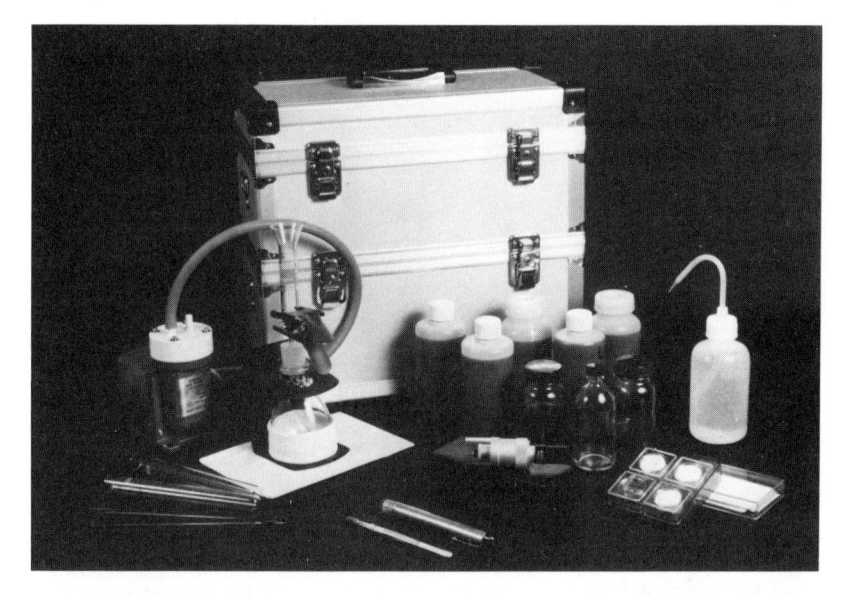

The Checker-Kit is suitable for lubricating oils, for contamination control of hydraulic systems and for solid particle debris monitoring. The kit consists of a small hand-held microscope, an electrically operated vacuum pump, membrane filter holder, membrane filters, necessary glassware, etc.

Range and sensitivity:
Generally 1 μm and upwards (ie. visible by optical microscope), using a 0·8 μm filter membrane.

Application:
The kit may be used on site where oil or hydraulic fluid samples can be taken. Portable, weight complete with electric pump 8 kg.

Cost:
Around £750 complete with electric pump.

Further details:
[WARNING — for latest models check supplier]
See pages 138 and 148.

CLIMET® CI-1000
LIQUID-BORNE
PARTICLE COUNTER

Manufacturer: Climet Instruments Co
UK Supplier: Gelman Sciences Ltd

---------- 12 ----------

Technique: OPTICAL – LIGHT OBSCURATION

Description:

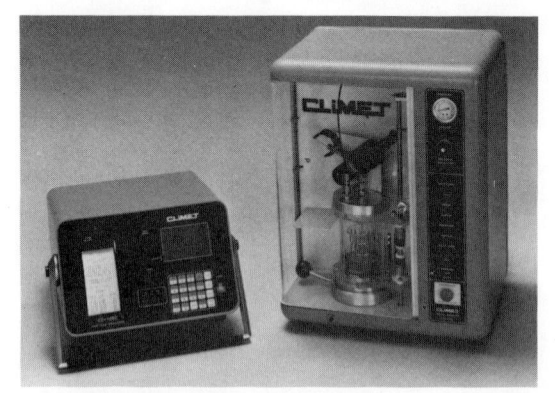

The CI-1000 consists of three modules — the sensor, the sampler and the electronic module. The sensor used is the Russell Sensor which offers significant gains in performance by the provision of a high intensity white light in the sensing zone; this yields a high signal-to-noise ratio with corresponding high accuracy and resolution. Two samplers cover a wide range of viscosity. The electronic module counts particles in 3000 size ranges simultaneously and then enables any choice of six size bands to be displayed (or printed).

Range and sensitivity:
Size range from 1 μm to 5mm (13 different sensors) with a dynamic range of 1:50 (eg. 2–100 μm), with a concentration maximum limit up to 44,000 particles/mL. Measuring zone is defined, eg for 5–250 μm it is 250 × 400 × 500 μm with a flow rate 100 mL/min; at the maximum flow of 700L/min the zone is 5 mm × 8 mm × 8 mm. Viscosity range 1–100 cSt. (The new RLLD laser diode Russell sensor covers 0·9 μm to 100 μm, calibrated with latex in water.)

Application:
Particulate contamination and debris monitoring in oils, fuels and hydraulic fluids in general industry, aerospace, etc.

Cost: From around £14000.

Further details: [WARNING — for latest models check supplier]
See pages 125 and 180. Ref: West 1990.

CLIMET® CI-1500
LIQUID QUALITY
TRANSDUCER

Manufacturer: Climet
Instruments Co
UK Supplier:
Gelman Sciences Limited

——————————— 13 ———————————

Technique: OPTICAL – NEPHELOMETRY/
TURBIDITY

Description:

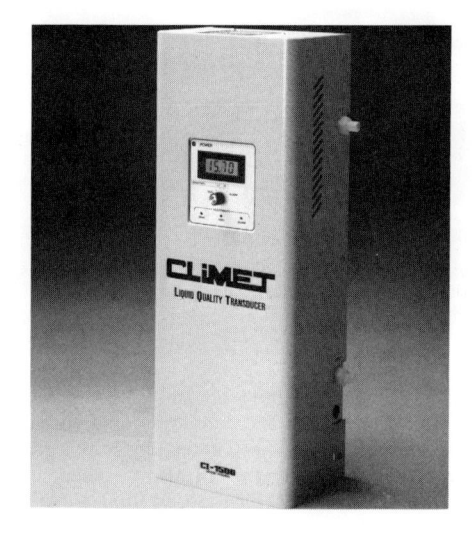

The CI-1500 uses laser diode light scattering at 180° to detect 0·01 μm particles in ultra clean liquids. It displays the turbidity value in mNTU's. It is an on-line or in-line monitor continuously updating the value for a very large range of flows.

Range and sensitivity:
Particles above 0·01 μm with the value of turbidity between either 0–20 mNTU, or between 0–200 mNTU. Sensitivity 0·02 mNTU. Flow rate from 100 mL/min to 10 L/min at a pressure up to 7 bar. Weight of unit 4·2 kg.

Application:
On-line and in-line measurement of ultra clean liquids, eg DI water and process fluids. Multiple units can be used at many locations around a plant with the outputs monitored with Climet CFMS software.

Cost:
Around £5000 per liquid quality transducer.

Further details:
[WARNING — for latest models check supplier]
See pages 127 and 155. Ref: Kreikebaum 1990.

CLIMET® CI-2000
LASER LIQUID-BORNE PARTICLE COUNTER

Manufacturer: Climet Instruments Co
UK Supplier: Gelman Sciences Ltd

14

Technique: OPTICAL – LIGHT SCATTER

Description:

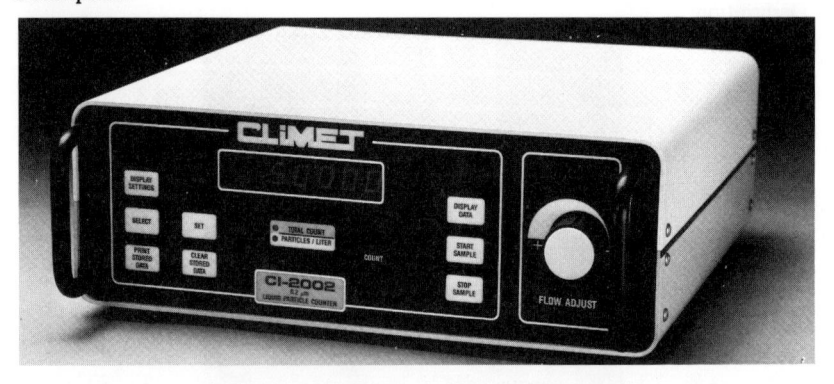

The CI-2000 is a battery operated portable fully self-contained particle counter for use in detection of sub micron particles. Two samplers are available, one for in-situ testing and the other for volumetric testing; both can be used on-line (at the suitable flow rates) and the in-situ can also be used in the off-line mode. Forward scatter at 10°–40° using a laser diode is used.

Range and sensitivity:
CI-2000 instruments count particles in two size bands simultaneously, either > 0.2 μm and > 0.5 μm, or > 0.3 μm and > 0.5 μm, depending on the model. A > 0.5 μm and > 1.0 μm version is also available to special order. In-situ versions sample 10% flow, volumetric versions sample 100% of the flow. Sample flow rates of up to 80 mL/min are obtainable with both versions. Maximum line pressure 7 bar. Concentration limits: 90,000 particles/mL.

Application:
On-line monitoring of DI water and process fluids and solvents in the semiconductor, aerospace, chemical industries, etc. Multiple units can be used around a plant with the data output logged with Climet CFMS software.

Cost:
Around £15000 to £25000.

Further details:
[WARNING — for latest models check supplier]
See pages 126 and 186. Ref: West 1990.

CM 20   Manufacturer & UK Supplier:
                                   UCC International Ltd.

─────────────────────── 15 ───────────────────────

Technique: OPTICAL – LIGHT OBSCURATION
Description:

The CM20 samples on-line without affecting the system operation. It can be operated direct from a UCC System 20 sensor which incorporates twin coupling ports, or be connected to a single sampling point (with the return taken to waste or the reservoir). A sample of fluid is drawn through the optical sensor which magnifies particle images that are projected onto a photo diode measuring the change in light intensity as each particle crosses a 25 μm optical window. Particles as large as 200 μm can pass through without blockage. Up to 300 test results can be stored or printed.

Range and sensitivity:
Particle sizes greater than 5, 15, 25 and 50 μm are separately counted and readouts given in actual counts, NAS 1638 or ISO 4406 codes as required. Viscosity range is 2 to 100 cSt with the CM20 connected to the System 20 on-line or up to 500 cSt with a single point sampler. Contamination range is ISO 7–24.

Application:
The CM20 weighs only 10 kg and is designed with maximum portability in mind when moving to any number of sampling points. Results can be downloaded to a computer via a software package "COMDAD" also produced by UCC.

Cost:
Around £6500 complete.

Further details:
[WARNING — for latest models check supplier]. See pages 125 and 180.

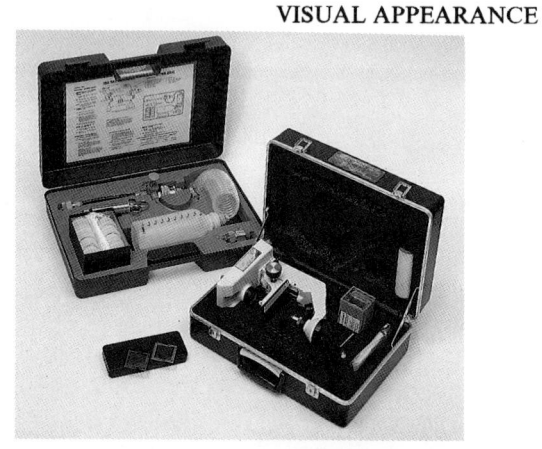

CONPAR®   Manufacturer & UK Supplier:
PARTICLE COMPARISON KIT   Howden Wade Limited
                                     Thermal Control Products

── 16 ──

Technique: VISUAL APPEARANCE
Description:

The Conpar condition monitoring system consists of three items — the Dynamic Sampling Set, the Particle Comparison Kit and the Contamination Control Module. The cased Sampling Set includes a sampling unit with a housing for one membrane filter; this unit is designed to be attached to a pressurised system in order to obtain a representative sample. A 3-way valve enables flow to be passed to waste initially before 100 mL is passed through the sampling monitor membrane. The membrane is made transparent and examined under the microscope and compared with a master slide, in order to determine the cleanliness level of the system.

Range and sensitivity:
Comparison slides are available for a range of standards as follows:
     Conpar   1H to 9H   DEF-STAN 05-42 400 to 100,000
     NAS 1638 1 to 12   (Equivalent ISO 4406 10/7 to 21/14)

Application:
Because of the large contaminant range available it is possible to cover from critical valve systems to more general hydraulics and fuels. Results can be achieved within 10 minutes from the time of sampling. The kit is fully portable, the Sampling Set weighing 4 kg and the Particle Comparison Kit (including the Conparscope microscope) 6·2 kg.

Cost:
Around £1500 for the Particle Comparison Kit complete with the Sampling Set and basic microscope, around £160 for each set of two Standard Slides (green-go and red-nogo).

Further details: [WARNING — for latest models check supplier]
See pages 103, 138, 148 and 282. Refs: Baker 1991, Goldsmith 1984 & 1986.

CONTAM-ALERT
Digital CONTAM-ALERT

Manufacturer: Diagnetics
UK Supplier: CHARN-MAC ENG-Ltd

Technique:

FILTER BLOCKAGE [PORE BLOCKAGE / FLOW DECAY]

Description:

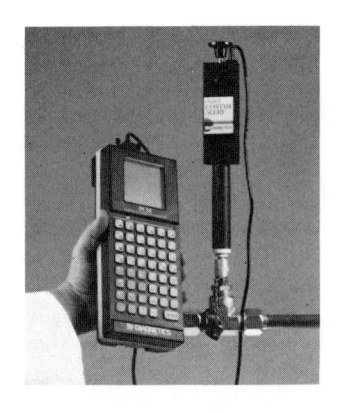

The Digital Contam-Alert (dCA) is a portable contaminant monitor which may be momentarily plugged into a field system as an on-line device, used with bottle samples, or (third version) used with reservoirs or gear boxes. The sensor employs the use of a photoetched screen through which fluid flows at constant pressure, screen pores gradually being covered by particles resulting in a flow restriction. The flow profile is measured and translated mathematically into an estimated particle size distribution. Backflushing of the screen is performed by hand. After some 250 operations calibration should be verified by user.

Range and sensitivity:
Three screen sizes are available (5 μm, 10 μm & 15 μm), covering a contamination range of ISO 4406 11/8 to 22/19. NAS 1638 and particle size distribution can also be displayed corresponding to a preset (or measured) distribution slope as determined by the test application. The sensor can accept fluid pressure in the range of 2 bar to 10 bar directly, or 0 bar to 210 bar with a suitable interface. Viscosity range 8 cSt to 150 cSt.

Application:
Hydraulic and lubricant systems. Depending on the fluid viscosity, pressure and cleanliness, test time 1–4 minutes.

Cost:
Around $(US) 10,000 to 13,000.

Further details: [WARNING — for latest models check supplier]
See pages 118, 158 and 196. Ref: Fitch 1991.

| CONTINUOUS DEBRIS | Manufacturer & UK Supplier: |
| MONITOR (CDM) — TECALERT | Ranco Controls Limited. |

Technique: MAGNETIC ATTRACTION [MAGNETIC FLUX PATH CHANGE]

Description:

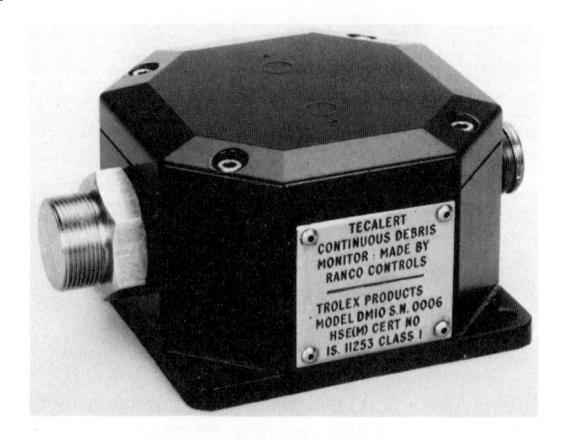

The objective of the CDM is to detect the build-up, against time, of ferromagnetic particles on a magnet. System fluid passes over a non-magnetic sensing shim under which a powerful permanent magnet resides on a spiral screw. When the magnet is near the shim, particles are trapped and detected by a Hall effect sensor in the magnet gap. At a set time, or when the quantity of particles becomes a certain value, the magnet is mechanically withdrawn releasing the debris ready for the next cycle.

Range and sensitivity:
Different ranges are possible. One is 100 mg to 600 mg for full scale reading, with a sensitivity of approximately 0.1mg with a capture efficiency from about 30% to 80%. In order to provide the maximum efficiency, flow over the CDM should be between 0·1 m/s to 0·5 m/s. The DM10 operates at a lower debris rate of up to 200 mg per cycle maximum with an 'unstick' time of 8 seconds.

Application:
In-line (low pressure) or on-line lubrication systems appropriate to such as gear boxes.

Cost:
Around £1300 for the DM10 CDM

Further details: [WARNING — for latest models check supplier]
See pages 122, 151 and 310. Refs: Bogue 1984, Mills 1985, Ranco Controls 1990.

COULTER® LS SERIES

Manufacturer: Coulter Corporation
UK Supplier: Coulter Electronics Limited

Technique: OPTICAL – FRAUNHOFER DIFFRACTION [LIGHT SCATTER]

Description:

Light is diffracted around a particle at angles inversely proportional to the size of the particle, ie. the smaller the particle the greater the angle of diffraction. Most particle sizers based on this technique have a single optical train for collecting and sensing the diffracted light, each over a limited size range; The COULTER® LS has multiple trains built in with a special binocular technology and, in the LS130, includes the exclusive Polarisation Intensity Differential Scattering (PIDS) optics which takes the size down to 0·1 μm in the one instrument.

Range and sensitivity:
0·1 μm to 800 μm in a single shot process, with Mie conversions below 0·5 μm.

Application:
Although a Dry Powder Module (DPM) is available for dry samples (in the LSP instrument), the LS models normally have the more common liquid sample module or small volume sample module. Analysis time is, on average, 60 seconds and covers between 72 and 100 channels of size.

Cost:

Further details:
[WARNING — for latest models check supplier]
See pages 124 and 168. Refs: Thomas 1990?, Bott & Hart 1990.

COULTER® MULTISIZER SERIES

Manufacturer & UK Supplier:
Coulter Electronics Limited

20

Technique: ELECTRICAL SENSING ZONE
(ESZ) [COULTER PRINCIPLE]

Description:

Particles in suspension are drawn through a small aperture by means of a vacuum source. An electrical current passing through the aperture between two electrodes, enables the particles to be sensed by the momentary changes in the electrical impedance as the particles pass through the aperture, since each particle displaces its own volume of electrolyte solution within the aperture itself. The voltage pulses generated (the magnitude of each pulse is essentially proportional to the volume of the particle which produced it) are amplified, counted and allocated into up to 256 size classes.

Range and sensitivity:
Using a range of aperture tubes, and an accuracy of sampling better than ± 0·5%, the following sizes can be covered

15 μm	0·4–9 μm	30 μm	0·6–18 μm
50 μm	1–30 μm	70 μm	1·4–42 μm
100 μm	2–60 μm	140 μm	2·8–84 μm
200 μm	4–120 μm	280 μm	5·6–168 μm
400 μm	8–240 μm	560 μm	11·2–336 μm
1000 μm	20–600 μm	2000 μm	40–1200 μm

Application:
Although restricted to electrically conducting fluids, it is suitable for high precision detection of particle volume.

Cost: Several models are available.

Further details: [WARNING — for latest models check supplier]
See pages 50, 116, 192 & 295.

COULTER® N4 SERIES

Manufacturer: Coulter Corporation
UK Supplier: Coulter Electronics Limited

21

Technique: OPTICAL – PHOTON CORRELATION SPECTROSCOPY (PCS)

Description:

The Brownian motion of particles is detected. This is the random motion, caused by diluent molecules colliding with the particles; the rate at which the particles diffuse being inversely proportional to particle size, the smaller the particle, the faster it diffuses. A laser beam is used to illuminate the particles and a sensor detects the intensity of scattered light fluctuations. The N4S and N4SD models have a fixed 90° scattering angle which is acceptable for many applications. The N4MD model allows a multi angle measurement which has advantages for mixed (masked) populations of particles, for smaller particles and for non-spherical or elongated particles.

Range and sensitivity:
0·003 μm to 3 μm for all models. Run time from 60 seconds up to several minutes, depending on the application. Scattering angles from 90° to 11°.

Application:
Although Brownian motion is normally used only for submicron analyses, this series is able to accurately detect up to 3 microns. This makes it suitable for a wide range of applications.

Cost:

Further details:
[WARNING — for latest models check supplier]
See pages 130 and 200.

381

| DANTEC PARTICLE DYNAMICS ANALYSER | Manufacturer: Dantec Elektronik [Denmark] UK Supplier: Dantec Electronics Limited |

22

Technique: OPTICAL – PHASE/DOPPLER SCATTER

Description:

A laser beam (typically He-Ne up to 5 watts) is used to illuminate the particles in suspension. In order to cover the detection of size, three detectors are used in one receiving optics which extends the size range to more than 40:1 for any given detector setting. The signal processor uses covariance techniques, similar to that used in the field of radar processing, to estimate the relative phase and frequency. Thus data can be recorded even in noisy environments.

Range and sensitivity:
The size range is much larger than most other methods, covering 0.5 μm to 10,000 μm (10 mm) with an accuracy typically 4%. Measuring distance is 310, 600, 1200 mm or even longer.

Application:
Particles, droplets and bubbles in sprays or combustion, or where particle transport needs an accurate non-intrusive measurement. Velocity, as well as size, can be simultaneously measured, even in noisy environments.

Cost:
From around £55,000

Further details:
[WARNING — for latest models check supplier]
See page 128.

DAWN® Manufacturer: Wyatt
Technology Corp.
UK Supplier: Optoken Instruments
Limited

———————— 23 ————————

Technique: OPTICAL– NEPHELOMETRY/
TURBIDITY [MULTI-ANGLE
LASER LIGHT SCATTERING]

Description:

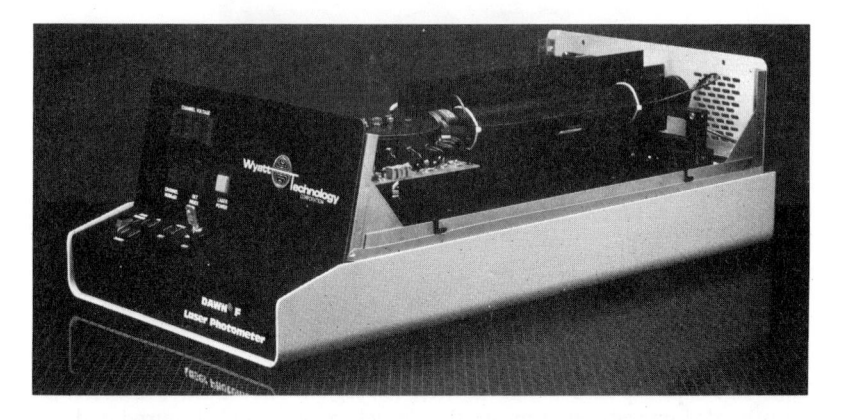

The DAWN® instruments have either 18 or 15 separate photodiode detectors to measure the laser scatter completely around the particle except direct reflection (180°). Fluid is placed in a central cuvette or a flow-through cell. This enables full scanning to be undertaken electronically at a much faster rate than mechanical means.

Range and sensitivity:
The instruments are for macro-molecular weight and size determination from a sample of a few millilitres. The Model F covers angles 5 to 175°, and the Model B from 23 to 128°.

Application:
Both models may be used in a static mode to make turbidity,as well as sedimentation rate studies. In Field Flow Fractionation (FFF) separation techniques, the Model F can be used as a detector, as well as for gel permeation/size exclusion chromatography.

Cost:
Around $US 50,000 (Model F), $US 30,000 (Model B)

Further details:
[WARNING — for latest models check supplier]
See pages 127 and 155.

DEBRIS TESTER

Manufacturer & UK Supplier:
Staveley NDT Technologies
Inspection Instruments (NDT) Ltd

Technique: INDUCTANCE

Description:

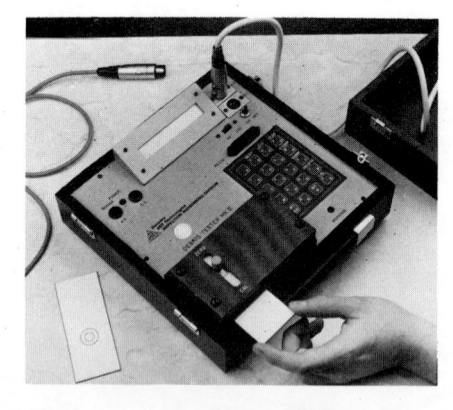

An unbalanced AC bridge principle is used, unbalance occurring when magnetic debris on a slide is placed within the detection region. The inductive measurement coil is located immediately beneath the slot where the sample is introduced, and the balance coil is positioned alongside. The construction of the Debris Tester only allows readings when the sample is in the correct place. The sample consists of a slide on which magnetic debris has been transferred, eg. using sticky tape on a chip plug removed from a magnetic chip detector (MCD). Internal calibration is possible.

Range and sensitivity:
Readings are given in Debris Test Units (DTU) covering a range of ferrous mass of 0·02 mg to 200 mg with an accuracy within 0·02 mg or 2%. Repeatability better than 1%. Automatic calibration on the Mk II, manual if required.

Application:
Two units are available, the MkII digital version and the older MkI analogue. Both are portable (weight about 6 kg). The MkII is able to store 98 previous readings and hence is suitable for carrying between test points where chip plugs can be removed and examined; results can be downloaded for trend analysis. Both Testers are suitable for routine laboratory MCD analysis.

Cost: Around £3000 (MkII) or £2500 (MkI)

Further details: [WARNING — for latest models check supplier]
See pages 121, 151, 300 and 305. Refs: British Coal 1984, Dickson 1991, Lewis 1984, Summerfield & Mathieson 1981.

ELZONE® Manufacturer: Particle Data Inc.
UK Supplier: Particle Data Ltd

— 25 —

Technique: ELECTRICAL SENSING ZONE (ESZ) [ELECTROZONE]

Description:

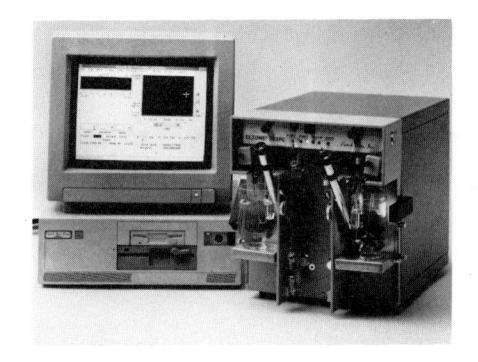

Particles are caused to flow, nearly almost individually, through a small liquid 'resistor' (sensing orifice). A current applied across the 'resistor' (between two electrodes) senses a change in impedance as the particles partially block the orifice during their transit. Pulses are generated in proportion to the displacement volume of the particles. The pulse spectrum is then amplified and scaled for particle size analysis.

Range and sensitivity:
The particle size range is 0·4 μm to 1200 μm, although an optional extension to 1900 μm is available. A patented variable logarithmic converter gives six ranges from 2·5:1 to 25:1 by particle diameter to give maximum accuracy and resolution over narrow or wide range distributions. Reproducibility is 1% by particle volume. Concentration is up to 1000 ppm and sensing rate between 1 and 2000 per second. 256 channels are available.

Application:
Although restricted to electrically conducting fluids, it is suitable for high precision detection of particle volume. Highly conductive metal particles can be analysed. The 180xy ELZONE® is a single unit, the 280pc ELZONE® is operated and analysed by separate PC.

Cost:
The 180xy around £15000 for the complete particle range, less for more restricted ranges. The 280pc depends on the pc used.

Further details: [WARNING — for latest models check supplier]
See pages 50, 116, 192 and 295. Ref: Berg 1991.

ENDECOTTS TEST SIEVES   Manufacturer & UK Supplier:
                             Endecotts Limited

Technique: SIEVING

Description:

Up to 8 full size or 17 half size sieves can be stacked on an Octagon sieve shaker. (This can be extended to 25 half height sieves on the EFL model.) Wet or dry sieving is available. The division of a sample into 10 identical smaller samples is achieved on the Rotary Sample Divider. Sieve diameters are available in 12 in, 300 mm, 8 in, 200 mm, 3 in or 100 mm.

Range and sensitivity:
Sieves are made and certified to international standards — BS410 1986, ASTM E11:81, ISO 3310/1 and ISO 3310/2. These cover a range from 20 microns (wire mesh) to 125 mm (perforated round or square hole). Smaller apertures are available down to 5 μm in electroformed nickel plate.

Application:
Normally sieving is used to qualify control material during production or to analyse the particle size distribution of a granularmetric sample, eg. ore, crystals, soil, sand, minerals, etc. However, sieving may be used to analyse quantities of particulate in fluids. Endecotts equipment is able to perform this in both the wet and the dry state, enabling particles of certain size to be examined. A comprehensive guide book, 'Test Sieving Manual', is available from Endecotts (£3.50).

Cost: Typical sieve cost around £30.

Further details: [WARNING — for latest models check supplier]
See pages 135 and 168. Ref: Endecotts 1977/1989.

FAS-CC100   Manufacturer & UK Supplier:
                                    Fairey Arlon Limited
——————————————— 27 ———————————————
Technique: VISUAL APPEARANCE
Description:

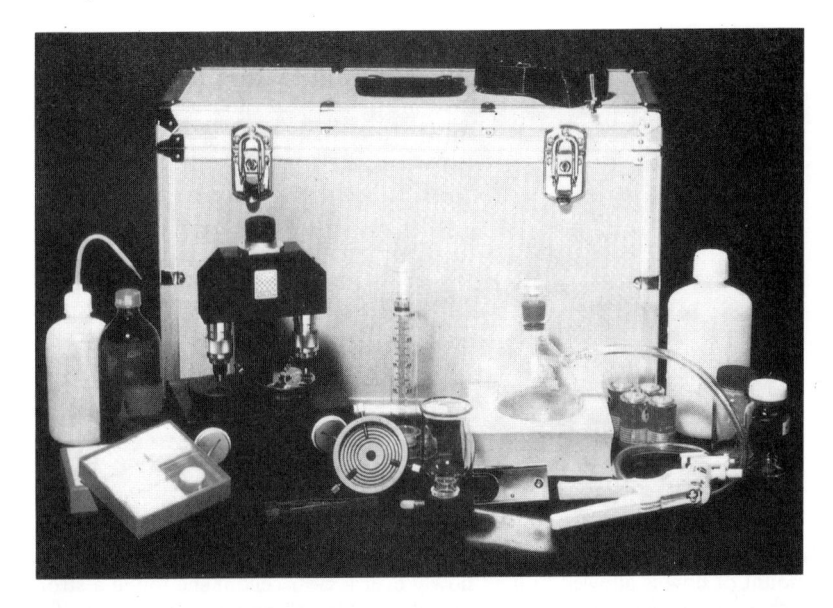

The FAS-CC100 kit is dedicated to contamination control of lubrication and hydraulic systems. The kit consists of a split image microscope with a set of 8 equivalent patches (dot matrices) of ISO 4406 hydraulic contamination Standards built-in, a hand operated vacuum pump, membrane filter holder, membrane filters (1·2 μm, 25mm dia), necessary glassware, universal camera adapter, etc.

Range and sensitivity:
Generally 1 μm and upwards (ie. visible by optical microscope).

Application:
The kit may be used on site where oil or hydraulic fluid samples can be taken. Portable, weight complete with microscope and hand operated pump 12·5 kg.

Cost:
Around £2120 complete.

Further details:
[WARNING — for latest models check supplier]
See pages 138 and 148.

FERROGRAPH REO 1   Manufacturer: VD AMOS
                          UK Supplier: Oilab Lubrication
                          Limited

28

Technique: MAGNETIC ATTRACTION
                          [FERROGRAPHY]

Description:

The Ferrograph REO 1 is a simpler version of the original Ferrograph, but yet includes the possibility of a further feature with the Densimeter. A measured amount of flow is allowed to flow slowly over a clear cellophane foil in a strong magnetic field. Ferrous debris is retained as a "trace" on the surface of the foil, larger debris first. The foil can then be analysed by microscope or by Densimeter. The REO 2 Densimeter can be used to determine the change of optical density along the foil (rather than just looking at the two extremes), and from this decide the type of wear that is occurring.

Range and sensitivity:
Generally ferromagnetic particles from 1 μm upwards are detected. Some non-ferrous particles are also trapped.

Application:
Primarily lubricating oils which may contain evidence of ferrous components wearing or fatiguing. Only used off-line from bottle samples.

Cost:
Around £1700 for Ferrograph alone, £2250 complete with Densimeter.

Further details:
[WARNING — for latest models check supplier]
See pages 122, 263 and 308.

FerroSCAN®   Manufacturer & Supplier: GasTOPS Ltd

29

Technique: MAGNETIC ATTRACTION [MAGNETIC CAPTURE]

Description:

This is a real time on-line monitoring technique. The fluid passes through a sensor which comprises of a sensor coil sandwiched between the pole faces of an electromagnet. When the electromagnet is energised, ferromagnetic particles in the fluid are attracted to the magnet poles which lie around the circumference of the pipe in which the fluid is flowing. The sensor coil is a component of an RF oscillator circuit, and the presence of particles close by causes a change in the inductance of the coil with a corresponding change in the natural frequency of the oscillator. After a trapping period (depending on the application) the electromagnet is de-energised, degaussed and the trapped particles are allowed to flush away before the cycle is repeated. The sensor imposes no restriction to normal fluid flow.

Range and sensitivity:
The sensor is sensitive to particles in the range of 1 μm to 1 mm. Fine debris (eg. 1 μm to 20 μm) is trended to provide an average concentration and baseline level for warning of excessive wear. Larger debris particles, eg. fatigue failure, are detected as superimposed "transient" spikes. Like many monitors, sensitivity is affected by viscosity, air, vibration and rapid temperature fluctuations, although under normal conditions these effects are small. Sensor output is linear for concentration levels up to 100 ppm for flow in the range 0·5 to 5 m/s, with an accuracy ± 1 ppm.

Application:
Flowing or pressurised lubrication systems.

Cost:
Around $US 3,500 to $US 4,000

Further details:
[WARNING — for latest models check supplier]
See pages 122, 151 and 312. Refs: Campbell 1991, Faulkner & MacIsaac 1991.

FLUID CONDITION		Manufacturer & UK Supplier:
MONITOR (FCM)		Lindley Flowtech Limited

30

Technique:				FILTER BLOCKAGE
Description:

The oil being tested is caused to flow through a high-precision filter mesh of a known size and number of pores. (This flow may either be produced by internal pump or from the machine system pressure.) Particles greater in size to the pores, are retained on the surface of the mesh and cause a partial blockage of the pores, with a rise in pressure drop across it. The rise in pressure drop is related to the number of particles trapped. Because an automatic reversal occurs, with backflushing, the instrument is continuous and includes a second size assessment automatically. Four counts are undertaken at each size before the average is displayed, usually well within 4 minutes.

Range and sensitivity:
Meshes provide measurement at 5 μm and 15 μm (viscosity range 5 cSt to 350 cSt, although extensions to this, either way, can be provided). The display provides a Standard Class such as ISO 4406 or NAS 1638, as well as the count of particles greater than each size.

Application:
Because the filter blockage principle is independent of particle type or of fluid opacity or mix, a wide range of machinery applications involving lubrication and hydraulic fluids is suitable.

Cost: Around £7000

Further details:
[WARNING — for latest models check supplier]
See pages 118, 196, 249 and 309. Refs: Raw & Hunt 1987, Wilkes & Hunt

FRITSCH ANALYSETTE 3/18　　　　Manufacturer: Fritsch GmbH
SIEVE/SHAKERS　　　　　　　　UK Supplier: Christison Scientific
　　　　　　　　　　　　　　　　Equipment Limited

Technique:　　　　　　　　　　　SIEVING [AUTOSIEVE]

Description:

A variety of sieves and shakers are available. Rotary sieve shakers (eg. Analysette 18) compare well with manual shakers. However, the Fritsch shaker also includes a secondary oscillating movement in the outer range which greatly improves the sieving effect and reduces time.

The Analysette 3 is a vibratory sieve shaker with variable speed as well as an intermittent mode. An additional feature is the special sieving head with three nozzles for wet sieving. Micro-precision sieves can be used on the vibratory shaker where a special control provides a lower amplitude but a frequency doubling.

Range and sensitivity:
Foil micro-sieves 5 μm to 100 μm with square holes expanded conically towards the bottom to avoid permanent blocking. Other sieves, including woven ones, are also available up to 200 μm. Up to 16 sieves may be stacked on the shakers.

Application:
The Analysette 18 is designed for coarse grained materials but is also suitable for fine grains; it replaces previously undertaken hand shaking. The Analysette 3 covers the smaller particles which may be in suspension.

Cost:
Around £1895 (Analysette 18), £1230 (Analysette 3)

Further details:
[WARNING — for latest models check supplier]
See pages 135 and 168. Ref: von Bernuth 1988b.

FRITSCH ANALYSETTE 20 SCANNING PHOTO-SEDIMENTOGRAPH	Manufacturer: Fritsch GmbH UK Supplier: Christison Scientific Equipment Limited

32

Technique: SEDIMENTATION [SCANNING PHOTO-SEDIMENTOGRAPHY]

Description:

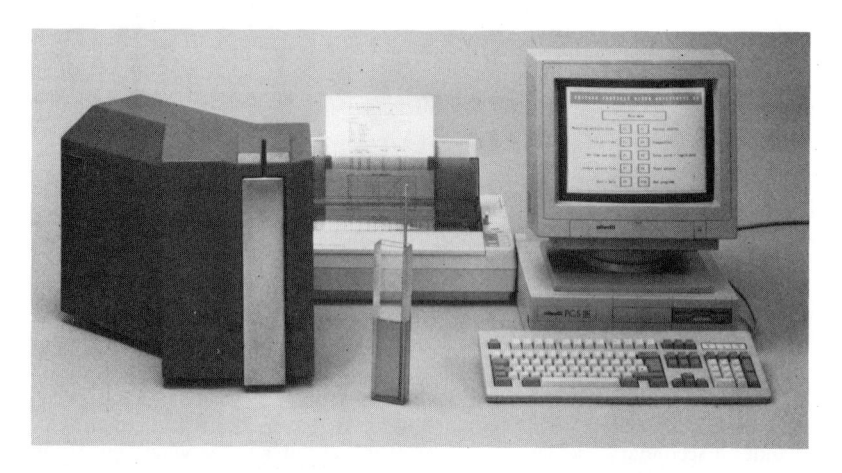

A larger than normal measuring vessel is used to eliminate interference of the sedimentation process by the vessel walls and base; the lower concentration of solids has an equally beneficial effect on the accuracy (eg. the larger particles do not drag the smaller ones down). The light required is provided by photodiodes (LED's) so that any disturbance of the sedimentation due to thermal convection is eliminated. A typical sampling time is of the order of 7 minutes.

Range and sensitivity:
Size range 0·5 μm to 500 μm. The highest accuracy is only achieved with particles greater than 2 μm because the absorption of light prevents clear results being attained.

Application:
Dry particles or debris in clear liquids

Cost:
Around £11,880

Further details:
[WARNING — for latest models check supplier]
See pages 134 and 168. Ref: von Bernuth 1988b.

FRITSCH ANALYSETTE 22 LASER PARTICLE SIZER	Manufacturer: Fritsch GmbH UK Supplier: Christison Scientific Equipment Limited
Technique:	OPTICAL – FRAUNHOFER DIFFRACTION [LASER DIFFRACTION SPECTROMETER, LASER PARTICLE SIZER]
Description:	

The Analysette 22 can analyse particles in suspension in a measuring cell or dry, in a dry feeding dispersion unit. A special feature of the laser employed is that it has a convergent beam; this allows a change in measuring range over the total scale to be achieved solely by moving the measuring cell (ie. no change of imaging lens). The time required for a complete distribution measurement is 2 minutes. Dispersion of the particles before measurement is achieved by an in-built ultrasonic bath.

Range and sensitivity:
Size range 0·16 μm to 1250 μm in a number of steps each with a dynamic range of 100 (eg. 5–500 μm). Dry debris as small as 0·5 to 1 g can be measured and wet fluid volumes of about 250 mL.

Application:
Process and quality control where particle distributions are required, primarily in mineral and synthetic raw materials. Particularly suitable for fast accurate measurements

Cost:
From around £18,400 to £43,000

Further details:
[WARNING — for latest models check supplier]
See pages 124 and 168. Ref: von Bernuth 1988a.

FULMER WEAR DEBRIS Manufacturer & UK Supplier:
MONITOR Fulmer Systems Ltd

34

Technique: WEAR (THIN FILM SENSOR)
Description:

A thin resistive metallic film (0·1 μm thickness, 50 ohms) is deposited on a ceramic substrate to form the sensor. This sensor is susceptible to erosion by all abrasive particles, metallic and non-metallic. The sensor includes a temperature compensation film and integral thermistor. Either the 'RW' mode may be used, for long term trend analysis, or the 'AB' for an instant reading of abrasivity; in each case the resistance change (ohms) of the sensor film is measured and compared with the compensation film. Flow from a system is accelerated through a 2 mm orifice and projected onto the wearing film at high velocity.

Range and sensitivity:
Sensitivity varies with shape and hardness of the particles as well as the fluid viscosity. Typically the Fulmer WDM is able to detect 1ppm for sands, catalytic fines, ferrous particles (35 μm to 150 μm). Resistance is measured approximately every 0·5 s; when it reaches 450 ohm, the sensor is changed.

Application:
A bypass connection needs to be made to a system, to allow an upstream pressure between 1·5 bar and 15 bar and a downstream pressure as close to atmospheric pressure as possible. Alternatively an auxiliary pump is fitted with a 1mm strainer to protect the pump and prevent blocking of the sensor housing.

Cost: Around £4000 hand-held or plant models. Replacement sensor £200.

Further details: [WARNING — for latest models check supplier]
See pages 139, 163, 249 and 272. Refs: Holmes 1985, Santilli 1989.

GALAI CIS-1, CIS-100, CIS-1000 & DSA-10	Manufacturer: Galai Production Limited
	UK Supplier: Roth Scientific Company Limited

Technique:	OPTICAL – TIME OF TRANSITION [and DYNAMIC SHAPE ANALYSIS]

Description:

The basic instrument uses a 5 mw He-Ne laser scanning in a circular motion by means of a rotating wedge prism and focussed down to a 1·2 μm spot. This 'spot' scans the sample volume at a constant speed detecting the dispersed particles. As each particle is individually bisected, interaction signals are generated and detected by a PIN photodiode. These signals are collected by a dedicated data acquisition card and analysed into 300 discrete size intervals. Shape analysis is provided by CCD video camera microscope. The acquired images, updated regularly, may be analysed.

Range and sensitivity:
Size range 0·5–1200 μm in two range modules; CIS-100 to 3600 μm. Concentration up to 10^9 particles per mL (1 μm particles). Resolution 0·3% full scale with a minimum of 0·2 μm. Image analysis with up to 256 grey levels with a resolution of 0·8 μm. Typical cycle time 30 seconds.

Application:
Laboratory, off-line, with the CIS-1, or on-line with the CIS-1000. Clear or opaque liquid down to 10% transparency, and hence suitable for some engine oils. The DSA-10 characterises shapes on slides.

Cost: From around £15,000 to £40,000 depending on system configuration.

Further details: [WARNING — for latest models check supplier]
See pages 131, 168 and 265. Ref: Karasikov & Krauss 1989.

HACH TURBIDIMETER

Manufacturer: Hach Europe S.A./B.V.
UK Supplier: Camlab Ltd.

36

Technique: OPTICAL – NEPHELOMETRY/TURBIDITY

Description:

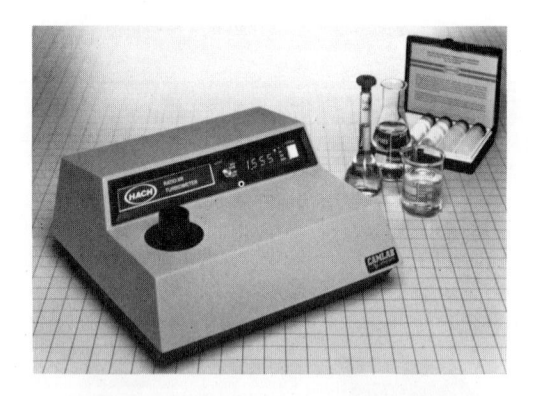

Hach turbimeters, using 90° scatter, cover a large range of use and hence operate in more than one extra technique to overcome some of the practical difficulties associated with in-line monitoring. For instance a bubble rejection circuit is employed which eliminates sudden spikes (this also rejects occasional large particles). In other models an optional degasing kit is available. The patented Ratio® turbidimeter uses forward transmitted and scattered light detection to reject the effect of sample colour. Display is in Nephelometric Turbidity Units.

Range and sensitivity:
Several models cover from the very low 0 (0·001 NTU) to 100 NTU up to 0 to 9999 NTU. with a resolution varying from 0·0001 NTU to 1·0 NTU even above 1000 NTU. Repeatability is generally ± 1% ± 1 digit.

Application:
Process and laboratory applications cover mainly water based situations, however, lubricants can be successfully catered for using the Ratio turbidimeter which compensates for the oil colour. The portable devices are generally around 6 kg in weight.

Cost:
From around £1000 to £3000 (laboratory instruments) and £2000 to £4500 (process instruments), depending on unit

Further details:
[WARNING — for latest models check supplier]
See pages 127 and 155.

HIAC/ROYCO® 8000 Series with Light Extinction Sensors	Manufacturer: HIAC/ROYCO Division of Pacific Scientific UK Supplier: HIAC/ROYCO Division of Pacific Scientific Limited

37

Technique: OPTICAL – LIGHT OBSCURATION [LIGHT EXTINCTION]

Description:

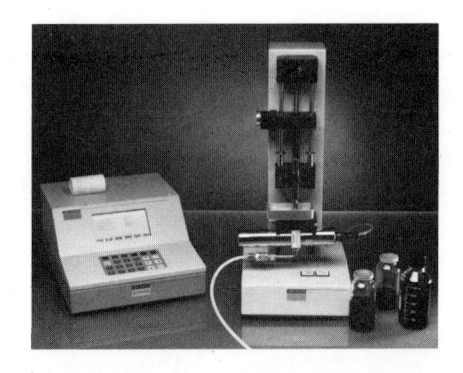

The HIAC/ROYCO is primarily an off-line instrument, although in certain filtration test systems it is used on-line. Samples of oil may either be in bottles (typically 250 mL bottle size) or in syringes taken directly from a system. Liquid is forced through the optical sensor at a fixed flow rate (appropriate to the calibration) and a count taken for a set liquid volume. The use of laser illumination has enabled the size range to be extended from earlier models (HIAC sensors for liquids commenced in 1962); the likelihood of blockage has also been reduced, and flow rates increased. The on-line use is where two sensors are attached, one either side of a filter, and the efficiency of filtration can be determined. Care has to be taken that the concentration levels are acceptable. Maximum on-line pressure is 70 bar.

Range and sensitivity:
Several sensors are available covering sizes from 0·5 μm to 2500 μm with concentration limits varying from 18000 particles per mL to 18 at 2500 μm. High concentration sensors are available. Viscosities up to 2500 cSt are possible.

Application:
Laboratory testing to a Standard. Filtration tests.

Cost: From around £10,000, typically around £17,000.

Further details: [WARNING — for latest models check supplier]
See pages 125, 180 and 199. Ref: Sommer et al 1991.

HIAC/ROYCO®
8000 Series
with Submicron Liquid Sensors

Manufacturer: HIAC/ROYCO
Division of Pacific Scientific
UK Supplier: HIAC/ROYCO
Division of Pacific Scientific Limited

Technique: OPTICAL – FORWARD REFLECTANCE [NEAR FORWARD ANGLE LIGHT SCATTER]

Description:

Laser diode submicron sensors have replaced the original He-Ne laser sensors. The laser diode is focussed to evenly illuminate the entire measuring volume. The seals and tubing and internal components used, make the instrument compatible with a wide range of fluids — from DI water to aggressive etches and acids used in the manufacture of semiconductors. Laser diode sensors offer all the advantages of the original He-Ne lasers, with the added advantages of being compact, long life, stable and robust.

Maximum line pressure is either 100 bar or 10 bar.

Range and sensitivity:
Several sensors are available covering sizes from 0·1 μm–20 μm with maximum concentration limits varying from 3500 to 30000 particles per mL. A combination of forward scatter and light obscuration, in one dual mode sensor, gives a range of 0·5 μm–350 μm.

Application:
Fine chemicals, semi-conductor, aerospace, pharmaceutical, etc.

Cost: From around £17,000

Further details: [WARNING — for latest models check supplier]
See pages 123 and 189.

HORIBA LASER SCATTERING Manufacturer: Horiba Ltd.
PSDA LA-900 UK Supplier: Horiba Instruments
 Limited

Technique: OPTICAL – FRAUNHOFER
 DIFFRACTION [LASER
 SCATTERING]
Description:

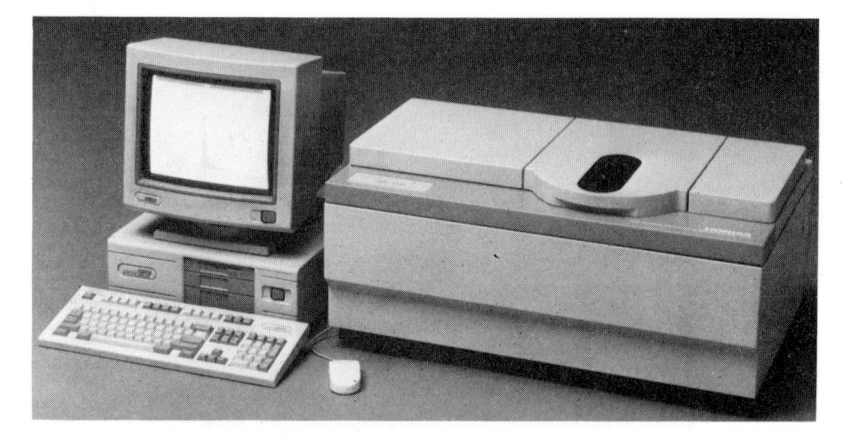

The LA-900 uses both Fraunhofer diffraction and Mie scatter theories and, by using a large diameter lens, is able to cover a significantly improved size range over a basic Fraunhofer device. A ring-shaped silicon photo-diode, ahead of the laser beam, is able to detect in 18 divisions; the smallest particles are detected using a tungsten lamp with scattering filters onto forward, side and back scatter detectors. Agitation is provided by mechanical agitation and ultrasonics. Time for a complete cycle is two minutes. A small fraction cell is provided where less than 10 mg/150 mL is available.

Range and sensitivity:
Size range 0·04 μm to 1000 μm. Quantity required 10 mg to 1 g for dispersing within 250 mL to 350 mL. Fraction cell takes 1 mg in 15 mL.

Application:
Most particulates, dry or in suspension. Emulsions and polymer latices are also possible.

Cost:

Further details:
[WARNING — for latest models check supplier]
See pages 124 and 168.

HORIBA PHOTOSEDIMENTATION Manufacturer: Horiba Ltd.
PSDA CAPA-700 UK Supplier: Horiba Instruments
Limited

Technique: SEDIMENTATION
[LIQUID PHASE
PHOTOSEDIMENTATION]

Description:

The CAPA-700 is a sophisticated desk-top instrument designed for rapid, accurate analysis using liquid phase photosedimentation. Both centrifugal and gravitational analysis is possible. The instrument also includes gradient analysis, two multi-mode analysis functions, 10 switch-selectable centrifuge speeds and multiple sedimentation distances. Size analysis may be presented on either a volume (equivalent to weight), area, length or number basis. Other important features include automatic self-diagnosis and automatic selection of the best test conditions.

Range and sensitivity:
Size range from 0·01 μm to 300 μm. Centrifuge speed from 300 to 10000 rev/min. Sedimentation distances from 10 mm to 40 mm for the gravitational mode (10–300 μm), and 5 mm or 10 mm for the centrifugal mode (0·01–30 μm).

Application:
Most particulates, dry or in suspension. Emulsions and polymer latices are also possible.

Cost:

Further details:
[WARNING — for latest models check supplier]
See pages 134 and 168.

HORIBA LIQUID PARTICLE COUNTER PLCA-520	Manufacturer: Horiba Ltd UK Supplier: Horiba Instruments Limited
Technique:	OPTICAL – LIGHT SCATTER [90° LASER SCATTER]

Description:

A 10 mW He-Ne laser is used to illuminate the particles as the liquid passes through the sensor. The laser is condensed vertically by a cylindrical lens and both edges are cut by a slit to locate the detection field in the centre of the oblate beam. The scattered light pulse is focussed through a pinhole at the 90° angle onto a photomultiplier tube. The pulse is allocated to up to six size channels.

Range and sensitivity:
Size range 0·2 μm to 10 μm. Liquid must be clear and contain no air bubbles. Sample required is between 200 mL and 1000 mL. Particle concentration less than 2000/mL.

Application:
Particulate in water or other clear liquids or chemicals.

Cost:

Further details:
[WARNING — for latest models check supplier]
See pages 126 and 186.

LASENTEC® LAB-TEC® and PAR-TEC®

Manufacturer: Laser Sensor Technology Inc.
UK Supplier: Belstock Controls

Technique:

OPTICAL – TIME OF TRANSITION [SCANNING LASER MICROSCOPE]

Description:

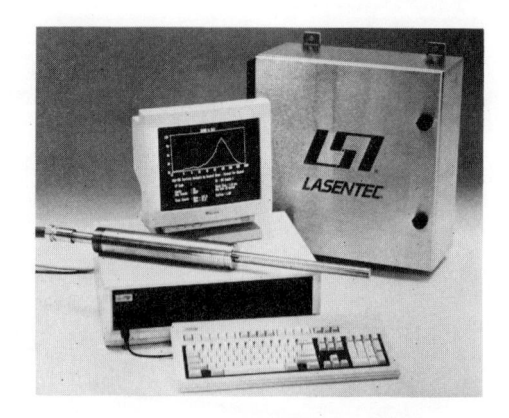

A single probe containing the laser and diode detectors is inserted into the liquid under test. The particles are detected in front of the sapphire window but only those which are in the focus of the laser beam are actually counted. The laser spot scans the sample volume (approximately 10 μm^3) at a constant 4500 rev/min detecting the dispersed particles. With the addition of mechanical stirring, or a magnetic stirrer, a very accurate distribution assessment can be achieved.

Range and sensitivity:
Size range from 1 μm to 1000 μm (the specific size range is determined by the computer when data has been acquired). Up to 70% concentration of particulates can be measured. The LAB-TEC model can be used with a 100 mL beaker, the PAR-TEC in-line. Counting concentration can be as high as 200,000 particles per second.

Application:
Due to the high concentrations possible, many applications are feasible without dilution.

Cost:
LAB-TEC around £14,000   PAR-TEC around £25,000

Further details:
[WARNING — for latest models check supplier]
See pages 131, 168 and 307. Ref: Hanseler & Preikschat 1987.

LIQUID CONTAMINATION MONITOR (LCM II)	Manufacturer & UK Supplier: Coulter Electronics Limited

Technique: FILTER BLOCKAGE [MESH OBSCURATION]

Description:

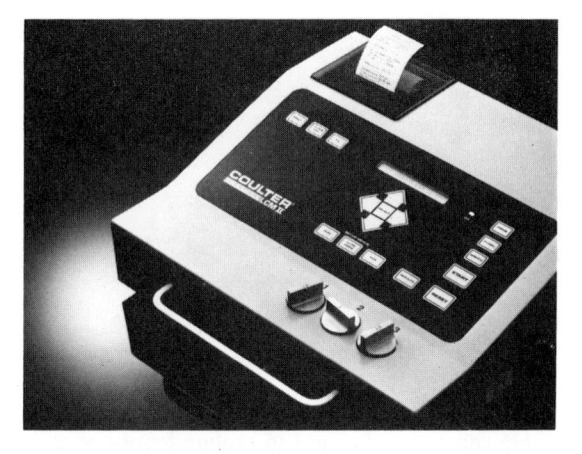

The oil, containing debris, is passed through three high-precision filter meshes, each with a known size and number of pores. Should any particle exceed the size of the pores, it is retained on the surface of the mesh. This reduces the open area of that mesh and increases the measured pressure drop across it. The rise in pressure drop is related to the number of particles trapped. Because an automatic reversal occurs, with backflushing, the instrument is continuous and normally provides an analysis within 90 sec. It can be programmed for up to 9,999 analyses

Range and sensitivity:
Meshes provide measurement at 5 μm, 15 μm and 25 μm, as well as a viscosity assessment (10 cSt to 500 cSt). The display provides a Standard Class such as ISO 4406, as well as the count of particles greater than each size. Count range at 5 μm is between 10^2 and 10^6 per mL of oil.

Application:
Because the filter blockage principle is independent of particle type or of fluid opacity or mix, a wide range of applications in lubrication and hydraulic machines is suitable.

Cost:

Further details:
[WARNING — for latest models check supplier]
See pages 118, 196 and 249. Ref: Cowan & Wenman 1990.

MAGNETIC CHIP COLLECTORS   Manufacturer & UK Supplier:
                               Muirhead Vactric Components
                               Limited

---44---

Technique: MAGNETIC ATTRACTION

Description:

The magnetic chip collector is designed to capture ferrous wear debris contained within lubricating oil, for visual inspection or detailed analysis. The device comprises a housing, permanently mounted in the lubrication system, and a removable magnetic probe that is positioned to expose the magnet to the circulating lubricant (normally in the scavenge or return line). The probe is usually held in place with a quick-release bayonet fitting and sealed by twin O-rings on the shank. When the probe is removed, a self-closing valve prevents loss of lubricant. Advice is available on the best position for fitting as regards oil viscosity, air/oil ratio, type of flow, etc. Swirl chambers are also available to assist collection.

Range and sensitivity:
Typical figures would be up to 14 bar within a temperature range from minus 70°C to plus 200°C. Ferrous debris of all sizes.

Application:
A simple and low-cost method of monitoring the wear rate of gears and bearings in critical components and systems, using pre-set time intervals. Typical examples would be those of fixed-wing aircraft or helicopters.

Cost:
From around £60 to over £200 depending on the size and complexity of the design. Some models incorporate safety features.

Further details:
[WARNING — for latest models check supplier]
See pages 10, 33, 122, 151, 300 and 311.

| MALVERN AUTOCOUNTER | Manufacturer & UK Supplier: |
| (ALPS 100L & 150H) | Malvern Instruments Ltd |

Technique: OPTICAL – LIGHT OBSCURATION [LIGHT EXTINCTION]

Description:

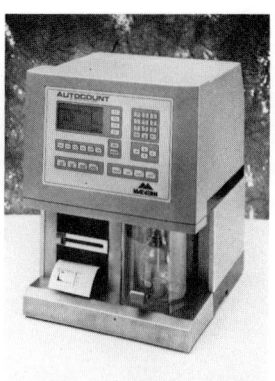

                100L   150H

Particular innovative characteristics of these models are the advanced white light source (tungsten), *in situ* calibration operation by the user and, in the case of the ALPS 150 H, a single self-contained unit incorporating an automatic bottle sampler. Output can indicate chosen Standard classes such as ISO 4406, NAS 1638, USP XXI. Individual particle counts at up to 50 sizes may be printed out for each test. Graphic and numerical displays are incorporated.

Range and sensitivity:
Size range is 2 μm to 150 μm with a dynamic range of 50 (eg. 2 μm–100 μm, 3 μm–150 μm). An extension to 5 μm to 250 μm is possible with an optional sensor. Volumetric accuracy is better than 1%. Viscosity range 1–300 cSt without dilution. Maximum particle concentration 47000 particles per mL.

Application:
Any homogeneous liquid within the viscosity band, although higher viscosities can also be tested with dilution. Suitable for oils, aqueous solutions and solvents. The Autocounters can be rapidly adapted for fully automated sampling by the addition of the Malvern Autosampler; on-line measurement can also be performed with a suitable flow controller.

Cost:

Further details: [WARNING — for latest models check supplier]
See pages 125 and 180.

MALVERN MASTERSIZER X   Manufacturer & UK Supplier:
                        Malvern Instruments Ltd

46

Technique: OPTICAL – FRAUNHOFER
                        DIFFRACTION

Description:

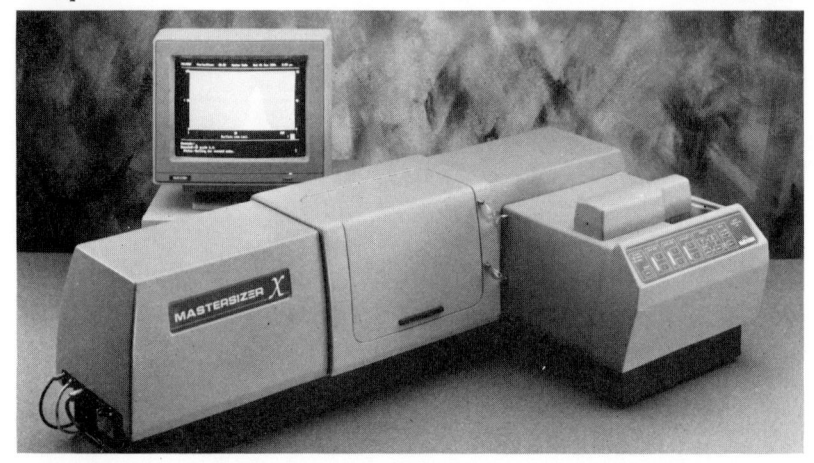

The MasterSizer X has an exceptionally large size range capability. It is able to cover the lower end of the range by means of three special innovations — an advanced single chip silicon photodiode array detector, the use of "reverse Fourier optics" to detect high angle scatter and an automatic precise positioning of the detector by servo motor. The larger end of the size range has been extended by more conventional means but with the accuracy of the servo motor automatically positioning the diode detector when the focal length of the lens is entered. Mie theory, standard optics and Fraunhofer approximations are used.

Range and sensitivity:
The total size range of the MasterSizer X is from 0.1 μm to 2000 μm. Up to 64 size classes are possible for each size band chosen (eg. 0.1–80 μm).

Application:
Both wet and dry dispersions can be analysed. Suitable for process plants and other applications where distributions over a wide size range are required.

Cost:
Note: Malvern Instruments also manufacture the Autosizer series based on photon correlation spectroscopy.

Check for latest details.

See pages 124 and 168. Ref: McFadyen 1989.

MATEC CHDF 1100 PSA   Manufacturer: Matec Applied Sciences Inc.
UK Supplier: ChemLab Scientific Products Limited

Technique: CAPILLARY HYDRODYNAMIC FRACTIONATION (CHDF)

Description:

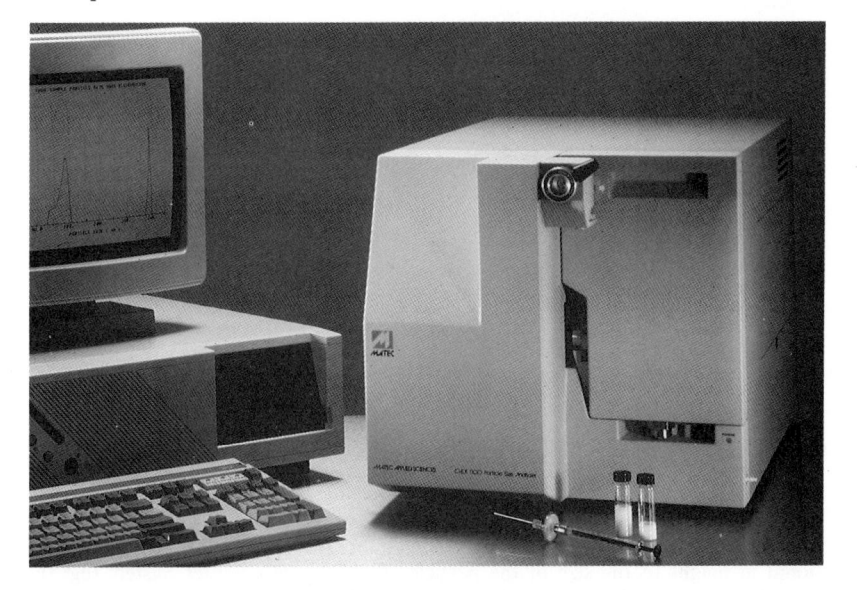

The technique is primarily a submicron method; resolution is therefore seen in % rather than absolute micron values. The technique is independent of particle density and hence a true distribution can be obtained even for mixtures. Analysis takes around 8 minutes and the resulting distribution is displayed both graphically and in tabular form. Distributions are either on size or on volume with averages, if required.

Range and sensitivity:
Standard size range is 0·015 μm to 1·1 μm. Particles with a more than 10% size difference are resolved. Sample size is 1 mL with a solids content in the region of 2–4% (prefiltering with a 5 μm filter is necessary).

Application:
Industrial latex samples and colloidal materials

Cost: Around £28,500
Further details:
[WARNING — for latest models check supplier]
See page 114. Ref: Ramos & Silebi 1990.

MET ONE LIQUID
PARTICLE COUNTERS and
ANALYSERS

Manufacturer: Met One Inc.
UK Supplier: Environmental
Monitoring Systems Ltd.

Technique: OPTICAL – FORWARD
REFLECTANCE [FORWARD
LIGHT SCATTERING]

Description:

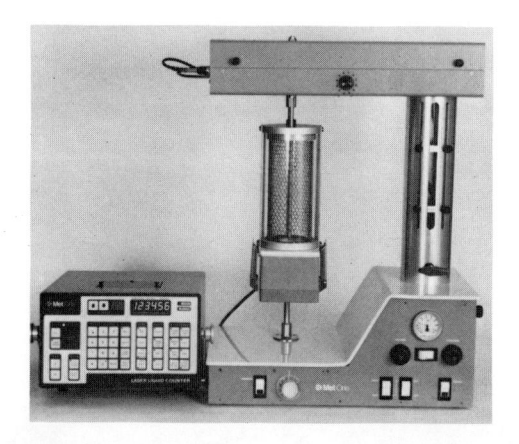

The Met One sensor uses a type of forward light scattering which requires only one solid state photodiode. As the particles in the liquid pass through the sensor (at the point of focus of the laser beam), they cause bursts of light energy to be projected. The photodiode converts these pulses into electrical energy, proportional in height to the size of the particle. The Particle Counter System (eg. the Model 250 Liquid Batch Sampling System) includes the sensor (eg. the Model 211) and a separate display (eg. the Model 214). A separate dual model (233) is available. A more inclusive advanced model (L9000 series) is also available.

Range and sensitivity:
Size range includes 6 bands from 0·5–25 μm to 5·0–600 μm with a display in six sizes. Coincidence is less than 10% at 3500 counts/mL in the standard sensor, and less than 10% at 10,500 for the high concentration sensor. Operating pressure is 8·5 bar but a high pressure option of 140 bar is available.

Application:
Off-line systems requiring particle assessments for contamination control or clean lubrication. The dual model is suitable for filtration efficiency testing.

Cost: From around £10,000

Further details: [WARNING — for latest models check supplier]
See pages 123 and 189.

MICRO PURE® MPS-D and MPS-3000 PARTICLE CONTAMINATION MONITORS

Manufacturer: Monitek Technologies Inc.
UK Supplier: Acal Auriema Limited

Technique: ULTRASOUND

Description:

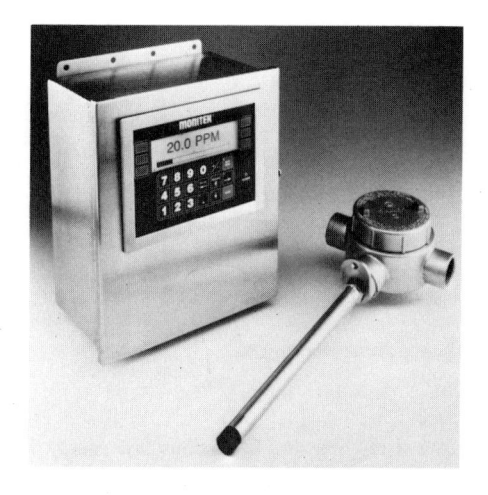

An acoustical transmitter is mounted in-line in the pipe wall without restricting the fluid flow. The transmitted beam is focussed to 22–38mm within the fluid and the return echo of the acoustical pulses is detected by the same sensor. As the echo magnitude is dependent on the size of the particle, the detection threshold can be adjusted to suit a particular process liquid. Backscatter (echo) is caused by liquid droplets, bacteria and air bubbles, as well as solid particles and fibres, but is independent of colour, density, opacity and flow rate.

Range and sensitivity:
Size range approximately 0·2 μm to 3 mm with a concentration range of either 0·01 ppm to 3% or 3–60% (based on ACFTD). Repeatability 1% of full scale, sampling rate 200 cycles/s.

Application:
As the sensor is fitted in-line it is particularly suited to process flows which involve acids and chemicals because no sampling is necessary. However, it is equally suitable for other types of system such as hydraulic fluids, lube oils and water. Off-shore use at 10 ppm is another application.

Cost: The MPS-D commences around £8500, the MPS-3000 at £30,000

Further details: [WARNING — for latest models check supplier]
See pages 137 and 164. Ref: Bruno 1982, Gaucher 1984.

MICROSCAN II PSA   Manufacturer: Quantachrome Corp.
                                 UK Supplier: Steptech
                                 Instrument Services Limited

Technique: SEDIMENTATION
Description:

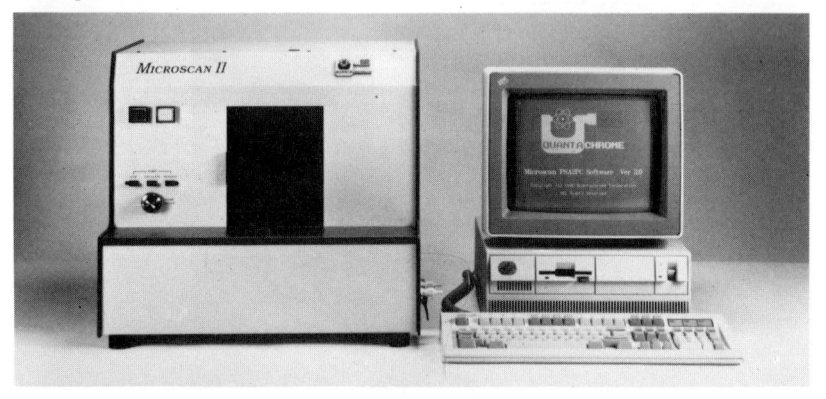

The Microscan II uses a moving non-hazardous low energy X-ray beam. As the particles settle under gravity in a liquid, a very narrow beam of X-rays scans the sample cell from bottom to top using an algorithm designed to minimize analysis time. Because the X-rays are attenuated in proportion to the mass of particles in the beam, the intrinsic data is a direct plot of particle size (diameter) versus mass. Data can be obtained as diameter versus number (%) or surface area. Sample dispersion (before commencement of the test) is by shear forces associated with circulation, although ultrasonics may also be used. Automatic bubble removal and system cleaning is also included.

Range and sensitivity:
Size range 0·1 μm to 300 μm with up to 300 data points. 7 discrete scan rates optimise analysis time without loss of resolution. Typical analysis time is 3 minutes. Graphical or tabular output giving histograms or cumulative curves for mass, surface area and particle count %, both logarithmic and linear.

Application:
Processes requiring control or determination of particle size distributions, including powdered and sintered metals, desiccants, ceramics, etc.
Cost:

Further details:
[WARNING — for latest models check supplier]
See pages 134 and 168. Ref: Quantachrome 1991.

MICROTRAC PSA
Series 9200 SRA and FRA

Manufacturer: Leeds & Northrup
UK Supplier: Leeds & Northrup Limited

Technique: OPTICAL – FRAUNHOFER DIFFRACTION

Description:

(The photograph shows the UPA and FRA together.)

UPA   FRA

The SRA (standard range analyser) and the FRA (full range analyser) both utilise forward Fraunhofer laser diffraction, but the FRA has an additional high angle diffraction detector module to extend the lower size limit (Mie scatter corrections are applied). The complete range is tested simultaneously. Samples may be presented wet or in powder form. The analyser allows output in graphical or tabular form covering volume %, surface area or other operator entered parameters. The units are modular and are suitable for desk top use.

Range and sensitivity:
The size range for the SRA is 0·7 μm to 700 μm, for the FRA it is 0·1 μm to 700 μm. Sample volume may vary between 13 mL (small volume cell) and 4 L (large volume recirculator), with an intermediate 250 mL (small volume recirculator). A dry feeder accessory provides 10–20 g/min.

Application:
Processes where the distribution of particles, fine or coarse, wet or dry, is required to be quickly and accurately determined, ie. on the same instrument.

Cost:
From around £22,000 to £30,000

Further details:
[WARNING — for latest models check supplier]
See pages 124 and 168.

MICROTRAC® PSA Manufacturer: Leeds & Northrup
Series 9200 UPA UK Supplier: Leeds & Northrup
 Limited

52

Technique: OPTICAL – PHOTON
 CORRELATION
 SPECTROSCOPY

Description:
(The photograph shows the UPA and FRA together.)

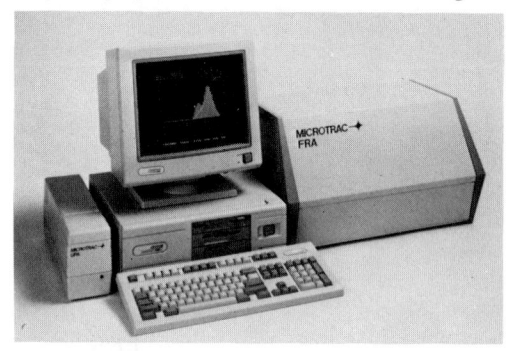

UPA FRA

The Microtrac UPA basically uses the principle of Brownian Motion of particles suspended in a fluid, however, the normal doppler shifts are compared with the waveform of an unshifted reference, using Fast Fourier Transform techniques. In addition the UPA measures the backscattered component of the light, combining reflected light with the backscattered light, to create a high-level signal strong enough to be fed directly to a photodetector. This enables a very wide range of concentrations to be analysed with minimum dilution.

Range and sensitivity:
Size range is 0·005 μm to 3 μm. The compact unit measures 152 mm high by 102 mm wide by 380 mm deep.

Application:
Processes where the distribution of fine particles in suspensions is required to be quickly and accurately determined, ie. on the same instrument with minimum dilution.

Cost:
Around £19,000

Further details:
[WARNING — for latest models check supplier]
See pages 130 and 200.

MILLIPORE®	Manufacturer: Millipore Corp.
FLUID CONTAMINATION	UK Supplier:
ANALYSIS KIT	Millipore (UK) Ltd

53

Technique: GRAVIMETRIC

Description:

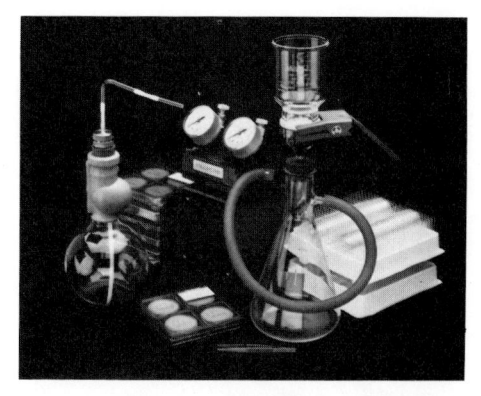

The special feature of the Millipore gravimetric test is the provision of matched pairs of membrane filters. These are of the same weight within 0·1 mg and can be used, one on top of the other, to determine the precise weight of debris per quantity of fluid without pre-weighing. All that is needed to be done is to determine the difference in weight between the two filters after the fluid has passed through them both. In order to complete each test as smoothly and rapidly as possible, the kit includes a vacuum pump with all the necessary glassware. Debris can later be examined under a microscope or spectrometrically if so required.

Range and sensitivity:
Either 0·45 μm or 0·8 μm membrane filters are used (47 mm diameter). The time for a test completion will vary with viscosity and level of debris. A high debris level will block the membrane quickly and reduce the rate of flow through it; the quantity of liquid to pass through can, therefore, be modified to suit the situation.

Application:
The technique is off-line only, but can be used for a very wide range of oils and fluids to assess the overall level of debris present in the liquid.

Cost:
Around £1200 including all glassware, a vacuum pump, petri slides, packs of membrane filters and various accessories.

Further details:
[WARNING — for latest models check supplier]
See pages 100, 119 and 143.

MILLIPORE® PATCH TEST KIT

Manufacturer: Millipore Corp.
UK Supplier: Millipore (UK) Ltd

54

Technique:
Description:

VISUAL APPEARANCE

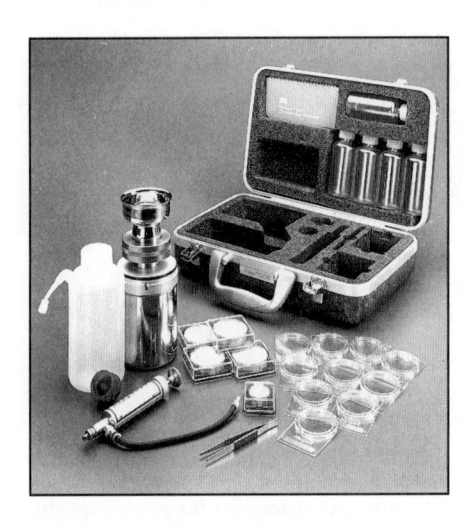

A rapid assessment of the general level of solid particulate contaminant in an oil is achieved with the Patch Test. A 5 μm membrane filter is used and, as required by the supplied standards book, 100 mL of oil is forced through it by vacuum. The membrane will be coloured to a certain density depending on the particulate present in the oil. However, the oil may also have a colouring effect; therefore, the standards provided come in different colour shades to compensate for any oil colouring effect. The supplied standards are for hydraulic fluids, but a user may make up his own standards to suit his situation.

Range and sensitivity:
Generally only particles greater in size to 5 μm will be detected. However, a smaller membrane pore size may be used in conjunction with specially prepared standard patches.

Application:
Off-line use only. Although primarily for hydraulic fluids, can also be used for fuels, lube oils, boiler water, etc.

Cost:
Around £1200 including all glassware, vacuum syringe, membrane filters, standards book and other accessories. A vacuum pump would add another £500 or so.

Further details:
[WARNING — for latest models check supplier]. See pages 100, 138 and 148.

MONITEK® TURBIDIMETER and CLAM® MONITOR

Manufacturer: Monitek Technologies Inc.
UK Supplier: Acal Auriema Limited

Technique: OPTICAL – NEPHELOMETRY/TURBIDITY

Description:

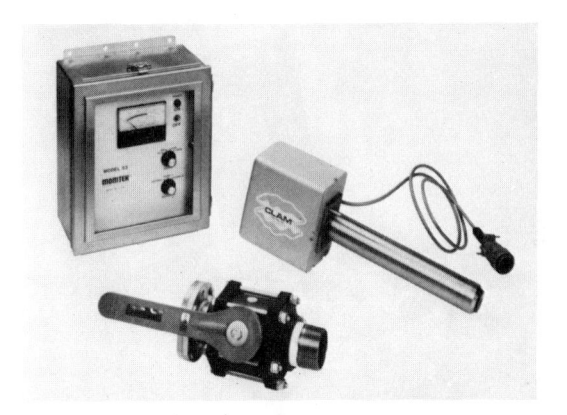

Monitek turbidimeters compare direct transmitted light with light which has been scattered at a narrow angle by particles. This ratioed forward scatter technique is independent of colour, particle size distribution and lamp intensity. It also has the advantage of having a known zero (ie. no light scatter) and being linear over the specified range. An electronic bubble rejection circuit is included. The CLAM® monitors also include a unique self-cleaning photoelectric sensing head, which ensures accurate, reliable reading and enables prolonged operation without costly cleaning or maintenance.

Range and sensitivity:
Range for the basic turbidimeter is from 0·1 ppm to 1000 ppm, but for the CLAM® it is from zero to 100,000 ppm (10%), calibrated for each particular system.

Application:
The basic turbidimeter is suitable for lube oil and other liquids. The CLAM® is more suitable for the investigation of suspended solids in industrial waste water and other fluids.

Cost:
From around £5000

Further details:
[WARNING — for latest models check supplier]
See pages 127 and 155.

| OILAB PORTABLE LUBRICANT & FUEL ANALYSIS KIT | Manufacturer & UK Supplier: Oilab Lubrication Limited |

Technique: VISUAL APPEARANCE

Description:

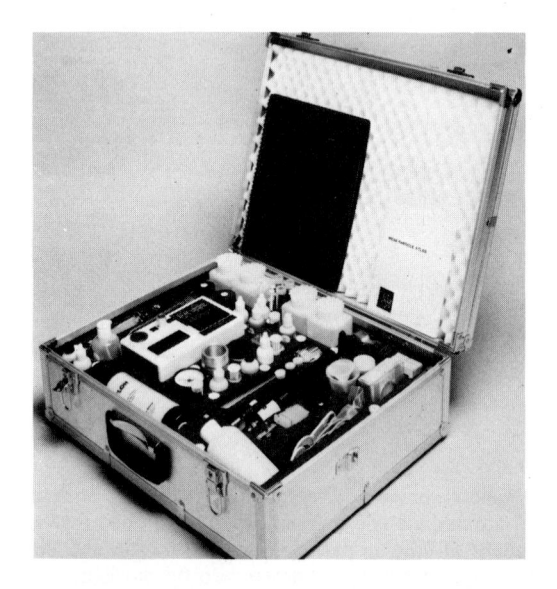

The Oilab kit is a multi-purpose oil analysis kit, able to determine several physical properties of an oil as well as the solid particle content. (Falling Ball viscometer, water content test, TAN test, TBN test, colour test, pour point test, density test, etc. are included.) The kit also includes a small 100× illuminated microscope, a sample pump, membrane filters, necessary plastic flasks, etc. The Wear Particle Atlas enables comparative debris work to be undertaken.

Range and sensitivity:
Generally 1 μm and upwards (ie. visible by optical microscope).

Application:
The kit may be used on site where oil or hydraulic fluid samples can be taken. Portable, weight complete with hand pump 12 kg.

Cost:
Around £2500 depending on how many functions required. Many optional extras are available such as photomicrography

Further details:
[WARNING — for latest models check supplier]
See pages 138 and 148. Ref: Newell 1991.

OILCHECK Manufacturer & UK Supplier:
UCC International Ltd

— 57 —

Technique: DIELECTRIC CONSTANT
Description:

The Oilcheck is a hand held portable oil monitor designed for use as a regular service monitoring instrument. The unit is automatically calibrated to zero by placing a few mL of fresh oil in the 'sensor well' and pressing the TEST button. After cleaning the well, a few mL of the test sample are placed in the well and the LO setting chosen for a reading. Numerical readings can be displayed and the green/red ring indicates the acceptability of an oil.

Range and sensitivity:
Range and setting (LO or HI) to be checked with system.

Application:
Fleet operators, garage mechanics and plant and equipment engineers.

Cost:
Around £260

Further details:
[WARNING — for latest models check supplier]
See pages 115 and 151.

OTSUKA LASER PARTICLE
ANALYSER LPA-3100

Manufacturer: Otsuka Electronics Co. Ltd.
UK Supplier:

Technique: OPTICAL – PHOTON CORRELATION SPECTROSCOPY and SEDIMENTATION

Description:

The LPA-3000/3100 is one of very few instruments which incorporate two quite independent techniques. In this way a much larger particle size range is possible. The smaller particle sizes are detected at 90° from Brownian movement scattering from a 5 mW laser beam, the larger particles are sensed by using gravity sedimentation with a LED light source. In the case of the sedimentation detection, a complete array of silicon photodiodes are used to observe the particles at different heights. The DLS-700 and ELS-800 units are also available for the small size particles and zeta potential measurements.

Range and sensitivity:
Size range from 0·003 μm to 5 μm (photon correlation spectroscopy) and 3 μm to 100 μm (sedimentation). Cell size 10mm square or cylindrical with a sample volume capacity of 2 mL/min.

Application:
Suitable for process control and all fields associated with fine particle measurement which may also involve some large particles. Emulsions, ceramics, paints, foods, etc.

Cost:

Further details:
[WARNING — for latest models check supplier]
See pages 130, 134 and 200.

PARKER PATCH TEST KIT Manufacturer & UK Supplier:
Parker Hannifin (UK) Ltd

Technique: VISUAL APPEARANCE
Description:

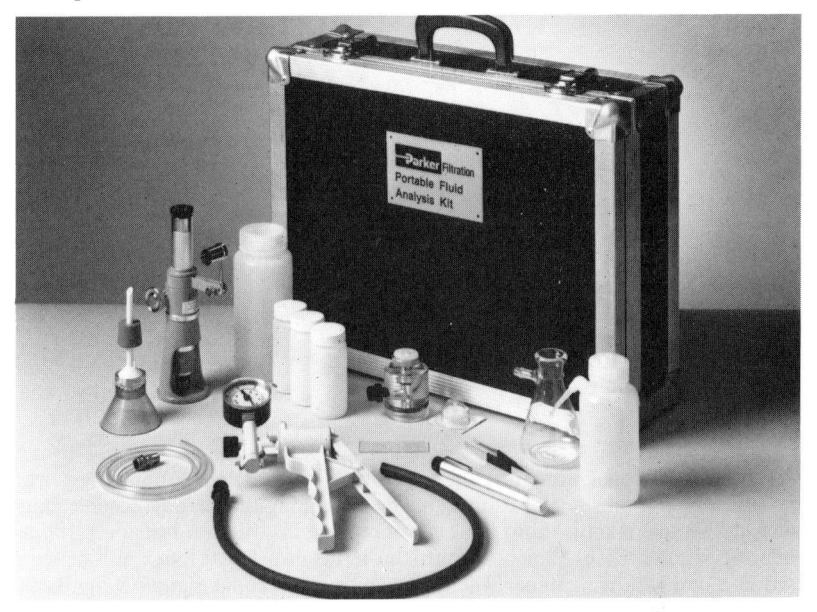

The Parker Patch Test kit is dedicated to contamination control of hydraulic systems and solid particle debris monitoring. The kit consists of a small hand-held illuminated microscope (60×), a hand operated vacuum pump, membrane filter holder, membrane filters, necessary glassware, plastic ware, etc. Comparison patches are also included in order to be able to estimate the contamination/cleanliness class of the fluid. Charts are provided to help in the analysis.

Range and sensitivity:
Generally 1 μm and upwards (ie. visible by optical microscope).

Application:
The kit may be used on site where oil or hydraulic fluid samples can be taken. Portable, weight complete 5·5 kg.

Cost:
Around £800 complete in hard body case

Further details:
[WARNING — for latest models check supplier]. See pages 138 and 148.

PARTICLE QUANTIFIER PQ90   Manufacturer: Swansea Tribology Centre
UK Suppliers: South — Analex Ltd
North & Midlands — Lindley Flowtech Ltd
International: Analex Ltd

Technique: MAGNETIC ATTRACTION [AC MAGNETOMETRY]

Description:

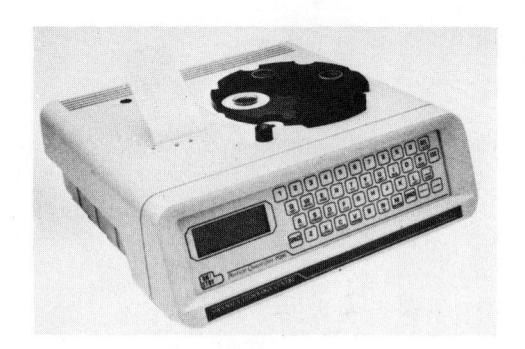

The PQ90 is designed to assess the quantity of ferromagnetic debris (eg. iron and nickel) in an oil sample. It does this by examining the ferromagnetic change when the sample is pulled into the sensing region. The full test lasts only 10 seconds per sample. Display or print out is in terms of a PQ90 Index and hence is used in a comparative manner where perhaps a 10% change signals a significant difference. Carbonaceous matter in the oil is not sensed. The debris contained in the oil can be presented to the instrument in situ within a plastic pot container (pot method), or as an oil-free debris laid in a specific pattern on a suitable substrate (RPD slide method) or on a membrane filter, as well as debris taken from magnetic plugs.

Range and sensitivity:
Particles detected in the range below 1 μm up to 2 mm.

Application:
All lubrication and hydraulic systems where oil samples can be reliably obtained for off-line tests. Industries such as mining, power generation, transport, construction, heavy engineering, aviation, etc.

Cost:
Around £4000. Very low running costs.

Further details:
[WARNING — for latest models check supplier]
See pages 122 and 151. Refs: Kwon et al 1987, Price et al 1987.

PHOTOMETRIC DISPERSION
ANALYSER PDA 2000

Manufacturer & UK Supplier:
Rank Brothers Ltd.

Technique:

OPTICAL – PHOTOMETRIC DISPERSION

Description:

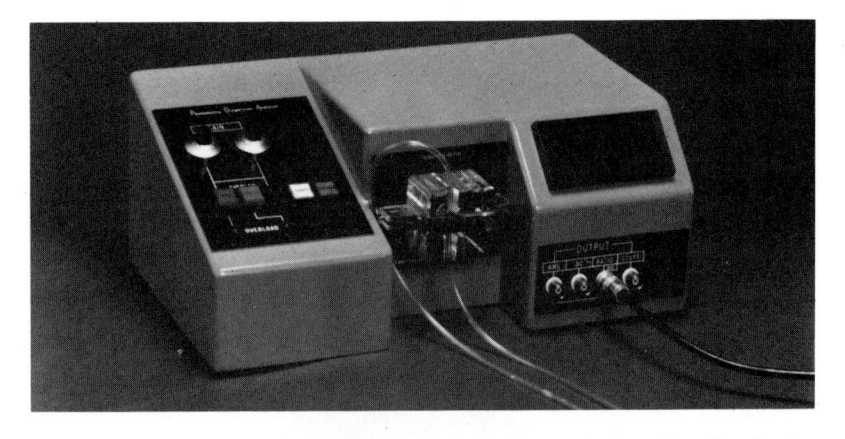

This is a simple and rugged instrument, yet extremely sensitive for flowing suspensions and emulsions. There are no orifices to clog and minimal problems associated with contamination of optical surfaces. The test cell incorporates a disposable plastic tube of either 1 mm or 3 mm bore depending on optical density (if high the 1 mm is used) and flow rate. The dc, rms and ratio outputs can be continuously monitored and digitally displayed. A limit switch provides some compensation for the presence of bubbles.

Range and sensitivity:
Particle size range is approximately 0.5 μm to 100 μm. Volume illuminated approximately 1mm^2. For particles around 2 μm, concentrations of the order of ppb can be detected; and for large particles even lower concentrations are possible. Upper limits of the order of several %. Flow rates from 2 mL/min to well over 20 mL/min. Compact with a weight of 4.5 kg

Application:
Selection of optimum flocculant dosages, control of dispersion and emulsification processes, assessment of the strength of aggregates (flocculations).

Cost:
Around £2900.

Further details:
[WARNING — for latest models check supplier]
See pages 129 and 166. Ref: Gregory & Nelson 1984.

PMS LBS-100

Manufacturer: Particle Measuring Systems Inc
UK Supplier: Particle Measuring Systems Europe Ltd

Technique: OPTICAL – LIGHT OBSCURATION

Description:

The IMOLV sensors are used with the Liquid Batch Sampler (LBS) to test and analyse bottle samples. As a bottle sampler the capacity of up to 1L makes the instrument particularly valuable for the larger sample, as well as small ones. Either a vacuum mode or the more usual pressure mode is available, enabling volatile liquids to be aspirated prior to the sample test.

Range and sensitivity:
The solenoid controlled operation of the LBS-100 is capable of handling liquids over a viscosity range from water (1 cSt) up to 50 cSt which covers most hydraulic and lubricating oils. The size ranges covered are 0·5 μm–60 μm, 2 μm–150 μm and 5 μm–300 μm, with maximum concentrations up to 10,000 particles/mL when oil is tested.

Application:
Hydraulic fluid contamination levels, debris levels in oils, water, solvents, etc.

Cost:
From around £10,000

Further details:
[WARNING — for latest models check supplier]. See pages 125 and 180.

PMS MICRO LASER PARTICLE
SPECTROMETER (μLPS)

Manufacturer: Particle
Measuring Systems Inc
UK Supplier: Particle Measuring
Systems Europe Ltd

Technique:

OPTICAL –
LIGHT OBSCURATION and
LIGHT SCATTERING

Description:

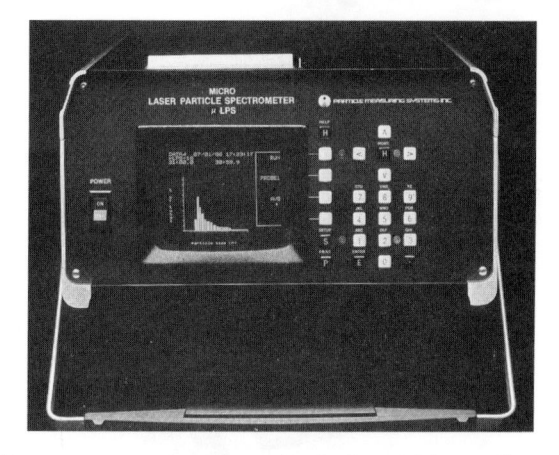

Two types of sensor are employed, one for 'volumetric' sampling which examines all the particles as they pass through a capillary tube (0·5 mm or 0·8 mm bore), the other is 'in-situ' where an optically defined area is chosen within the flow path. The in-situ type is ideal for sampling main process lines without the inherent blocking up of the sensor; in addition, as there is minimal pressure drop, the formation of confusing (counted) bubbles is less likely.

Range and sensitivity:
Volumetric sampling: size ranges from 0·2 μm–3 μm up to 0·5 μm–60 μm with concentrations up to 15,000 particles per mL (with a 90% confidence, ie. less than 10% coincidence). In-situ sampling: size ranges from 0·05 μm–0·2 μm up to 0·3 μm–5 μm with concentrations up to 10,000 particles per mL.

Application:
A variety of sensors and samplers which can connected to a single 'spectrometer', enables the PMS system to cover a wide range of applications from particle contamination and fine debris in oils and hydraulic fluids to deionised water.

Cost:
From around £8,500

Further details:
[WARNING — for latest models check supplier] See pages 126 and 180.

423

PMT 3120, PMT 2120   Manufacturer: Partikel-
PMT 3500, SKD-6   Messtechnik GmbH
                     UK Supplier:

Technique: OPTICAL – LIGHT OBSCURA-
                     TION [LIGHT BLOCKAGE]
Description:

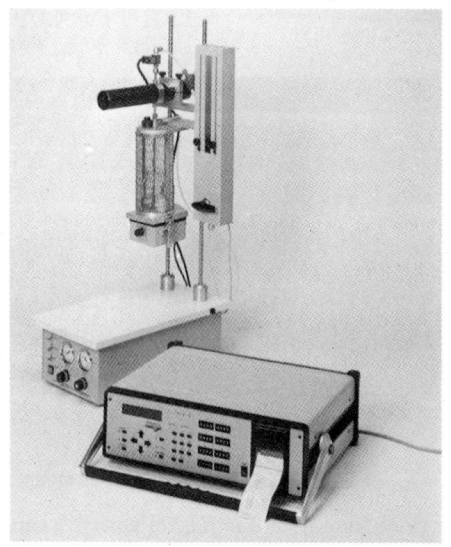

The PMT system may be used in-line, on-line or off-line (bottle sampling), although, by nature of the technique, the in-line measurement is limited to a certain particle concentration per volume. Three versions of the batch or bottle sampling samplers are available, covering viscous to much lighter liquids.

Range and sensitivity:
Total size range 1 μm–2000 μm with a dynamic range around 100 or even higher (eg. 1·5 μm to 500 μm). Special sensors at 80 μm–8000 μm and 200 μm–15,000 μm are also available. Display of size/number in up to 32 adjustable channels. Bottle samplers — standard 600 mL up to 4 bar, large 1·2 L at 5 bar, high 600 mL at 10 bar. Line pressures — standard 210 bar, optional 400 bar.

Application:
Contamination control of hydraulic oil, particle and debris counting in insulation fluids, air, water, solvents, etc. Multi-pass filter testing. Facility monitoring systems (over sensor control digitizers) with up to 256 measuring points.

Cost: From around 25,000 DM to 80,000 DM depending on the application.

Further details: [WARNING — for latest models check supplier]
See pages 125 and 180.

POLYTEC PARTICLE SIZER Manufacturer: Polytec GmbH
PSE-1500 UK Supplier: Laser Lines Ltd

Technique: OPTICAL – LIGHT SCATTER
Description:

A white light (tungsten halogen) source and a relatively large aperture in the optical system, are used to provide a very smooth curve of light intensity versus particle size. Such a near-linear relationship means sufficient resolution to allow 128 channel analysis over a particle dynamic range of 1:30. The measurement volume is defined by optical means, allowing particle detection on-line with little likelihood of the blocking up of the sensor. The scatter is detected at 90° by photodetector; the measuring volume being between 110-500 μm^3.

Range and sensitivity:
Size ranges from 1 μm–25 μm up to 2 μm–125 μm with maximum concentrations at 10^5 particles/mL down to 10^3. Velocity measurements from 0·1 m/s up to 10 m/s with an option to 20 m/s.

Application:
On-line and off-line sampling of process fluids. Contamination control. Measurement in emulsions, dusts, sprays and nozzles.

Cost:
Around £25,000

Further details:
[WARNING — for latest models check supplier]
See pages 126 and 186.

PURITY CONTROLLER Manufacturer: Hydac S.à.r.l.
RC 1000 UK Supplier: Flupac Ltd

Technique: OPTICAL – LIGHT OBSCURATION [LIGHT ABSORPTION]

Description:

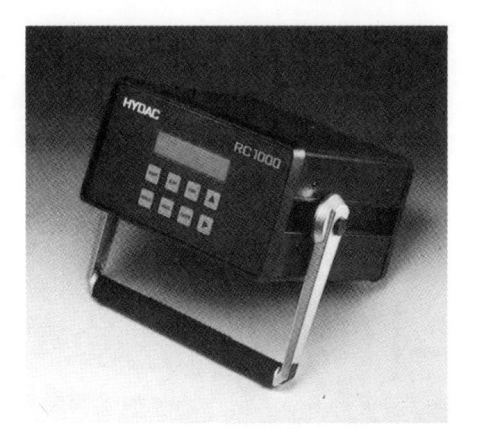

The rugged and compact construction means that the unit is portable and insensitive to vibration and high levels of contaminant. It is in two modules, electronic and hydraulic. Results are achievable within 60 seconds (at a flow rate of 100 mL/min). A 1mm diameter fluid path means that particles are unlikely to block up the sensor, and the sensor can be used on-line. The electronic light source does not deteriorate with time.

Range and sensitivity:
Size range 5 µm–100 µm at pressures 10 bar to 350 bar. Oil viscosities from 15 cSt–1000 cSt depending on line pressure (flow rate 50–150 mL/min). NAS contamination classes from 3 to 12 are measurable (alternatively ISO 4406 classes can be displayed).

Application:
On-line contamination control in hydraulic systems. Other on-line processes requiring particle level/count to be determined within the size range, eg. some wear debris .

Cost:
Around £7000 for complete system. Separate Sampling & Analysis Kit around £2000, Sampling Case (to take samples) around £900.

Further details:
[WARNING — for latest models check supplier]
See pages 125 and 180.

QUANTITATIVE DEBRIS MONITOR QDM®	67	Manufacturer: Vickers Inc. UK Supplier: Vickers Systems Limited TEDECO Division

Technique: MAGNETIC ATTRACTION

Description:
(Photograph shows QDM and control for an industrial gas turbine.)

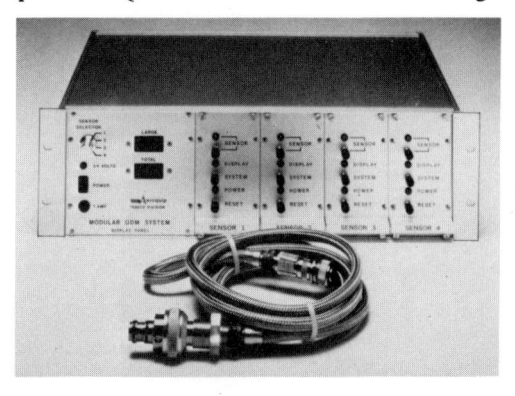

The QDM is fitted into the oil system at a point where debris is likely to be thrown toward the sensor (eg. at a bend or where a special cyclone unit is fitted) or settle out and be captured by the sensor. Ferrous debris in the flowing oil is magnetically captured by the QDM inductive sensor which provides voltage pulses proportional to the mass of captured particles. These pulses are amplified and transmitted to a conditioner which automatically classifies the signals into two categories. These relate to particles greater than 100 μg (approx. 500 μm size bearing spall flake) and greater than 800 μg (approx. 1000 μm flakes). Further processing allows trends and rates assessment.

Range and sensitivity:
Small chips (greater than 500 μm) and large chips (greater than 1 mm). Total particle counts over 10,000 possible; the processor displays total count and rate of count (eg. counts per hour), with adjustable warning levels. Maximum allowable pressure varies with fitting, but generally around 14 bar.

Application:
In-line lube systems, typically used on engines and gear boxes on helicopters and fixed wing aircraft.

Cost: Dependent on application and environment. A typical scavenge line system, as in photograph above, would cost around £7000.

Further details: [WARNING — for latest models check supplier] See pages 122, 151, 251 and 305. Refs: Astridge 1987, Belman 1991, Tauber 1983.

RION KL-01　　　　　　　　　　　Manufacturer: Rion Co. Ltd
　　　　　　　　　　　　　　　　　UK Supplier: Hawksley & Sons
　　　　　　　　　　　　　　　　　Limited

Technique:　　　　　　　　　　　OPTICAL – LIGHT OBSCURA-
　　　　　　　　　　　　　　　　　TION or LIGHT SCATTERING
　　　　　　　　　　　　　　　　　[LIGHT-SHIELDING or
　　　　　　　　　　　　　　　　　SCATTERING]
Description:

Four different sensors can be used with the RION liquid-borne particle measuring system, either on-line or off-line. The KS 60 series use light obscuration and the KS 58 light scatter. Six particle sizes can be chosen from the sensor range, all measured simultaneously. The mode can be chosen, ie. sample quantity (eg. 10 mL), particle count (up to the maximum concentration or 990,000)) or time (eg. 60 s for 10 mL, or 600 s for 100 mL, up to a maximum of 990 s). Once the setting has been made the instrument is fully automatic. Calibration varies depending on the sensor used as shown below.

Range and sensitivity:

Sensor	Liquid	Particle Size	Max. Concentration	Calibration
KS-60	Water/solvent	1–40 μm	2000 particles/mL	Latex
KS-61	Water/solvent	3–100 μm	800 particles/mL	Latex
KS-61AC	Oil	5–150 μm	800 particles/mL	ACFTD
KS-58	Pure water	0.3–10 μm	2000 particles/mL	Latex

Application:
A wide range of liquids from ultra pure water to hydraulic and lube oils. Mainly off-line due to optical characteristics.

Cost:

Further details: [WARNING — for latest models check supplier]
See pages 125, 180 and 186.

RION KL-20, KL-22		Manufacturer: Rion Co.Ltd
				UK Supplier: Hawksley & Sons
				Limited

Technique:			OPTICAL – LIGHT SCATTER
				[SIDEWAYS LIGHT
				SCATTERING]

Description:

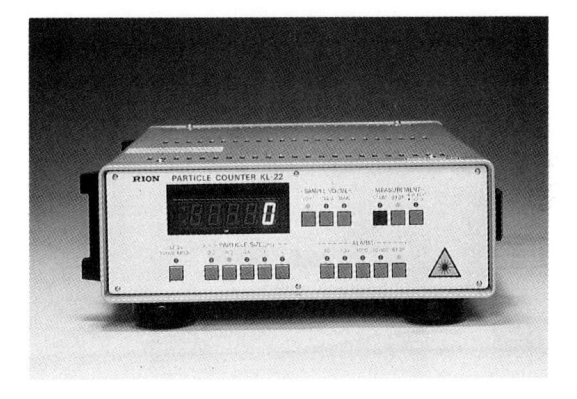

The sensor uses a laser diode and is constructed to be compatible with a range of liquids. On-line use is possible at a rate of 10 mL/min continuous, eg. 10 minutes would sample 100 mL, and a pressure up to 3 bar. Alarm levels can be set at cumulative particle counts (ie. 10, 100, 1000 and 10,000).

Range and sensitivity:
Five sizes are monitored — 0·2 μm, 0·3 μm, 0·5 μm, 1 μm and 2 μm. With a 95% confidence (5% coincidence) a particle count at a maximum of 1200 particles/mL is possible up to a total count of 99,999.

Application:
A wide range of liquids including ultra pure water, organic solvent, acid and alkali solutions, etc.

Cost:

Further details:
[WARNING — for latest models check supplier]
See pages 126 and 186.

ROTARY PARTICLE DEPOSITOR Manufacturer: Swansea
RPD Tribology Centre
UK Suppliers:
South — Analex Ltd
North & Midlands — Lindley Flowtech Ltd
International: Analex Ltd

70

Technique: MAGNETIC ATTRACTION

Description:

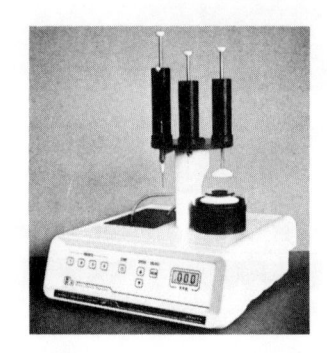

The basic idea of the RPD is akin to ferrography where particles are deposited in different striations depending on their size. However, unlike the conventional ferrograph, the RPD uses a rotating substrate (32 mm square) on which a measured volume of the oil is slowly deposited by pipette. Beneath the substrate a cylindrical magnet assembly, comprising two rare earth magnets, provides the magnetic attraction. The substrate is held to the magnetic face by means of suction. The magnet/slide combination can rotate at different speeds. The whole operation of (a) Sample deposition (typical speed 70 rev/min), (b) Washing with solvent (150 rev/min) and (c) Drying (200 rev/min), takes 5–10 minutes.

Range and sensitivity:
The prepared slide usually has three rings, with the inner ring containing all size of ferrous particle (perhaps 1–2000 μm), the intermediate 1–50 μm, and the outer ring less than 10 μm. Some non-ferrous debris is also deposited.

Application:
Lube oils where a rapid means of depositing the debris is required, ie. for microscope or image analysis. The excellent dispersion facilitates such analysis.

Cost: Around £4000, with some consumables. Very low running cost.

Further details: [WARNING — for latest models check supplier]
See pages 122, 240 and 263. Refs: Kwon et al 1987, Price & Roylance 1984, Williams 1988.

SediGraph 5100   Manufacturer: Micromeritics®
                             Instrument Corp.
                             UK Supplier: Micromeritics Ltd

Technique: SEDIMENTATION
Description:

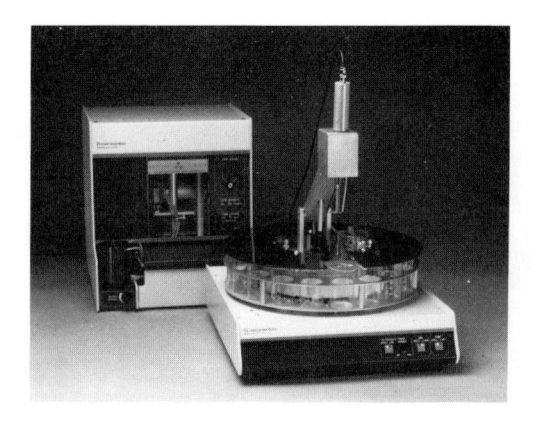

The 5100 uses a finely collimated beam of low energy X-rays and a detector to determine the distribution of particle sizes in the cell containing the sedimentation liquid. The source and collector remain stationary and the cell moves vertically between them. Sample dispersion is provided mechanically, by means of a variable speed magnet or by ultrasonics. Bubbles are automatically detected and relentlessly eliminated by tilt and de-aeration. The particles must be more absorptive of X-rays than the sedimentation liquid. The SediGraph results agree closely with the BCR standard material specifications as regards traceability. The MasterTech 51, with an ultrasonic probe and sample carousel, is also shown.

Range and sensitivity:
Size range 0·1 μm to 300 μm, although particles outside this range are also accounted for. Software allows addition of sieve data up to 1000 μm. 50 mL of dispersed sample is used.

Application:
All inorganic powders including pigments, metal powders, minerals, metal oxides, ceramics, etc.

Cost:

Further details:
[WARNING — for latest models check supplier]
See pages 134 and 168.

SPECTREX® SPC-510 LASER PARTICLE COUNTER SYSTEM	Manufacturer: Spectrex Corp. UK Supplier:

72

Technique:	OPTICAL – FORWARD REFLECTANCE [NEAR-ANGLE LIGHT SCATTERING]

Description:

The Spectrex counter uses a scanning laser beam detection of particles from 0·5 μm to 100 μm. Scatter caused by particles in the rapidly rotating 1mW He-Ne laser beam is picked up by a secondary lens which detects only forward scatter between 6° and 19°. In this angular range the forward reflectance is dependent on the size of the particle. There is an automatic rejection of the out-of-focus pulses. The liquid being tested is normally in-situ within the sample bottle; as the analysis occurs within a central volume, dirt on the bottle walls is out of focus.

Range and sensitivity:
The laser beam reflectance detects particles from 0·5 μm and greater, and sizes them between 0·5 μm and 100 μm. 17 1 μm steps are possible from 0·5–17 μm, and 16 5 μm steps thereafter up to 100 μm. The data is collected in 30 seconds with a maximum concentration of 999 particles/mL. Testing of only 5 mL is possible with the small vial attachment.

Application:
Direct particle counting within bottled samples of lube oil, fuel, hydraulic fluid, ocean water, drinking water, etc. (Excessive concentrations may be diluted.) The automode program permits "Sedimentation against time" studies. Users include the military, aerospace, semi-conductor and water treatment plants.

Cost: Around $US 17,500

Further details: [WARNING — for latest models check supplier]
See pages 123 and 189. Refs: Clayton 1980?, Underwood et al 1986.

SPI-WEAR®	Manufacturer: Spire Corp.
	UK Supplier:

Technique: RADIOACTIVATION [SURFACE LAYER ACTIVATION]

Description:

SPI-WEAR can be used in two ways for detecting the wear of a previously surface activated component. It may either detect the wear by direct measurement of a change in radio activity (SLA marker technique), or it can observe the radio activity of the lube oil which 'washes' the component (Tracer technique).

Range and sensitivity:
The amount of wear detectable is approximately 1% of the total activation depth, thus for a depth of 50 μm, 0·5 μm can be detected. Activation from 1 μm to 1mm is possible covering the sensitivity range of 0·01 μm to 10 μm.

Application:
The technique is for in-line use for any components which can be suitably activated. Even plastics can be implanted by the use of special foils, but most analysis concerns metals. Highly selective components can be tested. Used for automotive and critical aerospace components.

Cost:
Around $US 35,000 (computer driven laboratory system); $US 9,000 (Handheld model, less sensitive)

Further details:
[WARNING — for latest models check supplier]
See pages 132 and 252. Ref: Blatchley & Sioshansi 1990.

SYMPATEC HELOS PSA	Manufacturer: Sympatec GmbH
UK Supplier: Sympatec GmbH UK

Technique:	OPTICAL – FRAUNHOFER DIFFRACTION

Description:

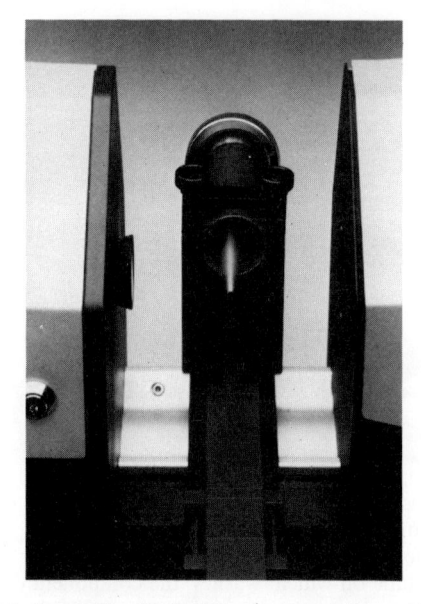

The HELOS (He-Ne Laser for Optical Spectrometry) provides seven measuring ranges to cover a very wide particle size distribution, both off-line and on-line. It is a modular design with the central unit housing the entire optics. 31 semicircular rings form the multielement detector. An ultrasonic bath provides dispersion with the help of two agitators. 5 mW laser.

Range and sensitivity:
Size ranges from 0·1 μm–35 μm, up to 0.5 μm–2625 μm. Ranges being formed by means of lenses of different focal lengths. 31 logarithmically divided particle size classes. Measuring time from 1 s to 10,000 s (approx. 2 hours), with diffraction pattern scanned 400 times/s.

Application:
Suspensions requiring assessment of particle distribution over a wide size range, either off-line or on-line.

Cost: From around £25,000

Further details: [WARNING — for latest models check supplier]
See pages 124 and 168.

WEAR PARTICLE ANALYZER MODEL 56

Manufacturer: Tribometrics Inc
UK Supplier: Oilab Lubrication

Technique: MAGNETIC ATTRACTION [HIGH GRADIENT MAGNETIC SEPARATION]

Description:

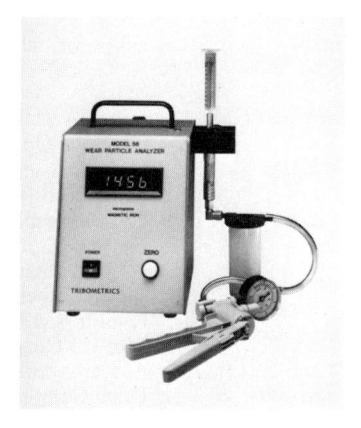

A small sample of oil is drawn through a special magnetic filter, called a FIRON® filter, composed of a bed of steel fibres which become magnetised when the FIRON® filter is inserted in the instrument. Particles are both captured because of magnetic attraction to the fibres and the filter restriction. Different grades of FIRON® filters are available to cover different applications. The Model 56 Wear Particle Analyzer displays the equivalent value of micrograms of iron metal from the increase in magnetic flux. Both analogue and digital versions are available. Particles are easily recovered for microscopic analysis by backflushing after the FIRON® filter is removed.

Range and sensitivity:
Particle size captured: 95% or greater of magnetic cast iron and steel particles 0·5 μm and larger, the majority of non-magnetic particles 10 μm and larger. FIRON® filters are available with different fibre sizes and packing densities to preferentially capture particles of different size and handle fluids of different viscosities.

Application:
All oil or fluid wetted wearing components, such as in engines, transmissions, gearboxes and hydraulic systems. Off-line use, portable (weighs 4·5 kg).

Cost: Around $US 6000

Further details: [WARNING — for latest models check supplier]
See pages 122, 151 and 308. Refs: Jones & Larkin 1988, Lewis 1991.

SECTION 2
IMAGE ANALYSERS FOR PARTICLE DETECTION AND ANALYSIS

All analysis systems are supplied with the necessary software, and in some cases that is all (unless requests are made for additional hardware from other manufacturers, tailored for customers specific requirements). However, other analysers are complete specials from the manufacturer, in their own consoles. There are still other systems which lie in between these two extremes with wired cards (eg. for image capture — 'frame grabbers'), boxed modules, etc. The initial description outlines what is the basic supply.

The analysers are identified by their usual names or, where that is not clear, by the manufacturer's name.

IMAGE ANALYSIS SYSTEM	MANUFACTURER
1. Cue-2/3/4/VS	Olympus Optical (Europa) GmbH
2. Hamamatsu C1172 & DIPS	Hamamatsu Photonics K.K.
3. IBAS	Kontron Bildanalyse GmbH
4. Juliet II®	The Shapespeare Corp.
5. Keyance VH-5900	Keyance Corp.
6. Magiscan®	Applied Imaging Ltd (formerly Joyce-Loebl)
7. MicroEye, MicroScale	Digithurst Ltd
8. Moritex Scopeman	Moritex Corp.
9. Optomax V, VIDS V and PROTOS	Ai Cambridge Ltd (formerly AMS)
10. Quantimet 500, 520 and 570	Leica Cambridge Ltd
11. Seescan	Seescan Ltd

A typical image analysis system (Digithurst Limited)

1. CUE 2/3/4, CUE-VS   Manufacturer: Olympus®
                              Optical Co. (Europa) GmbH
                              UK Supplier:
                              Olympus Optical (UK) Ltd

IMAGE ANALYSIS

Supplied:
  Software alone for desk-top computers, or
  Hardware built-in functions, CCD camera, etc.

Special Features:
Cue-2 and Cue-4 are designed for black/white analysis, Cue-3 for colour (distinguishing 16·7 million colours). All cover 512 × 512 pixels × 8 (or 12) bits, 256 grey scale. Cue-VS includes a high resolution of 1024 × 1024 pixels. Cue-4 includes many advanced functions such as FFT. 'One touch' program can be put in by customer. Up to 35 different parameters can be measured on each display, eg. area, perimeter, convexity, c.g., aspect ratio, shape factor, hole area, etc.

Cost:
Cue-2 software around £2500, System (less microscope) around £15,000, Cue-3 — £26,000, Cue-4 — £42,000, Cue-VS — £11,000.

Further details (check supplier):
See pages 48, 120 and 243.

---

2. Hamamatsu C1172   Manufacturer: Hamamatsu
Particle Counter and   Photonics K.K. Systems Div.
DVS-1000 & DVS-3000 systems UK Supplier:
                            Hamamatsu Photonics UK Ltd

IMAGE ANALYSIS

Supplied:
  Hardware modules including C1172 Particle Counter and video cameras. The DVS real-time video image quality improvement systems (units with remote control).

Special Features:
The C1172 counts up to 30,000 particles in real time. Average of 8 counts displayed continuously, area of count and size selectable. The DVS image enhancement systems enable real-time signals to be greatly improved — particularly valuable when there is low S/N, low illumination or low contrast. A number of novel features are included, such as background subtraction, moving average, 20 patterns of look-up table (LUT) and pseudo-colour display. Image memory size up to 688 × 512 (16 bit).

Cost:
(approx) C1172 — £10,000, DVS-1000 — £6000, DVS-3000 — £9000

Further details (check supplier):
See pages 48, 120 and 243.

3. IBAS　　　　　　　　　　　Manufacturer: Kontron
　　　　　　　　　　　　　　　Elektronik GmbH
　　　　　　　　　　　　　　　UK Supplier:
　　　　　　　　　　　　　　　Kontron Elektronik Limited

IMAGE ANALYSIS ─────────────────────────

Supplied:
　　Software, and hardware modules (controlled by an IBM® or compatible).

Special Features:
Particle size analysis with a range of image processing and parameter measurement tools. Analysis and display up to 1280 × 1024 pixels utilising floating point array processor and host processor. Image acquisition from B/W, Real Colour analogue and digital cameras. Digital connection to SEM is available. Frame grabber version VIDAS and Semi-Automatic system VIDEOPLAN can be upgraded to full IBAS at any time.

Cost:
From around £8000 to £45,000

Further details (check supplier):
See pages 48, 120 and 243.

4. JULIET II®　　　　　　　　Manufacturer: The Shapespeare
　　　　　　　　　　　　　　　Corporation
　　　　　　　　　　　　　　　UK Supplier:
　　　　　　　　　　　　　　　Brookhaven Instruments Limited

IMAGE ANALYSIS ─────────────────────────

Supplied:
　　Processor module including colour VGA display and considerable software.

Special Features:
Each particle image is measured in a standardized configuration to determine the profile centroid and perform a principle axis transformation before analysis. More than 50 particle features can be determined including classical size features (eg. area, perimeter, convex perimeter, etc), classical shape (eg. aspect ratio, roundness, Feret's H/V, etc) and morphology (eg. macro and micro roughness, etc).

Cost:
Around £14,000 (complete system), £7500 "technology package" for customers own processor hardware.

Further details (check supplier):
See pages 48, 120 and 243.

5. Keyance Monitor Microscope Manufacturer: Keyance Corp
VH-5900 UK Suppliers: Camlab Limited

IMAGE ANALYSIS ────────────────────────────
(Hand-held)

Supplied:
Portable microscope probe, controller module and optional monitor.

Special Features:
Up to 1000 times magnification with up to 30 times greater depth of field than an optical microscope, due to the integral white light illumination head; enables even hand-held analysis to be undertaken. Six lens sizes are available — 20×, 50×, 100×, 200×, 500× and 1000×. (A lens of 50× magnification has a depth of field of 6.5mm, the 1000× a depth of field of 0·05mm.) Resolution better than 350 TV lines horizontal and vertical (up to over 420 TV lines vertical on one model).

Cost:
Around £5500 for the controller, £1200 for each lens and £2000 for the probe.

Further details (check supplier):
See pages 48, 120 and 243.

───

6. Magiscan® and Manufacturer & UK Supplier:
Mini-Magiscan® Applied Imaging International
 Limited (formerly Joyce-Loebl)

IMAGE ANALYSIS ────────────────────────────

Supplied:
Software (GENIAS®) and hardware process modules. Monitors, microscope controller with high resolution CCD camera, and printer also included in complete package.

Special Features:
Extensive software package covers numerous applications including particle sizing, metallurgy and lubrication. Image processor uses 512 × 512 pixels (8 bit) with 256 grey scales. Programmable real-time frame averager for pre-processing noisy images. Integrated system for storing 72,000 images. SEM connection possible.

Cost:
From around £20,000.

Further details (check supplier):
See pages 48, 120 and 243.

7. MicroEye and Manufacturer & UK Supplier:
MicroScale Digithurst Limited

IMAGE ANALYSIS ─────────────────

Supplied:
    Software and control cards for easy installation in a PC providing image capture, compression, analysis and output.

Special Features:
720 × 512 pixels resolution (24 bit colour video images). With RGB palette 16.7 million colours can be distinguished. 3-D profiles and surface visualisation. Numerous features analysed on particles.

Cost:
Around £4000 for the MicroEye image capture card, Around £7000 for the colour image MicroScale including MicroEye. Software alone around £1000–£1500.

Further details (check supplier):
See pages 48, 120 and 243.

───────────────────────────────

8. Moritex Scopeman Manufacturer: Moritex Corp.
    UK Supplier:
    Finlay Microvision Co. Ltd.

IMAGE ANALYSIS ─────────────────

(Hand-held)

Supplied:
    Hand-held camera probe with a range of lenses. Probe includes built in illumination. Hardware interface to standard video equipment, or self-contained. Many accessories also available.

Special Features:
The range of 13 fixed lenses covers from 0·7×–800× with a resolution to over 380 TV lines. Two zoom lenses provide 5×–40× and 35×–210×. Depth of field is from 200 mm down to 0·2 mm (800×). Both contact and non-contact lenses are available. Two interfaces are available — the MS-503 provides intensity of illumination and colour control for direct connection to a video monitor, the MS-501 is a self contained unit with its own 8 mm VTR and 6 inch monitor with built-in text superimposition. A low-cost Micro Scopeman and a built-in PC Scopeman are also available.

Cost:
From around £2700 basic) to over £14000 (MS-501)

Further details (check supplier):
See pages 48, 120 and 243.

9. OPTOMAX V,   Manufacturer & UK Supplier:
VIDS V and PROTOS   Ai Cambridge Limited
                                            (formerly, Analytical Measuring
                                            Systems)

IMAGE ANALYSIS

Supplied:
    Complete self contained image analyser (Optomax V) or video frame store and graphics card (VIDS V). Protos field specific image analyser (colony counter).

Special Features:
The Optomax V is a single desk-top unit with integral 9 inch monitor. Two independent 256 grey level detectors with a resolution of 704 × 560 pixels (8 and 32 bit). A high resolution camera and colour monitor are also available. The VIDS V software is similar, available on a floppy disc. Field and feature specific parameters covered include, the usual features, plus perimeter, frame area, intercepts, orientation, longest dimension, etc.

Cost:
From around £10,000 to £30,000

Further details (check supplier):
See pages 48, 120 and 243.

---

10. QUANTIMET 500, 520 and 570   Manufacturer & UK Supplier:
                                                   Leica Cambridge Ltd

IMAGE ANALYSIS

Supplied:
    A range of powerful processing and analysis systems providing hardware, software and documentation for immediate use with a microscope. Extensive application packages for contaminants in hydraulic oil, fibres analysis and engine wear debris analysis.

Special Features:
The Q500 is a low-cost, bench top image analyser which is operated via a Windows interface, and includes a wide range of grey and binary image processing functions. Microscope automation of stage and focus is available with real-time shading correction and image averaging. The top-of-the-range Q570 incorporates a powerful range of mathematical morphology functions, and is capable of true-colour analysis. With the Stereoscan range of SEM's, high resolution and chemical analysis capability is added.

Cost:
A wide range depending on requirements.

Further details (check supplier):
See pages 48, 120 and 243.

11. SEESCAN IMAGE ANALYSERS

Manufacturer & UK Supplier:
Seescan Plc

IMAGE ANALYSIS

Supplied:
   Cameras, video printers, microscopes, software and a range of hardware processor modules.

Special Features:
The menu driven programs are written in an image processing language, incorporating features of BASIC and 'C'. This enables a high rate of processing. Field and feature measurements include the usual functions, up to 8 grey scale band areas within the measuring frame (displayed using pseudo colours), intensity, centre of mass, shape factor ($4\pi a/p^2$), elongation, optical density, etc.

Cost:
From around £5000 to £50,000 depending on software and hardware required. A special package is available conforming fully to BS3406, the standard for particle sizing.

Further details (check supplier):
See pages 48, 120 and 243.

SECTION 3
SERVICES AND LABORATORIES

Services can be divided into two categories, those which provide a full oil analysis including spectrometric analysis, and those which specialise in contamination control and undertake particle counts with the possibility of some visual observations (and, maybe, the basic physical properties of the oil). Some indication of the type of service is given, but no mention is made of actual equipment used because this is constantly being updated.

The normal mode of operation is for a company to discuss the type of machinery involved with the service agency — in particular, the types of failure or contamination which could be a nuisance and the possible places for obtaining samples from the system. The service agency will then supply pre-cleaned bottles, a means of obtaining the sample and suitable labelling. On receiving the bottle, the agency then examines (immediately, or after a few days?) the oil and notifies the company (immediately, or by fax or post?) of what it is able to find and the possible consequences to the machine. Costs could vary from £5 to £75 per sample depending on the analysis required and what sort of report (colour photographs ?) is requested.

The following are examples of Full Oil Analysis, including spectrometric analysis on wear debris, and contamination counts. Please note the brief comments are in no way comprehensive; full brochures, and costs, should be obtained from the companies:

CENT   Century Oils
(Trend service given with immediate warning when excessive levels imply a serious fault developing)

INDIC-8   BP Oil UK Limited
(Industrial 8-stage equipment monitoring service with full customer liaison and trend indications)

PALL CONTAMINATION ANALYSIS   Pall Industrial Hydraulics Ltd
(Offers particle counting, sizing and contamination analysis in support of Pall Customers worldwide.)

PAR-TEST® FLUID ANALYSIS   Parker Hannifin plc
(Analysis of up to 27 wear metals, water, viscosity, particle count, etc. Fast turn-round, recommendations.)

PREDICT   BP Oil
(As Indic-8 plus a full identification of all particles by ferrography and microscopy. Premium service.)

REDWOOD OILSCAN   SGS Redwood Ltd
(Standard analysis or "Ferrographic Analysis". Both give trends and recommendations. Fast response.)

SAFEGUARD   Castrol (U.K.)Ltd
(Full customer liaison and immediate response with serious problems. Normal reply within 24 h of receipt.)

SCIENTIFIC SERVICES OIL ANALYSIS British Rail Research
(Nation wide service provided with a typical sample turn-round of 24 hours and faster service when needed.)

SPECTRO TRIBOLOGICAL LABORATORY   Spectro Inc.
(This is a complete laboratory facility, with training, which can be supplied to a company for in-house use.)

TRIBO PREDICT   Condition Monitoring Services
(A range of services including full analysis. Times and reports to suit customers. Full liaison.)

WEARCHECK   Wearcheck Laboratories
(Results mailed to customer within 36 hours, with trend analysis — within 24 hours if serious. Floppy disc.)

Example of visual reporting
(Wearcheck Laboratories)

The following are examples of specialist particle count analysis, including wear debris and contamination counts:

FES, INC   FES, Inc., USA
(Specialist fluid diagnosis as well as particle counts and visual examination. Contamination control advice.)

HOWDEN WADE LTD
(Gravimetric analysis of fuels, solid particle size & counts for oil and hydraulic fluid samples. Standards.)

INDUSTRIEBEDARF REHM   Germany
(Manufacturer and supplier of precleaned bottles to ISO 3722 and 5884. Laboratory automatic particle counting.)

MONITION   Monition Ltd
   (Oil sampling and system testing on-site. Trend analysis undertaken. Other monitoring available.)

QUANTACHROME Analytical Laboratory   Quantachrome Corp. USA
   (Particle size from 1 μm to 300 μm measured; also roughness factors by sedimentation.)

SKF LUBRICATION OIL ANALYSIS SERVICE SKF (UK) Service
   (Two levels provided, ie. Basic — viscosity, water, TAN, wear debris: Full — particle sizes and analysis.)

The Index and Address lists also include laboratories and establishments which will undertake a small amount of analysis for local industry, or for those using their other facilities, eg. oil companies.

APPENDIX 2

Listing of Addresses & Functions

The description given, only concerns that part of the activity which relates to wear debris analysis or particle detection; in other words, most companies would be involved in much more than that described.

Note that the telephone and fax numbers given, are those to be used within each country; international codes are additional.

NAME & ADDRESS

DESCRIPTION

A^3 GmbH — see end of list

Acal Auriema Limited
Industrial Division
442 Bath Road
Slough, Berks, SL1 6BB   UK
Tel: 0628 604353   *Fax:* 0628 603730

UK SUPPLIER: Micro Pure and Monitek equipment.

Advanced Polymer Systems
3696 Haven Avenue
Redwood City
CA 94063   USA
Tel: (415) 366-2626   *Fax:* (415) 365-6490

MANUFACTURER: Polymer standards for the calibration of nephelometric instruments, etc.

AEA Technology (Harwell)
Thin Layer Activation Unit
B477 Harwell Laboratory
Oxon, OX11 0RA   UK
Tel: 0235 432885   *Fax:* 0235 433029

SERVICE: Irradiation of specific parts for testing with TLA.

Aerometrics Inc.
894 Ross Drive
Unit 105
Sunnyvale
CA 94089   USA
Tel: (408) 745-0321   *Fax:* (408) 745-6379

MANUFACTURER: Phase doppler particle analyser.

[see Laser Lines]

AHEM — see British Fluid Power Association (BFPA)

Ai Cambridge Limited
London Road
Pampisford
Cambridge
CB2 4EF   UK
Tel: 0223 834420   *Fax:* 0223 835050

MANUFACTURER: Image analysers and automated microscopy products, eg. Optomax, VIDS Quickstep and Protos.

A.I.M. München Vertriebs GmbH
Ganghoferstraβe 21
8031 Gernlinden/Maisach
Germany
Tel: 08142 18005   *Fax:* 08142 18008

SUPPLIER: Oil sampling kits and precleaned bottles for off-line samples.

Analex Limited
P.O. Box 14
Pangbourne
Reading
Berks
RG8 7EB   UK
Tel: 0491-875500   *Fax:* 0491 875300

SUPPLIER: PQ and RPD (UK South and international distributor)
SERVICE: Maintenance and condition monitoring.
(See Condition Monitoring Services.)

Analytical Measuring Systems (AMS) — see Ai Cambridge Limited

Applied Imaging International Limited
Dukesway
Team Valley
Gateshead
NE11 0PZ   UK
Tel: 091 482 2111   *Fax:* 091 482 5249

MANUFACTURER: Image analysers (Magiscan), sub-micron particle size analysers (disc centrifuge).
(Formerly Joyce Loebl)

Applied Research Laboratories (ARL)
24911 Avenue Stanford
Valencia
CA 91355   USA
Tel: (805) 295-0019   *Fax:* (805) 295-0419

MANUFACTURER: Spectrometric instrumentation including ICP-AES and DCP-AES.

Applied Research Laboratories (ARL)
En Vallaire
1024 Ecublens
Switzerland
Tel: 21 691 1515   *Fax:* 21 691 1531

MANUFACTURER: Spectrometric instrumentation including ICP-AES and XRF.

Asesoramiento Técnico en Lubricación Industrial
(F.E. Lantos, J.C. de Lantos & E. Lantos)
A. Del Valle 971
1638 Vicente López
Buenos Aires, Argentina
Tel: (1) 791-9525   *Fax:* (1) 791-9525

SERVICE: Consultancy in wear and tribology. Laboratory testing and analysis, including spectrometry and SEM. Fuel analysis.

Bahco AB, Sweden — see George Fischer Foundry Systems Inc.

Baird Europe B.V.
Produktieweg 30
P.O. Box 81
NL-2380 AB Zoeterwoude
Netherlands
Tel: 071-413151   *Fax:* 071-414899

MANUFACTURER: Multielemental oil analysers.

Bassaire Limited
Duncan Road
Swanwick
Southampton
SO3 7ZS   UK
Tel: 0489 885111   *Fax:* 0489 885211

MANUFACTURER: Clean air cabinets, for assembly of cleaned components and fine particle filtration.

Bath Fluid Power Centre — see Fluid Power Centre, University of Bath

Belstock Controls
10 Moss Hall Crescent
Finchley, London
N12 8NY   UK
Tel: 081-446 8210   *Fax:* 081-446 6991

UK SUPPLIER: Particle size analysers (Bromley Instruments and Lasentec).

BHR Group
British Hydromechanics Research
  Group Limited
Cranfield, Bedford
MK43 0AJ   UK
Tel: 0234 750422   *Fax:* 0234 750074

SERVICE: Courses and conferences organised. Equipment and system testing. Debris analysis.

G. Bopp & Company Limited
Block 'C'
115 Brunswick Park Road
New Southgate
London
N11 1LJ   UK
Tel: 081-368 9989   *Fax:* 081-368 5229

UK SUPPLIER: Woven wire cloth and mesh, suitable for sieving and fine filtration.

BP OIL UK Limited
BP House
Breakspear Way
Hemel Hempstead
Herts
HP2 4UL   UK
Tel: 0442 232323   *Fax:* 0442 225509

SERVICE: Oil condition monitoring for the automotive industry, general engineering (Indic-8) and ultra-critical capital intensive applications (Predict).

British Fluid Power Association (BFPA)
235/237 Vauxhall Bridge Road
London
SW1V 1EJ   UK
Tel: 071 233 7044   *Fax:* 071 828 1917

SERVICE: Numerous publications to assist in improving the reliability of fluid power systems.

British Rail Research
P.O. Box 2
London Road
Derby
DE2 8YB   UK
Tel: 0332 386619   *Fax:* 0332 386720

SERVICE: Analysis of used engine oil using spectrometry and other techniques. Physical and dissolved gas analysis of transmission oils.

Bromley Instruments
2 Union Road
Croydon
Surrey
CR0 2XU   UK
Tel: 081 683 3080   *Fax:* 081 689 9085

Brookhaven Instruments Corp.
750 Blue Point Road
Holtsville, New York 11742   USA
Tel: (516) 758-3200   *Fax:* (516) 758-3255

Brookhaven Instruments Limited
Chapel House
Stock Wood
Worcs
B96 6ST   UK
Tel: 0386 792727   *Fax:* 0386 792727

Brunel
The University of West London
The Brunel Centre for
   Manufacturing Metrology
Uxbridge, Middx.
UB8 3PH   UK
Tel: 0895 274000   *Fax:* 0895 232806

California Measurements Inc
150 East Montecito Avenue
Sierra Madre
CA 91024   USA
Tel: (818) 355-3361   *Fax:* (818) 355-5320

Camlab Limited
Nuffield Road
Cambridge
CB4 1TH   UK
Tel: 0223 424222   *Fax:* 0223 420856

Carlo Erba Strumentazione
Strada Rivoltana
20090 Rodano (Milan), Italy
Tel: (2) 950591/9588161

Castrol (U.K.) Limited
Industrial Division
Pipers Way
Swindon, Wilts
SN3 1RE   UK
Tel: 0793 512712   *Fax:* 0793 486083

MANUFACTURER: Particle size
analyser (Bromley Instruments
sedimentation PSA).

[see Belstock Controls]

MANUFACTURER: Particle size
analysers.
[see below]

UK SUPPLIER: Brookhaven
particle size analysers,
Duke Scientific particle
size standards, Shapespeare
Corp. image analysers.

SERVICE: Courses in Condition
Monitoring including subjects
such as predictive maintenance
and fluid contaminant
measurement.

MANUFACTURER: Clean room
air particle collector
for SEM and EDX analysis -
sizes from 0.05 μm – > 2 μm.

UK SUPPLIER: McVan Analite
and Hach turbidity meters.
Keyance hand-held video
microscope and monitor.

MANUFACTURER: Elemental
analysers.
[see Fisons]

SERVICE: Safeguard sampling
kits and analysis of oils,
including transformer oils
and dissolved gas analysis.

Century Oils Limited
P.O. Box 2
New Century Street, Hanley
Stoke-on-Trent
ST1 5HU   UK
Tel: 0782 202521   *Fax:* 0782 202073

SERVICE: Oil and debris analysis.

CHARN-MAC-ENG-Ltd
Peel House Gate
Stocks Lane, Luddenden
Halifax, West Yorks HX2 6SP   UK
Tel: 0422 884252   *Fax:* 0422 885913

UK SUPPLIER: Diagnetics Contam-Alert II portable contaminant monitor.

ChemLab Scientific Products Limited
Construction House
Grenfell Avenue, Hornchurch
Essex RM12 4EH   UK
Tel: 04024 76162   *Fax:* 04024 37231

UK SUPPLIER: Matec CHDF 1100 particle size analyser.

Christison Scientific Equipment Ltd
Albany Road, Gateshead
NE8 3AT   UK
Tel: 091 477 4261   *Fax:* 091 490 0549

UK SUPPLIER: Fritsch sieves and particle size analysers.

CILAS — see Compagnie Industrielle des Lasers

CJC Napier Limited
Enterprise City
Unit 37 Green Lane
Spennymoor, Co. Durham
DL16 6JF   UK
Tel: 0388 420721   *Fax:* 0388 420718

MANUFACTURER: Off-line filtration units and oil transfer equipment.

Clayton Instruments Limited
26 Moorlands Estate
Metheringham, Lincoln
LN4 3HX   UK
Tel: 0526 322670   *Fax:* 0526 322669

UK SUPPLIER: Elesa hydraulic accessories, including filler caps for oil tanks/reservoirs.

Climet Instruments Co.
P.O. Box 1760
Redlands
CA 92373   USA
Tel: (714) 793-2788   *Fax:* (714) 793-1738

MANUFACTURER: Liquid nephelometer and particle size analysers.
[see Gelman Sciences]

CMS International
16A Sovereign Street
Bedford View
P.O.Box 418, Isando 1600
South Africa
Tel: 011 622-3625   *Fax:* 011 615-7437

SUPPLIER: Predictive maintenance software.

Compagnie Industrielle des Lasers
Route de Nozay
BP 27
91460 Marcoussis, France
Tel: (1) 64 54 48 25 *Fax:* (1) 69 01 37 39

MANUFACTURER: CILAS
Granulometer (Fraunhofer) and
CILAS 1064 (laser scattering)
particle sizers.

Condition Monitoring Services
P.O. Box 14
Pangbourne, Reading
Berks, RG8 7EB UK
Tel: 0491 875500 *Fax:* 0491 875300

SUPPLIER: Predictive
maintenance software.
[see Analex]

Cormon Ltd
Cormon House
South Street
Lancing, W.Sussex
BN15 8AJ UK
Tel: 0903 766861 *Fax:* 0903 763192

SERVICE: Provide expertise
for TLA monitoring.
[see AEA Technology]

Coulter Electronics Ltd
Northwell Drive
Luton, Beds
LU3 3RH UK
Tel: 0582 491414 *Fax:* 0582 490390

MANUFACTURER: Scientific
equipment including the
LS series, Multisizer
series, N4 series and LCM II.

CSI Europe — see end of list

Dantec Elektronik
Medicinsk og Videnskabeligt
  Maleudstyr A/S
Tonsbakken 16-18
DK 2740 Skovlunde, Denmark
Tel: 4542 84 2211 *Fax:* 4542 84 6346

MANUFACTURER: Phase
Doppler particle sizing
equipment (sprays and
particles in air).
[see Dantec Electronics]

Dantec Electronics Limited
Techno House
Redcliffe Way, Bristol
BS1 6NU UK
Tel: 0272 291436 *Fax:* 0272 213532

UK SUPPLIER: Dantec Phase
Doppler particle sizing
equipment (sprays and
particles in air).

Diagnetics Inc
5410 S. 94th E. Avenue
Tulsa
OK 74145 USA
Tel: (918) 664-7722 *Fax:* (918) 664-7724

MANUFACTURER: Portable
contaminant monitor
(digital Contam-Alert).
[see CHARN-MAC-ENG]

Digithurst Limited
Newark Close, Royston
Herts SG8 5HL UK
Tel: 0763 242955 *Fax:* 0763 246313

MANUFACTURER: Image analysis
systems and software.

Duke Scientific Corp.
1135D San Antonio Road
P.O. Box 50005
Palo Alto
CA 94303   USA
Tel: (415) 962-1100   Fax: (415) 962-0718

MANUFACTURER: Certified particle size standard spheres.
[see Brookhaven Instruments]

Dyno Particles A.S.
Svelleveien
P.O. Box 160
N-2001 Lilleström, Norway
Tel: 6 81 70 01   Fax: 2 31 76 10

MANUFACTURER: Dynospheres® calibration kits. Standard particles.

ELESA® s.p.a
1-20052 Monza (MI)
via Pompei, 29
Italy
Tel: 039 83 22 81   Fax: 039 83 63 51

MANUFACTURER: Filler and breather caps (with filters) for tanks and reservoirs.
[see Clayton Instruments]

Endecotts Limited
Lombard Road
Morden Factory Estate
London SW19 3BR   UK
Tel: 081-542 8121   Fax: 081-543 6629

MANUFACTURER: Test sieves and shakers.

Environmental Monitoring Systems Ltd
G11 Mayford Business Centre
Smarts Heath Road
Woking, Surrey
GU22 0PP   UK
Tel: 0483 722463   Fax: 0483 740462

UK SUPPLIER: Met One particle counters.

Fairey Arlon Limited
Fareham Industrial Park
Fareham
Hants
PO16 8XG   UK
Tel: 0329 826161   Fax: 0329 825758

MANUFACTURER: Filtration products, Fluid Analysis System FAS-CC100 contamination control and F.A.S.T. Filtration Education & Videos.

Fawcett Christie Hydraulics Ltd
Sandycroft Industrial Estate
Chester Road
Deeside, Clwyd
CH5 2QP   UK
Tel: 0244 535515   Fax: 0244 533002

MANUFACTURER: Oil/air separators for contamination control of hydraulic systems.
UK SUPPLIER: Mahle Pi C 9000 mobile particle counter.

FES Inc
5111 N. Perkins Road
Stillwater
OK 74075   USA
Tel: (405) 743-4337   Fax: (405) 743-2012

SERVICE: Consultancy. Lubrication/wear tests. Contamination control and fluid diagnosis.

Finlay Microvision Co. Limited
Finlay House
Southfields Road
Kineton Road Industrial Estate
Southam, Warwicks
CV33 0JH   UK
Tel: 0926 813043   *Fax:* 0926 817186

UK SUPPLIER: Moritex hand-held video microscope.

Fisons Instruments
Sussex Manor Park
Gatwick Road
Crawley, West Sussex
RH16 2RP   UK
Tel: 0293 561222   *Fax:* 0293 561980

UK SUPPLIER: ARL, Carlo Erba Instruments and Kevex spectrometric instrumentation.

Fisons Instruments
24911 Avenue Stanford
Valencia, CA 91355   USA
Tel: (800) 551-8741   *Fax:* (805) 295-0419

MANUFACTURER: ARL spectrometers.

Fisons Instruments Inc.
32 Commerce Center
Cherry Hill Drive
Danvers, MASS 01923   USA
Tel: (508) 777-8034   *Fax:* (508) 777-0678

MANUFACTURER: VG Elemental spectrometers.

Fluid Power Centre
The University of Bath
Claverton Down
Bath, Avon
BA2 7AY   UK
Tel: 0225 826371   *Fax:* 0225 826928

SERVICE: R & D facility, consultancy and courses in fluid power (covering contamination and its control). M.Sc. and Diploma.

Flupac Limited
Woodstock Road
Charlbury
Oxon OX7 3EF   UK
Tel: 0608 811211   *Fax:* 0608 811259

UK SUPPLIER: Hydac Purity Controller, Sampling and Analysis kits.

Fritsch GmbH
Industriestraβe 8
6580 Idar-Oberstein
Germany

Tel: 06784 70-0   *Fax:* 06784 70-11

MANUFACTURER: Laboratory instruments including fraunhofer diffraction particle sizer, seiving and centrifugal analysers. [see Christison]

Fulmer Systems Limited
c/o I.P. Hunt, Barn Close
Old Lane, Farthinghoe
Northants NN13 5NZ   UK
Tel: 0295 710414

MANUFACTURER: Wear Debris Monitor using the abrasive action of particles on a thin conducting film.

Galai Production Ltd
Industrial Zone
10500 Migdal Haemek
Israel
Tel: (972) 6-543369 *Fax:* (972) 6-540891

MANUFACTURER: Particle size and shape analyser.
[see Roth Scientific]

GasTOPS Limited
1011 Polytek Street
Gloucester, Ontario K1J 9J3
Canada
Tel: (613) 744-3530 *Fax:* (613) 744-8846

MANUFACTURER: FerroSCAN® in-line ferromagnetic debris sensor using inductance changes.

Gelman Sciences Limited
10 Harrowden Road
Brackmills, Northampton
NN4 0EZ   UK
Tel: 0604 765141 *Fax:* 0604 761383

MANUFACTURER: Membrane filters and associated glassware.
UK SUPPLIER: Climet particle analysers.

George Fischer Foundry Systems Inc.
407 Hadley Street
P.O. Box 40, Holly
Michigan 48442   USA
Tel: (313) 634-8251 *Fax:* (313)634-2181

UK SUPPLIER: Bahco (Dietert) micro particle centrifugal classifier.

Hach Company
P.O. Box 389, Loveland
Colorado 80539   USA
Tel: (303) 669-3050 *Fax:* (303) 669-2932

MANUFACTURER: Turbidity meters.
[see Camlab]

Hach Europe S.A./N.V.
B.P. 229
B-5000 Namur 1, Belgium
Tel: 32 81 44.53.81 *Fax:* 32 81 44.13.00

MANUFACTURER: Turbidity meters.
[see Camlab]

Hamamatsu Photonics K.K.
Systems Division
812 Joko-cho
Hamamatsu City
431-32 Japan
Tel: 0534-35 1569 *Fax:* 0534-35 1574

MANUFACTURER: Image processing and measuring systems.
[see Hamamatsu Photonics UK]

Hamamatsu Photonics UK Limited
Lough Point
2 Gladbeck Way
Enfield, Middx
EN2 7JA   UK
Tel: 081-367 3560 *Fax:* 081-367 6384

UK SUPPLIER: Hamamatsu image processing and measuring systems.

Harwell — see AEA Technology

Hawksley & Sons Limited
Marlborough Road, Lancing
Sussex, BN15 8TN UK
Tel: 0903 752815 *Fax:* 0903 766050

UK SUPPLIER: Rion particle counters and associated equipment.

Hiac Royco World Headquarters
Division of Pacific Scientific Company
Technical Service Park
2431 Linden Lane
Silver Spring, Maryland 20910 USA
Tel: (301) 495-7000 *Fax:* (301) 495-0478

MANUFACTURER: Optical particle counters.
[see Pacific Scientific]

Hitachi Scientific Instruments
Nissei Sangyo Co. Ltd
C.P.O. Box 1316
Tokyo, Japan
Tel:

MANUFACTURER: Spectrometers including microwave induced plasma MS element analyser. SEMs.

Hitachi Scientific Instruments
Merlin Building, Ivanhoe Road
Hogwood Industrial Estate
Hogwood Lane, Finchampstead
Wokingham, Berks RG11 4QQ UK
Tel: 0734 328632 *Fax:* 0734 328779

UK SUPPLIER: Spectrometers.

Holmbury Limited
P.O. Box 107
Tonbridge, Kent
TN11 8EZ UK
Tel: 0892 870933 *Fax:* 0892 870886

MANUFACTURER: Hydraulic hose quick release couplings with flat faces.

Horiba Ltd.
Miyanohigashi
Minami-ku
Kyoto, Japan
Tel: 81 75-313-8123 *Fax:* 81 75-321-5725

MANUFACTURER: Particle size distribution analysers, Particle counters.
[see Horiba Instruments]

Horiba Instruments Limited
Harrowden Road
Brackmills, Northampton
NN4 0EB UK
Tel: 0604 765171 *Fax:* 0604 765175

UK SUPPLIER: Particle size distribution analysers, Particle counters.

Howden Wade Ltd
Thermal Control Products
Crowhurst Road
Brighton
East Sussex
BN1 8AJ UK

Tel: 0273 506311 *Fax:* 0273 557123

MANUFACTURER: Conpar membrane filter sampling kit, comparison microscope and standard contamination slides.
SERVICE: Fluid cleanliness assessment by microscope count and gravimetric techniques.

Hunt, Trevor M.
50 Kingsholm Road
Westbury-on-Trym
Bristol, BS10 5LH   UK
Tel: 0272 507194   *Fax:* 0272 507194

SERVICE: Consultancy in wear
debris analysis and other
condition monitoring.
Lectures and course notes.

Hydac S.à.r.l.
Zone Industrielle du Held
Rue R. Schumann
57350 Stirling-Wendel, France
Tel: 87 85 85 50   *Fax:* 87 85 90 81

MANUFACTURER: Portable
on-line Purity Controller RC 1000.
Sampling and Analysis kits.
[see Flupac]

Hydraulic System Products Ltd
Monckton Industrial Estate
Monckton Road
Wakefield, West Yorks
WF2 7AL   UK
Tel: 0924 364748   *Fax:* 0924 290450

MANUFACTURER: HSP hydraulic
sampling points and pipe fittings.

Hydrotechnik GmbH
Holzheimer Straβe 94-96
D-6250 Limburg 1
Germany
Tel: 06431 4004-0   *Fax:* 06431 45308

MANUFACTURER: Minimess high
pressure sampling points and
probes.
[see Hydrotechnik UK]

Hydrotechnik UK Ltd
Unit 4
55A Yeldham Road
Hammersmith, London
W6   UK
Tel: 081-741 9934   *Fax:* 081-741 9935

UK SUPPLIER: Portable debris
check kit CCK4, including
filter membrane. Minimess test
points, fittings, etc.

Industriebedarf Rehm
Alte Dorfstraβe 151
2841 Lembruch/Dümmer See
Germany

Tel: 05447 629   *Fax:* 05447 1515

MANUFACTURER/SUPPLIER:
Precleaned sample bottles (ISO
5884) for particle detection in liquids.
Sampling kits for
spectrometric oil analyses.

Insitec Measurement Systems
2110 Omega Road, Suite D
San Ramon
CA 94583   USA
Tel: (510) 837-1330   *Fax:* (510) 837-3864

MANUFACTURER: PCSV in-line
particle counting, sizing and
velocimetry by light scatter
(primarily for air streams).

Institut de la Filtration et des
  Techniques Séparatives liquide-solide
Rue Marcel Pagnol
47510 Foulayronnes–Agen, France
Tel: 53 95 83 94   *Fax:* 53 95 66 95

SERVICE: Laboratory or on-site
testing of oils and filtration
systems. Consultancy.

Instruments S.A. UK Limited
2-4 Wigton Gardens
Stanmore, Middx
HA7 1BG   UK
Tel: 081-204 8142   *Fax:* 081-204 6142

Jobin-Yvon
Division d'Instruments S.A.
16-18 rue du Canal
B.P. 118
91163 Longjumeau Cedex, France
Tel: (1) 69 09 34 93   *Fax:* (1) 69 09 07 21

UK SUPPLIER: Jobin-Yvon spectrometers (ICP).

MANUFACTURER: Spectrometers including ICP.
[see Instruments S.A. UK]

Joyce-Loebl Limited — see Applied Imaging International Limited

Kevex Instruments
(A VG Instruments Group Company)
355 Shoreway Road
P.O. Box 3008
San Carlos
CA 94070-1308   USA
Tel: (415) 591-3600   *Fax:* (415) 595-4217

MANUFACTURER: Spectrometers including the X-ray microfluorescence analyser, combining XRF and SEM.
[see Fisons]

Kevex Instruments Inc
24911 Avenue Stanford
Valencia
CA 91355   USA
Tel: (805) 295 0019   *Fax:* (805) 295-0419

MANUFACTURER: Spectrometric analysers including XRF.
[see Fisons]

Keyance Corporation
2-13 Aketa-cho
Takatsuki
Osaka 569
Japan
Tel: 0726-84 2233   *Fax:* 0726-84 0444

MANUFACTURER: Hand-held video microscope.
[see Camlab]

Kleentek Limited
Sinagawa Techno Bloc 4F 2-7-7
Higash-Ohi
Shikagawa-ku
Tokyo
Japan
Tel: (03) 3740-4141   *Fax:* (03) 3740-4966

MANUFACTURER: Kleentek Electrostatic liquid cleaner.
[see United Air Specialists (UK) Limited]

Koflo Corp.
309 Cary Point Drive
Cary
Il 60013   USA

Tel: (312) 426-2211   *Fax:* (312) 426-2601

MANUFACTURER: In-line static liquid mixers, to improve debris mixing before taking samples of oil.
[see Pump & Packaging]

Kontron Elektronik GmbH
Breslauer Straβe 2
D-8057 Eching
Munich, Germany
Tel: 089 31901-1   *Fax:* 089 31901-311

Kontron Elektronik Limited
Blackmoor Lane, Croxley Centre
Watford, Herts, WD1 8XQ   UK
Tel: 0923 245991   *Fax:* 0923 54118

MANUFACTURER: IBAS image analysers and SEM image processing.
[see Kontron Elektronik]

UK SUPPLIER: IBAS image analysers.

Labcaire Systems Limited — see end of list

Lantos, F.E. — see Asesoramiento Ténico en Lubricación Industrial

LASENTEC®
Laser Sensor Technology Inc
P.O.Box 3912
Bellevue
WA 98009   USA
Tel: (206) 881-7117   *Fax:* (206) 881-8964

MANUFACTURER: In-line particle size analyser (PAR-TEC® and LAB-TEC®).
[see Belstock Controls]

Laser Lines Limited
Beaumont Close
Banbury, Oxon
OX16 7TQ   UK
Tel: 0295 267755   *Fax:* 0295 269651

UK SUPPLIER: Polytec particle size analysers, Aerometrics phase doppler particle analyser.

Leeds & Northrup
Sumneytown Pike
North Wales
PA 19454   USA
Tel: (215) 699-2000   *Fax:* (215) 699-3702

MANUFACTURER: Particle size analysers (Microtrac).
[see Leeds & Northrup Ltd]

Leeds & Northrup Limited
Wharfdale Road
Birmingham, B11 2DJ   UK
Tel: 021-706 6171   *Fax:* 021-706 4824

UK SUPPLIER: Particle size analysers (Microtrac).

Leica Cambridge Limited
Clifton Road, Cambridge
CB1 3QH   UK
Tel: 0223 411411   *Fax:* 0223 210692/
(Image Analysis – 0223 412776)

MANUFACTURER: Quantimet imageanalysers. Stereoscan 360 SEM.

Lindley Flowtech Limited
385 Canal Road
Frizinghall
Bradford
West Yorks
BD2 1AX   UK
Tel: 0274 530066   *Fax:* 0274 530078

MANUFACTURER: Fluid Condition Monitor (FCM).
SERVICE: Training in wear debris monitoring.
UK SUPPLIER: (UK North – PQ and RPD).

Machinery Monitoring
(G.H. Mills)
22 Yukon Terrace
Westwood, E. Kilbride
Glasgow, Scotland, G75 UK
Tel: 03552 23255

SERVICE: Consultancy in machine monitoring including wear debris monitoring.

Mahle Industriefilter
Knecht Filterwerke GmbH
Werk Ühringen
Postfach 1309
Schleifbachweg 45
7110 Ühringen, Germany
Tel: 07941 67-0 *Fax:* 07941 67-429

MANUFACTURER: Pi C 9000 mobile contamination particle counter 10-315 bar, 1-1000 cSt, full NAS 1638 range (5 μm up) and ISO 4406 standards classes.
[see Fawcett Christie]

Malvern Instruments Limited
Spring Lane South
Malvern, Worcs
WR14 1AT UK
Tel: 0684 892456 *Fax:* 0684 892789

MANUFACTURER: Laser diffraction particle size analysers.

Matec Applied Sciences Inc
75 South Street
Hopkinton
MA 01748 USA
Tel: (508) 435-9039 *Fax:* (508) 435-5289

MANUFACTURER: Capillary hydrodynamic fractionation particle size analyser.
[see Chemlab Scientific]

McVan Instruments Pty Ltd
58 Geddes Street
Mulgrave 3170
P.O. Box 298, Mulgrave North
Victoria, Australia 3170
Tel:

MANUFACTURER: Analite turbidity meter.
[see Camlab]

Metachem Diagnostics Limited
29 Forest Road
Piddington, Northampton
NN7 2DA UK
Tel: 0604 870370 *Fax:* 0604 870194

UK SUPPLIER: Seradyn calibration particles.

Met One Inc
481 California Avenue
Grants Pass, OR 97526 USA
Tel: (503) 479-1248 *Fax:* (503) 479-3057

MANUFACTURER: Laser diode liquid borne particle counters.
[see Environmental Monitoring]

Micromeritics Instrument Corp.
One Micromeritics Drive
Norcross
GA 30093-1877 USA

Tel: (404) 662-3660 *Fax:* (404) 662-3696

MANUFACTURER: Sedigraph x-ray detection of particle sedimentation.
SERVICE: Analysis of samples.
[see Micromeritics below]

Micromeritics Limited
4, The Ringway Centre
Eddison Road
Houndmills, Basingstoke
Hants
RG21 2YH   UK
Tel: 0256 844094   *Fax:* 0256 842431

UK SUPPLIER: Sedigraph x-ray detection of particle sedimentation.

Millipore (UK) Limited
The Boulevard
Blackmoor Lane
Watford, Herts
WD1 8YW   UK
Tel: 0923 816375   *Fax:* 0923 818297

UK SUPPLIER: Millipore membrane filters, glassware etc for gravimetric and patch tests and the preparation of microscope membrane slides.

Millipore Intertech
P.O. Box 255
Bedford
MA 01730   USA

Tel: (617) 275-9200

MANUFACTURER: Membrane filters and associated holders, all glassware required for the preparation of debris samples for microscope examination. [see Millipore (UK)]

Mills, George H. — see Machinery Monitoring

Monitek Technologies Inc.
1495 Zephyr Avenue
Hayward
CA 94544   USA
Tel: (415) 471-8300   *Fax:* (415) 471-8647

MANUFACTURER: Turbidity devices (Monitek). Micro Pure ultrasonic particle detectors. [see Acal Auriema]

Monition Limited
Bolsover Enterprise Park
P.O. Box 10
Station Road, Bolsover
Chesterfield S44 6BD   UK
Tel: 0246 825212   *Fax:* 0246 827999

SERVICE: Predictive maintenance with oil sampling and other condition monitoring methods. On-site service.

Moritex Corp.
Meisei Building
8-9 Sakuragaoka-cho
Shibuya-ku, Tokyo 150
Japan
Tel: 03-476-1021   *Fax:* 03-476-1698

MANUFACTURER: Hand-held video microscope imaging system. [see Finlay Microvision]

Muirhead Vactric Components Limited
Oakfield Road
Penge
London SE20 8EW   UK
Tel: 081 659 9090   *Fax:* 081 659 9906

MANUFACTURER: Wide variety of magnetic chip plug debris collectors.

Nicolet Instruments Ltd
Budbrooke Road
Warwick, CV34 5XH UK
Tel: 0926 494111 *Fax:* 0926 494452

UK SUPPLIER: FT-IR spectrometers.

NIKAT Associates
(C. Nicholls)
Meyer House
42 City Road
Chester CH1 3AE UK
Tel: 0244 313202 *Fax:* 0244 320224

SERVICE: Consultancy in condition monitoring including vibration. Regular courses in condition monitoring techniques.

Novasina AG
Talstrasse 35-37
CH-8808 Pfäffikon/SZ
Switzerland
Tel: 055 47 65 65 *Fax:* 055 47 62 62

SUPPLIER: McVan Analite turbidity meter.

Oilab Lubrication Limited
Sutherland House
31 Sutherland Road
Wolverhampton
WV4 5AR UK
Tel: 0902 334106 *Fax:* 0902 333010

UK SUPPLIER: Portable Oilab analysis kits, including debris checks. Ferrograph REO 1 and Densimeter REO 2. Model 56 Wear Particle Analyzer.

Olympus Optical Co.(Europa) GmbH
Postfach 10 49 08
Wendenstraβe 14-16,
D-2000 Hamburg 1
Germany
Tel: 040 2 37 73-0

MANUFACTURER: Cue-2/3/4/VS image analysis systems.

Olympus Optical Co. (UK) Limited
2–8 Honduras Street, London
EC1Y 0TX UK
Tel: 071 253 2772 *Fax:* 071 251 6330

UK SUPPLIER: Image analysers, microscopes and associated equipment.

Optokem Instruments Limited — see end of list

Otsuka Electronics Co. Ltd
3-26-3, Shodai-Tajika
Hirakata
Osaka, 573 Japan
Tel: 81 720-55-8550 *Fax:* 81 720-55-9100

MANUFACTURER: Laser particle size analysis systems.

Outokumpu Electronics Oy
P.O. Box 85
SF-02201 Espoo
Finland
Tel: 3580 4211 *Fax:* 3580 4212614

MANUFACTURER: XRF elemental analysers.

**Oxford Instruments –
Industrial Analysis Group**
19/20 Nuffield Way
Abingdon, Oxon OX14 1TX   UK
Tel: 0235 532123   *Fax:* 0235 535416

MANUFACTURER: XRF elemental analyser for wear metals.

Pacific Scientific Limited
HIAC/ROYCO Division
11 Manor Courtyard
Hughenden Avenue
High Wycombe, Bucks
HP13 5RE   UK
Tel: 0494 473232   *Fax:* 0494 472566

UK SUPPLIER: HIAC/ROYCO particle counters.

Pall Industrial Hydraulics Limited
Europa House
Havant Street
Portsmouth
PO1 3PD   UK
Tel: 0705 753545
Fax: 0705 297647/831324

SERVICE: Filtration products and consultancy in contamination control for hydraulic, lubrication, fuel and coolant fluid systems.

Parker Hannifin plc
Filter Division
Peel Street
Morley
Leeds
LS27 8EL
Tel: 0532 537921   *Fax:* 0532 527815

UK SUPPLIER: Filtration products, LCM II contamination monitor and Patch Test kit.
SERVICE: Par-Test® laboratory analysis of fluids, and on-site analysis as required.

Partech (Electronics) Limited
Eleven Doors
Charlestown,
St. Austell
Cornwall
PL25 3NN   UK
Tel: 0726 74856   *Fax:* 0726 68850

MANUFACTURER: Suspended solids monitoring equipment including turbidimeters. Portable, laboratory and industrial units particularly for water.

Particle Data Inc.
P.O. Box 265
111 Hahn Street
Elmhurst
IL 60126   USA
Tel: (312) 832-5653   *Fax:* (312) 832-5686

MANUFACTURER: ELZONE® particle size analyser to determine the volume of the particle in a conducting fluid.

Particle Data Limited
2 Goose's Foot Estate
Kingstone,
Hereford
HR2 9HY   UK
Tel: 0981 250479   *Fax:* 0981 251434

UK SUPPLIER: ELZONE® particle size analyser to determine the volume of the particle in a conducting fluid.

Particle Measuring Systems Inc.
1855 South 57th Court
Boulder
Colorado 80301   USA
Tel: (303) 443-7100   *Fax:* (303) 449-6870

Particle Measuring Systems Europe Ltd.
Nene House
Drayton Way
Drayton Fields Industrial Estate
Daventry, Northants
NN11 5EA   UK
Tel: 0327 301400   *Fax:* 0327 301323

Partikel-Messtechnik GmbH (PMT)
Carl-Zeiss-Strasse 11
D–7250 Leonberg-Gebersheim
Postfach 1516
Germany
Tel: 0 7152 51008   *Fax:* 0 7152 51676

Perkin-Elmer Limited
Maxwell Road
Beaconsfield, Bucks
HP9 1QA   UK
Tel: 0494 676161   *Fax:* 0494 678324

Polytec GmbH
D-7517
Waldbronn-KA
P.O. Box 1140
Germany
Tel: 07243 604-0   *Fax:* 07243 69944

Pump & Package Limited
The Old Manse
20 King Street
Desborough
Northants
NN14 2RD   UK
Tel: 0536 762674   *Fax:* 0536 761973

Quantachrome Corp.
5 Aerial Way
Syosset
NY 11791   USA
Tel: (516) 935-2240   *Fax:* (516) 935-2194

MANUFACTURER: Optical particle analysers (PMS).
[see Particle Measuring Systems Europe]

UK SUPPLIER: Optical particle analysers (PMS).

MANUFACTURER: Optical particle counters and systems(PMT).

MANUFACTURER: Spectrometric instrumentation including FT-IR.

MANUFACTURER: Optical particle sizer.
[see Laser Lines]

UK SUPPLIER: Koflo in-line static liquid mixers, to improve debris mixing before taking samples of oil, and blending of chemicals for water treatment.

MANUFACTURER: Microscan II particle size analyser.
Supplier of particle technology instrumentation, services and reference materials.
[see Steptech Instr. Services]

Quantachrome GmbH
Eismannsberger Straβe 34
8901 Eurasberg
Germany
Tel: 08208 1841   *Fax:* 08208 1471

(See above)

Ranco Controls Limited
401 Southway Drive
Southway Industrial Estate
Plymouth
Devon
PL6 6QT   UK
Tel: 0752 737166   *Fax:* 0752 739246

MANUFACTURER: Continuous Debris Monitor DM10 for ferrous debris on-line.

Rank Brothers Limited
High Street
Bottisham
Cambridge
CB5 9DA   UK
Tel: 0223 811369   *Fax:* 0223 811441

MANUFACTURER: Photometric Dispersion Analyser.

Rion Co. Ltd.
20-41 Higashimotomachi 3-chome
Kokubunji
Tokyo 185
Japan
Tel: 03-379 2352   *Fax:* 03-370 4828

MANUFACTURER: KL series particle counters.

[see Hawksley & Sons]

Roth Scientific Company Limited
Roth House
12 Armstrong Mall
The Summit Centre
Southwood, Farnborough
Hants
GU14 0NR   UK
Tel: 0252 513131   *Fax:* 0252 543609

UK SUPPLIER: Galai analysers for size and shape.

Seescan Plc
Unit 9
25 Gwydir Street
Cambridge
CB1 2LG   UK
Tel: 0223 460004   *Fax:* 0223 460116

MANUFACTURER:
Image analysers.

Seiko Instruments Inc.
6-31-1, Kameido
Koto-ku
Tokyo
136 Japan
Tel: 033-637-1393   *Fax:* 033-638-1279

MANUFACTURER: Element monitor XRF analyser.
[see Stormage Scientific]

Seradyn Inc
P.O. Box 1210
Indianapolis
IN 46206   USA
Tel: (317) 266-2000   *Fax:* (317) 266-2991

SGS Redwood Limited
Rosscliffe Road
Ellesmere Port
South Wirral, Cheshire
L65 5AS   UK
Tel: 051-355 4931   *Fax:* 051-356 3253

Shapespeare Corp.
128½ East Washington Street
Iowa City, Iowa 52240   USA
Tel: (319) 337-9463

Sheffield University
Department of Mechanical and
   Process Engineering
P.O. Box 600
Mappin Street
Sheffield
S1 4DU   UK
Tel: 0742 768555   *Fax:* 0742 753671

SKF (U.K.) Limited
Sundon Park Road
Luton, Beds
LU3 3BL   UK
Tel: 0582 490049   *Fax:* 0582 504203

SKF (U.K.) Limited
Bradbourne Drive
Tilbrook, Milton Keynes
MK7 8BJ   UK
Tel: 0908 838383   *Fax:* 0908 838403

Sonofloc
Separationstechnologien GmbH
Am Kanal 27
A-1112 Wien, Austria
Tel: 1-743521   *Fax:* 1-748833

**Southampton Institute of
   Higher Education**
East Park Terrace
Southampton
SO9 4WW
Tel: 0703 229381   *Fax:* 0703 222259

MANUFACTURER: Calibration particles, Latex spheres.
[see Metachem Diagnostics]

SERVICE: Oilscan oil analysis including wear metals and particle count. Also water TAN, TBN, flash point, etc. Provision of detailed diagnosis.

MANUFACTURER: Juliet II® image analyser.

SERVICE: Courses in condition based maintenance management, covering several subjects including wear debris analysis and its application to industrial machines.
Information — Dr. R. Firoozian.

SERVICE: Research into debris effects on bearings.

SERVICE: Lubrication oil analysis at two levels of information. Both include wear debris analysis.

Manufacturer: Ultrasonic filtration units.

SERVICE: Post graduate course in Systems Monitoring and Diagnosis (1 year), covering a range of topics including wear debris analysis.

Spectrex Corp.
3580 Haven Avenue
Redwood City, CA 94063   USA
Tel: (415) 365-6567   *Fax:* (415) 365-5845

MANUFACTURER: Laser particle counters.

Spectro Analytical U.K. Limited
Fountain House
Great Cornbow
Halesowen, West Midlands
B63 3BL   UK
Tel: 021-550 8997   *Fax:* 021-550 5165

UK SUPPLIER: Spectro OES for oil analysis.

Spectro Inc
160 Ayer Road
Littleton
MA 01460   USA
Tel: (508) 486-0123   *Fax:* (508) 486-0030

MANUFACTURER: Oil analysers including combined Arc Emission and ICP.

Spire Corporation
One Patriots Park
Bedford
MA 01730-2396   USA
Tel: (617) 275-6000   *Fax:* (617) 275-7470

MANUFACTURER: Surface Layer Activation equipment (SPI-WEAR®).

Staveley NDT Technologies
Inspection Instruments Division
18 Buckingham Avenue
Slough, Berks
SL1 4QB   UK
Tel: 0753 76216   *Fax:* 0753 821038

MANUFACTURER: Portable Debris Tester.

Steptech Instrumentation Service Ltd
Steptech House
Primrose Lane
Arlesey, Bedfordshire
SG15 6RD   UK
Tel: 0462 733566   *Fax:* 0462 733909

UK SUPPLIER: Quantachrome Microscan II particle size analyser.

Stormage Scientific Systems Limited
2 Fleetsbridge Business Centre
Upton Road
Poole, Dorset
BH17 7AA   UK
Tel: 0202 666426   *Fax:* 0202 666413

UK SUPPLIER: Seiko Instruments including elemental monitor.

Sulzer (UK) Limited
Sulzer-Chemtech Division
Westmead
Farnborough, Hants
GU14 7LP
Tel: 0252 544311   *Fax:* 0252 377636

MANUFACTURER: Flow mixers for fitting in laminar and in turbulent flow.

Swansea — Department of Mechanical Engineering
University College of Swansea
Singleton Park
Swansea
SA2 8PP   UK
Tel: 0792 295534   *Fax:* 0792 295701

SERVICE: Courses in tribology and machinery condition monitoring. Product development. Oil and debris analysis.

Swansea Tribology Centre
The Innovation Centre
University College of Swansea
Singleton Park
Swansea, SA2 8PP   UK
Tel: 0792 295369   *Fax:* 0792 295613

SERVICE: Product development, eg. RPD, PQ. Oil and debris analysis.
[see Analex and Lindley Flowtech]

Sympatec GmbH
System-Partikel-Technik
Burgstätter Straße 6
D-3392 Clausthal-Zellerfeld
Germany
Tel: 05323 717-0   *Fax:* 05323 717-229

MANUFACTURER: Laser diffraction particle size analysers.
[see below]

Sympatec GmbH
Unit 38
Bury Business Centre, Kay Street
Bury, Lancs
BL9 6BU   UK
Tel: 061 705 1878   *Fax:* 061 763 1887

UK SUPPLIER: Laser diffraction particle size analysers.

Tedeco — see Vickers Inc.

Thermal Control Products — see Howden Wade

Tribometrics Inc.
2475 4th Street
Berkeley
CA 94710   USA
Tel: (510) 540-1247   *Fax:* (510) 527-7247

MANUFACTURER: Model 56 Wear Particle Analyzer.
[see Oilab Lubrication]

UCC International Limited
P.O. Box 3
Thetford, Norfolk
IP24 3RT   UK
Tel: 0842 4251   *Fax:* 0842 3702

MANUFACTURER: Portable optical particle counter CM20 and Oilcheck oil analyser.

United Air Specialists (UK) Limited
Heathcote Way
Heathcote Industrial Estate
Warwick
CV34 6LY   UK
Tel: 0926 311621   *Fax:* 0926 315986

MANUFACTURER: Checker Kit oil analysis equipment using membrane filters (cased).
UK SUPPLIER: Kleentek electrostatic liquid cleaner.

Universities — see name of city/town

Varian Australia Pty Limited
679 Springvale Road
Mulgrave, Victoria
Australia 3170
Tel: (3) 560 7133   *Fax:* (3) 560 7950

MANUFACTURER: Spectrometers including AA and ICP-OES.
[see Varian]

Varian Limited
28 Manor Road
Walton-on-Thames
Surrey
KT12 2QF   UK
Tel: 0932 243741   *Fax:* 0932 228769

UK SUPPLIER: Spectrometers including AA and ICP-OES.

VD AMOS
Technicko obchodni agentura
FOR EKO
Post Box 209
746 01 Opava
Czechoslavakia
Tel: 0653 217370
Fax: 069 448606/443487

MANUFACTURER: Ferrograph REO 1 and Densimeter REO 2.
[see Oilab Lubrication]

VG Elemental
Ion Path Road Three
Winsford, Cheshire
CW7 3BX   UK
Tel: 0606 861022   *Fax:* 0606 552588

MANUFACTURER: Elemental mass spectrometers including ICP-MS.
(See also Fisons Instruments)

Vickers Inc.
Tedeco Division
24E Glenolden Avenue
Glenolden
PA 19036   USA
Tel: (215) 583-9400   *Fax:* (215) 583-3985

MANUFACTURER: Magnetic chip detectors (electrical, self-sealing), QDM® and Lubriclone® mixing chamber.
[see Vickers Systems]

Vickers Systems Limited
TEDECO Division
Larchwood Avenue
Bedhampton
Havant
Hants
PO9 3QN   UK
Tel: 0705 487260   *Fax:* 0705 492400

UK SUPPLIER: TEDECO chip plugs and detectors, QDM® and Lubriclone®.

Wearcheck Laboratories
Llandudno
North Wales
LL30 1SA   UK
Tel: 0492 581811   *Fax:* 0492 585290

SERVICE: Oil and debris analysis, including trends and report within 24 hours of serious condition.

Wearcheck (Pty) Limited
P.O. Box 731
16 School Road
Pinetown 3600
South Africa
Tel: (031) 7014291   *Fax:* (031) 728854

SERVICE: Oil analysis and maintenance consultancy.

Whatman LabSales Limited
St. Leonard's Road
20/20 Maidstone
Kent
ME16 0LS   UK
Tel: 0622 674821   *Fax:* 0622 682288

UK SUPPLIER: Membrane filters and associated laboratory equipment.

Wyatt Technology Corp.
820 East Haley Street
P.O. Box 3003
Santa Barbara
CA 931130   USA
Tel: (805) 963-5904   *Fax:* (805) 965-4898

MANUFACTURER: Laser light scattering instrument for absolute molecular characterization (Dawn®).
[see Optokem Instruments]

A^3 GmbH
Analytik Applikation Apparatebau
Josef-Beyerle-Straße 7
D-7252 Weil der Stadt 1, Germany
Tel: 07033 6014   *Fax:* 07033 6781

MANUFACTURER: BS series particle counters based on light scatter.
[see Labcaire Systems Ltd]

Computational Systems Inc
835 Innovation Drive
Knoxville
TN 37932   USA
Tel: (615) 675-2110   *Fax:* (615) 675-3100

MANUFACTURER: CSi OilView Portable Oil Analyzer, based on impedance/capacitance of sample.
[see CSI Europe]

CSI Europe
Innovation House, 23 Llwyni Drive
Connah's Quay
Deeside, Clwyd, CH5 4NJ   UK
Tel: 0244 822115   *Fax:* 0244 823100

UK SUPPLIER: CSi OilView Analyzer giving particle and ferromagnetic content, and change in oil chemistry.

Labcaire Systems Limited
15 Hither Green
Clevedon
Avon, BS21 6XU   UK
Tel: 0275 340033   *Fax:* 0275 341313

SUPPLIER: A^3 BS series particle counters.

Optokem Instruments Limited
Unit 25, Deeside Enterprise Centre
Rowleys Drive
Shotton, Clwyd, CH5 1PP   UK
Tel: 0244 836753   *Fax:* 0244 836754

UK SUPPLIER: Wyatt Technology Corp light scattering instruments.
SERVICE: Analysis of samples.

INDEX

The abbreviations for chemical and spectrometric apparatus are listed in Chapter 9 from page 205.

The names of companies (given in capitals) are shown abbreviated, ie. the company designation 'Ltd', 'Corp', GmbH, etc. has been omitted for clarity. Main references, which may involve more than one page, are highlighted in **bold** (the first page number only is given).

A
A^3 (ANALYTIK APPLIKATION APPARATEBAU), 126, 469
ABEX DENISON, 34
Absorption
  Air — *see* Air
  Atomic — *see* Atomic absorption
  Debris — *see* Debris
  Light — *see* Light
ACAL AURIEMA, 409, 415, 446
Acid dilution method, 222, 223
Acoustic emission, 3, 9
Actiprobe [Cormon], 132, 252, 306, **361**
Additives — *see* Oil
Addresses, 446ff
ADVANCED POLYMER SYSTEMS, 293, 446
A.E.A.TECHNOLOGY (Harwell), 361, 446
Aerodynamic shape factor — *see* Shape
AEROMETRICS, 128, **362**, 446
Agglomeration, 17, 42, 85, 116, 192, 266, 284
AHEM — *see* BFPA
Ai CAMBRIDGE, 441, 446
A.I.M. MÜNCHEN VERTRIEBS, 99, 100, 447
Air
  Absorption in oil, 26
  Breather, intake, 28, 40, 73, **76**, 276, 318

Bubbles, 26, 121, 135, 165, 183, 187, 191, 192
  Effect on counters, 126, 197, 230
  Elutriation, **175**, 295, 365
  Measurement, 26
Airborne particles, dust
  Ingestion into system – *see* Debris ingestion
  Monitoring, 68, 129, 181, 184, 208, 209, 266
Air Cleaner Fine Test Dust (ACFTD), 13, 50, 184, 199, 284, 289, **291**, 296
Aircraft engine monitoring
  Fixed wing, general, 10, 12, 63, 65, 93, 108, 153, 209, 214, 219, 232, 274, 286, 299, 301, 311
  Helicopter, 52, 57, 63, 93, 251, 305
Aerospace, 105, 185, 277, 297
ALPS [Malvern], 125, 184, **405**
ANALEX, 420, 430, 447
Analite [McVan], 127, 155, **158**, **363**, 449
Analysette [Fritsch], 124, 134, 135, 168, 169, 172, **391**, **392**, **393**
Appearance — *see* Visual appearance
Applications, 14, 115ff, **140**
APPLIED IMAGING INTERNATIONAL, 134, 168, 172, **173**, 364, 439, 447
APPLIED RESEARCH LABORATORIES, 223, 227, 228, 230, 231, 307, 447

Arc/Spark, 218
ASESORAMIENTO TÉCNICO EN LUBRICACIÓN INDUSTRIAL, 22, 59, 447
Ashing rotating disc electrode (ARDE), 21, **222**
Aspect ratio (length to width) — *see* Shape
Assembly, 29
Atomic absorption (AA), 211, 212, **213**, 311
Atomic emission (AE), 104, 213, **218**, 223, 311
Atomic fluorescence, 227
Atomic number, 210ff
Attenuated total reflectance, 216
Autocorrelation spectroscopy — *see* Photon correlation spectroscopy
Autocounter [Malvern] — *see* ALPS
Automated identification, 246
Automobile monitoring (road vehicles), 14, 98, 276, 304, 308
Autosizer [Malvern], 130, 202, 406

B
β-ratio (beta ratio), 73
BAHCO, 135, 168, **175**, 365
BAIRD, 207, 218, 219, 220, 223, 225, 227, 447
BASSAIRE, 67, 448
BATH FLUID POWER CENTRE — *see* FLUID POWER CENTRE
Bath Tub (distribution), 287
Bearings
  Design, 9, 39, 105
  EHL — *see* Lubrication
  Failures, fatigue, pitting, spalling, 10, 12, 27, 30, 39, 208, 209, 275, 300, 310, 311, 312, 318
  Life (see also Wear), 8, 9, 23, 56, 57, 284
  Lubrication (see also Lubrication, Oil), 4, 11, 104
  Monitors, 4, 9ff, 274
BELSTOCK CONTROLS, 367, 402, 448

BFPA — *see* BRITISH FLUID POWER ASSOCIATION
BHR GROUP, 55, 136, **158**, 276, **366**, 448
BI-90, BI-DCP [Brookhaven], 130, 134, 168, 172, 200, **368**, **369**
Bibliography, 330ff
Biological growth — *see* Microbial contamination
Blocking, **61**, 126, 136
BOPP, 174, 448
Boltzmann's constant, 201
Borescope — *see* Endoscopy
Bottles — *see* Sampling
BP OIL, 443, 448
BRITISH COAL (National Coal Board), 36, 154, 302, 305, 310, 384
BRITISH FLUID POWER ASSOCIATION, 66, 70, 277, 448
BRITISH RAIL, 301, 302, 306, 307, 444, 448
BROMLEY INSTRUMENTS, 134, 168, 172, **173**, **367**, 448, 449
BROOKHAVEN INSTRUMENTS, 130, 134, 168, 172, **201**, 293, 368, 369, 438, 449
Brownian motion (see also Photon correlation spectroscopy), 47, **82**, 111, 114, 130, 200
BRUNEL UNIVERSITY, 328, 449
Bubbles — *see* Air
Built-in — *see* Debris

C
Calibration, 40, 49, 182, 267, 284, **289**, 322
Calibration particles — *see* ACTFD, Latex particles
CALIFORNIA MEASUREMENTS, 68, 69, 176, 449
CAMBRIDGE INSTRUMENTS — *see* LEICA CAMBRIDGE
CAMLAB, 363, 396, 439, 449
CAPA-700 [Horiba], 134, 168, 172, **400**
Capillary hydrodynamic fractionation, **114**, 407

CARLO ERBA, 236, 449
CASPA, 247
CASTROL (UK), 333, 444, 449
Catalytic fines, 59
Cavitation, 8, 12
CCK 4 Checker Kit [Hydrotechnik], 138, 148, **370**
CDM — *see* Continuous debris monitor
CENT [Century Oils], 274, 275, 443
Centrifugal filter — *see* Filters
Centroid — *see* Shape
CENTURY OILS, 274, 443, 450
CHARN-MAC-ENG, 377, 450
CHDF 1200 [Matec], 114, **407**
Checker Kit [UAS], 138, 148, **371**
CHEMLAB SCIENTIFIC PRODUCTS, 407, 450
Chip plug — *see* Magnetic chip detector/collector
Choice (of monitor), 3, 9, 106, **140**, **320**
CHRISTISON SCIENTIFIC EQUIPMENT, 391, 392, 393, 450
Chunk shape — *see* Shape
CI 1000 [Climet], 125, 180, **372**
CI 1500 [Climet], 127, 155, **157**, **373**
CI 2000 [Climet], 126, 186, **374**
CILAS, 451
Circularity — *see* Shape
CIS [Galai], 131, 168, 266, 267, **395**
CJC NAPIER, 71, 450
CLAM [Monitek], 127, 155, **415**
CLAUSTHALL UNIVERSITY — *see* SYMPATEC
CLAYTON INSTRUMENTS (see also ELESA), 450
Clean cabinet, clean room, 64, 67, 70, 95
Cleanliness — *see* Contamination control
Clearances, 12, 60, 61, 159, 277, 284
CLIMET INSTRUMENTS, 125, 126, 127, 155, 157, **184**, 372, 373, 374, 450
CM20 [UCC International], 125, 180, **185**, **375**

CME, 315
CMS — *see* CONDITION MONITORING SERVICES
CMS INTERNATIONAL, 450
Coal (see also Mining), 41, 42, 62, 147
Coincidence (electronic, optical), 180, 196
Collection of debris (see also Sampling), **79**
Colliery — *see* Coal, BRITISH COAL
Colorimetric process, 211
Colour, 30, 40, 41, 42, 150, 178, 245, 246, 257ff, 294
Combustion systems, 22, 129, 204, 306
Commissioning, 28, 324
Compatibility, 5, 12, 20, 21, 103
Component casualties, 315ff
Compounds (monitoring), 237
COMPUTATIONAL SYSTEMS, 469
Computer aided particle analysis — *see* CASPA
COMRAD, 334
CONDITION MONITORING SERVICES, 444, 451
Conpar [Howden Wade], 103, 138, 148ff, 240, 277, 282, 376, 455
Consultants, 329, 447, 456, 459, 461
Containers — *see* Bottles, Reservoirs
Contam-Alert — see Digital Contam-Alert
Contamination
 Classes (levels), 16, 59, 65, 67, 70, 94, 95, 271, **276**, 310
 Control, cleaning, 29, **54**, 63, **276**
 Types (see also Debris), 19, 22ff, 54
Continuous debris monitor (CDM) [Ranco Controls], 122, 151, **152**, 310, **378**
Control spool, 12, 59, 61, 317
Coolant, Cooler, Cooling, 15, 24, 208, 209, 301, 317
CORMON, 132, 306, 361, 451
Corrosion, 8, 22, 23, 26, 39, 209
Cost — *see* Economics

COULTER ELECTRONICS, 116, 117, 118, 124, 130, 170, **195**, 199, 201, 249, 266, 379, 380, 381, 403, 451
Coulter principle — *see* Electrical sensing zone
Counters, 49, 142, 178, **179**, 295ff
Coupling, 7, 29, 78, 94
Crackle test — *see* Water content
Crack testing, 3
Crank case, 208
Crane, 12, 16, 308
CSI EUROPE, 469
CU (Concentration Unit) — *see* Nephelometry
CUE [Olympus Optical], 437
Curl shape — *see* Shape

D

DANTEC ELECTRONICS, 128, **382**, 451
Dawn [Wyatt Technology], 127, 155, **383**
Dead end, 29, 81, 94
Dean & Stark — *see* Water content
Debris
    Absorption — *see* Ingestion below
    Built-in, 16, 28, 29, 250
    Collection, **79**
    Density, 48, 82, 83, 114, 131, **147**
    Generation, 7, 16, 28, 29, **30**, 249, 251
    Hardness, 30, 58, 163, 239, 246, 249
    Identification, **36**, **239**, 254
    Implanted, 28, 302
    in Filters, 105
    Ingestion (Absorption, Ingression), 16, 28, 29, **31**, 44, 50, 63, 78, 82, 208, 209, 313
    Mixing, **85**, 319
    Movement (Migration), 86, 88, 90
    Origin, 7, 19, 40, 43, **51**, 53, 204, 239, 271
    Shape — *see* Shape
    Size — *see* Size
Debris Tester [Staveley NDT], 93, 121, 151, **154**, 300, 301, 306, **384**

Deburring, 44
Degradation of fluid, **19**
Design (see also Bearings, Lubrication, Reservoir Terotechnology), 77
Density
    Fluid, 26, 83
    Particle — *see* Debris
    Ratio (meter), 82, 262, 263, 388
Detection of particles, **64**, **106**, 285
DIAGNETICS, 118, 158, **160**, 200, 377, 451
Dielectric constant, permittivity, **115**, 151, 417
Diesel Engine, 4, 24, 37, 39, 209, 256, 302, 304
Diesel oil (see also Fuel), 22, 59, 84, 226
Diffraction — *see* Fraunhofer diffraction, Light
Digital Contam-Alert [Diagnetics], 118, 158, **161**, 196, 200, 249, **377**
DIGITHURST, 436, 440, 451
Diluent, dilution of (see also Solvent)
    Fuel, 284, 301, 304
    Grease, 9, 104
    Oil, 25, 117, 123, 183, 195, 202, 215, 225, 226, 227, 255, 260, 262
Dimension of debris — *see* Size
Dip stick, 71, 98
Direct current plasma (DCP), **227**
Disaster, 1, 55, **62**, 76, 271, 287, 310, 318, 323, 325
Disc centrifuge — *see* Sedimentation
Disposal, 316, 324
Distilled water, DI water, 95, 128, 373, 374, 398, 423, 428, 429
Distribution of sizes (see also Bath Tub distribution), 50, 110, 111, 115, 119, 120, 124, 131, 132, 142, **167**, 182, 184, 193, 196, 291, 292
DM10 — *see* Continuous debris monitor
Doppler shift, 111, 128, 202
Drag coefficient, 81, 134

DSA (Dynamic Shape Analyser), **395**
DTU — *see* Nephelometry, Debris Tester
DUKE SCIENTIFIC, 293, 452
Dynamic light scattering — *see* Photon correlation spectroscopy
Dynamic state monitoring, 3
DYNO PARTICLES, 293, 294, 452

E

Economics, 2, 8, 10, 62, 64, 298ff, 315ff, 322ff
Edinburgh, University of, 153
Education — *see* Training
EDX — *see* Energy dispersive X-ray
Efficiency (see also Filters), 13, 27, 32, 59, 89, 135
Electrical, **151**
  Conductance, **116**, 151, 155
  Resistance — *see* ESZ
Electrical sensing zone (ESZ), 50, **116**, **192**, 295, 380, 385
Electrolysis, 28
Electrophoretic mobility, 111
Electrostatics (see also Filter), 76, 82
Elements (see also Gases, Metals, Non-metals), **204**, **207**
ELESA, 77, 452
Elutriation — *see* Air elutriation
Elzone [Particle Data], 116, 194, 196, **385**
Emission — *see* Atomic emission
ENDECOTTS, 135, 168, **174**, **386**, 452
Endoscopy, 305
Energy consumption, 3
Energy dispersive X-ray (EDX), 39, 228, **234**, 243, 269, 270
ENVIRONMENTAL MONITORING, 408, 452
Equivalent spherical diameter — *see* Size
Erosion, 59
Etching, 12, 38, 135
Ethylene propylene — *see* Seals
Evidence (removed, or erroneous), 16, 17, 19
Experience, **298**, 318

F

FAIREY ARLON, 138, 148, 387, 452
FAS-CC110 [Fairey Arlon], 138, 148, **387**
Fatigue — *see* Bearings, Wear
FAWCETT CHRISTIE, 76, 452, 459
FCM [Lindley Flowtech], 63, 118, **199**, 309, 310, **390**
Feret's diameter — *see* Size
Ferrography, 4, 9, 51, 104, 122, 240, 251, **254**, 303, 304, 305, 308, **388**, 443
Ferrokinetic stage (microscope), 149, 241
FerroSCAN [GasTOPS], 122, 151, **152**, 312, **389**
Ferrous
  Compatibility, 21
  Debris (Iron element), 9, 39, 147, 208, 211, 232, **250**, 275
  Monitoring (see also Ferrography), 11, 89, 116, **121**, **122**, **151**, 241, 250
FES, 444, 452
Fettling, 44
Fibre shape — *see* Shape
Filling — *see* Oil transfer
Filterability, 33ff
Filter blockage technique, **118**, 136, 158ff, **196**, 249, 377, 390, 403
Filter membrane — *see* Membrane filter
Filters & filtration (air, gas), 40, 67, 275, 313
Filters & filtration (oil, liquid), 17, **32**, 73, 105
  Clogging indicator, 70, 74, 75, 313
  Efficiency, Rating, 17, 32, 33, 57, **73**, 299
  Media:
    Fibres (cellulose, etc), 74
    Mesh, 118, 135, 136, 174ff, 196ff, 297
    Warp, weft, 135, 155, 174
  Types:
    Blocking, 60, 74, 75
    Centrifugal, 60, 90
    Electrostatic, 76, 91
    Magnetic, 33, 74, 75, 435
    Ultrasonic, 74, 91

FINLAY MICROVISION, 440, 453
FISONS (see also ARL, CARLO ERBA, KEVEX and VG ELEMENTAL), 453
Fire resistant fluids, 4, 21, 23
Fitting, 7, 321
Flake shape — *see* Shape
Flocculation, 130
Flow, 8, 13, **80**, 114, 118
  Laminar, 67, 80, 81, 82
  Mixing, 81, 85, 87, 98, 319
  Turbulent, 80, 81, 82, 84, **85**, 93, 165, 167
Flow field flow fractionation, 114
Fluid cleanliness — *see* Contamination
Fluid cloudiness, 26
Fluid Condition Monitor — *see* FCM
FLUID POWER CENTRE, 85, 97, 199, 277, 328, 453
Fluid sampling — *see* Sampling
Fluid state monitoring, 3, 8ff
Fluorescence — *see* Atomic fluorescence
FLUPAC, 426, 453
Flushing, 16, 28, 29, 61, 62, 68, 69, 70, 250, 310, 311
Foaming, 20, 27
Fork lift truck, 312, 320
Forward reflectance, 110, **123**, **189**, 398, 408, 432
Forward scatter — *see* Fraunhofer diffraction
Fourier lens, 169, 244
Fourier transform IR (FT-IR), 216
Fraunhofer diffraction, 110, **124**, **168**, 184, 379, 393, 399, 406, 411, 434
Fretting — *see* Wear
Friction, 6, 58, 81, 87
FRITSCH, 124, 134, 135, 168, **169**, 175, 391, 392, 393, 453
Fuel dilution, 284, 301
Fuel, 22, 59, 60, 208, 209
FULMER SYSTEMS, 139, **163**, 249, 272, **394**, 453

G
GALAI PRODUCTION, 131, 168, 176, **265**, 395, 454
Gamma rays — *see* Radioactivation
Gases, 225
  Acetylene, 215
  Argon, 229
  Chlorine, 27
  Helium, 230
  Hydrogen, 208, 210
  Nitrous oxide, 218
  Oxygen, 25, 208, 210
Gaseous contamination, 19, **26**
GASTOPS, 122, **152**, 312, 389, 454
Gaussian distribution, 169
GBC, 213, 223
Gears, gear boxes (see also Bearings, Wear), 4, 5, 6, 14, 15, 52, 57, 96, 100, 208, 209, 251, 274, 302, 304, 306, 310, 317
GELMAN SCIENCES, 372, 373, 374, 454
Generators, 305
GEORGE FISCHER FOUNDRY SYSTEMS, 175, 365, 454
GLACIER METAL, 12
Granulometer [CILAS], 451
Gravimetric analysis, **119**, **143**, 149, 284, **413**
Grease, 4, 9, 42, **104**, 208, 209, 318, 319
Grey scale (image analysis), 120, 244, 245, 437ff
Grinding, 39, 52, 66

H
HACH, 127, 155, **156**, **396**, 449, 454
HAMAMATSU PHOTONICS, 437, 454
Hall effect, 151
Hand-held scanner (video), 248, 439, 440
Handling, 65
Hardness of particles — *see* Debris
Harwell — see A.E.A. TECHNOLOGY
HAWKSLEY & SONS, 428, 429, 455

Health monitoring, 1ff
Heat effect on ferrogram slide, 258
Heat treatment, 30
Helicopter — see Aircraft engines
Helos [Sympatec], 124, 168, 170, **434**
HFA, HFB, HFC, HFD, 24
HIAC/ROYCO [Pacific Scientific], 123, 125, **181**, 189, 199, **397**, **398**, 455
HITACHI SCIENTIFIC INSTRUMENTS, 213, 236, 455
HOLMBURY, 78, 455
HORIBA INSTRUMENTS, 124, 126, 134, 168, 170, 172, **187**, 235, 399, 400, 401, 455
Hose — see Pipe
HOWDEN WADE, 103, 138, 148, **149**, 240, 376, 444, 455
HSP — see HYDRAULIC SYSTEM PRODUCTS
HYDAC, 125, **184**, 426, 456
HYDRAULIC SYSTEM PRODUCTS, 93, 94, 456
Hydrodynamic chromatography — see Capillary hydrodynamic fractionation
HYDROTECHNIK, 94, 138, 148, 370, 456

I
Iatrogenic, 9, 323
IBAS [Kontron], 438
Identification — see Debris, Microscopy
IFTS — see INSTITUT DE LA FILTRATION
Image analysis, 48, **120**, **243**, 267, 268, 395, **436**, 446
IMOLV [PMS], 125, **422**
Indic [BP], 443
Inductance, **121**, 151, 384
Inductively coupled plasma (ICP), 212, 218, **223**, 238, 307
Industrial levels (contamination), 55, 274, 283, 287
INDUSTRIEBEDARF REHM, 99, 100, 444, 456

Infra-red (IR), 110, 151, 158, **216**
Ingression — see Debris
In-line, On-line, Off-line, 17, 79, **107**, 140
INSITEC MEASUREMENT SYSTEMS, 124, 168, **171**, 456
Insolubles, 275, 276, 284
Isotope — see Radioactivation
Installation, 324
INSTITUT DE LA FILTRATION ..., 456
Instruments (see also Choice), 357ff
INSTRUMENTS S.A., 457
Intensity kits, 147, 149
IPA, 95, 210
Irradiation — see Radioactivation
ISO — see Standards

J
Jacks, 287
Jamming of components, 16, **60**
JOBIN YVON, 223, 226, 227, 457
JOYCE-LOEBL — see APPLIED IMAGING INTERNATIONAL
Juliet [Shapespeare], 438

K
Karl Fischer — see Water content
Karlsruhe — see POLYTEC
Kerosene, 225, 227
KEVEX, 228, 233, 234, 457
KEYANCE, 248, 439, 449, 457
KLEENTEK, 91, 456
KL Series [Rion], 125, 126, 185, 189, **428**, **429**
KNECHT FILTERWERKE — see MAHLE
KOFLO, 87, 457
KONTRON ELEKTRONIK, 438, 458
Kurtosis, 303

L
L_{10} (bearing life), 23
LA-900 [Horiba], 124, 168, **399**

LABCAIRE SYSTEMS (see also A³ GmbH), 469
Labelling (bottles), 97
Laboratories — *see* Oil analysis, Services
LAB-TEC [Lasentec], 131, 307, 308, **402**
Lacquers, 28
Lambert's law (turbidity), 156
LASENTEC (Laser Sensor Technology), 131, 168, **176**, 307, 402, 458
Laser diffraction — *see* Optical diffraction
LASER LINES, 362, 425, 458
Latex spheres, 40, 289, 290, 293, 294
LBS [PMS], 125, 185, **422**
LCM II [Coulter Electronics], 118, **199**, **403**, 462
Leak, leakage
  Air, 26, 276
  Coolant, 24, 208, 209, 301
  Oil, chemicals, 3, 9, 14, 55, 72, 275, 317
LEEDS & NORTHRUP, 124, 130, 168, 170, **202**, 411, 412, 458
LEICA CAMBRIDGE, 441, 458
Level (oil), 9, 72, 76
Level (contamination) — *see* Contamination classes
Light (see also Optical), 109
  Obscuration, Extinction, 110, **125**, **180**, 196, 372, 375, 397, 405, 422, 424, 426, 428
  Scatter, 109, 111, **126**, **186**, 374, 401, 423, 425, 428, 429
LINDLEY FLOWTECH, 118, **199**, 249, 309, 390, 420, 430, 458
Liquid — *see* Fluid, Oil, Water
Liquid Contamination Monitor — *see* LCM II
Load, 8, 317
Location of monitor, 17, 32, 299, 316, 319ff
LONGLANDS COLLEGE, 328
LPA [Otsuka Electronics], 130, 134, 203, **418**
LS Series [Coulter Electronics], 124, 168, 170, **379**

Lubrication (see also Oil)
  Design, 14, 32, 316
  Elastohydrodynamic lubrication, 7
  Monitoring, 14, 300ff
  Standards, 284ff
  Wear effects, 5, 8, 37, 56, 287
Lumpiness — *see* Shape

M
MACHINERY MONITORING, 459
Machine tool, 14, 65
Magiscan [Applied Imaging International], 439, 447
Magnetic attraction monitors, 89, 121, **122**, **151**, 378, 388, 389, 404, 420, 427, 430, 435
Magnetic chip detector/collector, 10, 11, 13, 33, 62, 79, 89, **92**, 116, 121, 151, 300, 301, 311, 384, **404**
Magnetometry, 122, 152
MAHLE, 459
Maintenance, 1, 2, 28, 72, 73, 316, 321, 324, **325**
MALVERN INSTRUMENTS, 124, 125, 130, 168, 171, **184**, 202, 405, 406, 459
MARCHWOOD LABORATORIES, 152, 162
Marine, 14, 65, 209, 226
Marker (tracer), 132, 226, 252, 302
Martin's diameter — *see* Size
Mass spectrometry, **236**
Mastersizer [Malvern], 124, 168, **171**, **406**
MATEC APPLIED SCIENCES, 114, 407, 459
McVAN INSTRUMENTS, 127, 155, 363, 449, 459
Medical monitoring, 2, 3, 106
Membrane filter, 34, **100**, 119, 142, 143, **148**, 162, 230, 239ff, 295
Mesh — *see* Filter media
Mesh obscuration — *see* Filter blockage
METACHEM DIAGNOSTICS, 293, 459
Metals (see also Gases, Non-metals)
  Aluminium, 21, 58, 147, 208, 210, 225, 230, 275, 301, 302

Metals—*contd.*
  Antimony, 208
  Barium, 208, 225
  Brass, 21, 41, 58, 147, 209
  Bronze, 20, 21, 41, 147, 209
  Cadmium, 211, 258
  Calcium, 34, 208, 209, 211, 218
  Chromium, 39, 41, 104, 208, 209, 211, 258, 269, 275, 301
  Cobalt, 211, 226
  Copper, 21, 39, 42, 58, 208, 209, 211, 232, 258, 275, 301, 302
  Gold, 147, 211, 258, 302
  Iron — *see* Ferrous, Steel below
  Lead, 21, 147, 208, 209, 211, 258, 275
  Magnesium, 208, 209, 210, 218, 258
  Manganese, 39, 208, 209, 211, 302
  Mercury, 195, 211
  Molybdenum, 208, 209, 211, 258
  Nickel, 104, 208, 209, 211, 258
  Platinum, 211
  Potassium, 230
  Silver, 20, 209, 211, 232, 258, 302
  Sodium, 209, 210, 230, 275, 301
  Steel, 29, 41, 204, 240, 250, 258
  Tin, 208, 209, 211, 258, 275
  Titanium, 209, 211, 225, 230, 232, 258
  Tungsten, 208, 211
  Uranium, 211, 230
  Vanadium, 211
  White metal, 30, 39, 305
  Zinc, 208, 209, 210, 211, 218, 258
  Zirconium, 211, 226
MET ONE, 123, **190**, **408**, 459
Microbial contamination, 22
MicroEye, MicroScale [Digithurst], 440
Micro laser particle spectrometer [PMS], **423**
MICROMERITICS INSTRUMENT, 134, 168, 172, 431, 459, 460
Micron (definition), 7, 29
Micro Pure [Monitek], 137, **164**, **409**, 446
Microscan [Quantachrome], 134, 168, **173**, **410**
Microscopy (see also SEM), **241**
  Counting, Size, 48, 184, 284, 295
  Hand-held, 149, 248, 370, 371, 419, 439, 440
  Particle identification, 36, 138, 149, 240, 257
  Range, Types, 149, 175, **241**
Microtrac [Leeds & Northrup], 124, 130, 168, **170**, 202, **411**, **412**
Microwave induced plasma, 238
Mie scatter, 11, 126, 331
Migration — *see* Debris movement
MILLIPORE, **96**, **100**, 119, 138, **144**, 148ff, 240, 284, 413, 414, 460
Mineral oil — *see* Oil type
Minimess [Hydrotechnik], 94, 161
Mining (see also Quarrying), 14, 262, 304, 306, 308, 310, 313, 321
Mixer — *see* Flow
Mobile, 65, 85, 86, 320
Model 56 — *see* Wear Particle Analyzer
MONITEK TECHNOLOGIES, 124, 127, 137, 155, **164**, 409, 415, 446, 460
MONITION, 445, 460
Monitor choice — *see* Choice
Monochromator, 214, 218
MORITEX, 248, 440, 460
Movement of particles — *see* Debris
MPS Series [Monitek], 137, **409**
MTBF (Mean Time Between Failures), 160, 312
MUIRHEAD VACTRIC COMPONENTS, 92, 93, 122, 151, 404, 460
Multisizer [Coulter Electronics], 116, 192, 295, **380**
Multi-trend, 273

N

N4 Series [Coulter Electronics], 130, 201, **381**
National Coal Board — *see* BRITISH COAL
Nebulization, 225, 227, 229

Nephelometry/Turbidity, 25, 81, 111, **127**, **155**, 363, 373, 383, 396, 415, 446
NICOLET INSTRUMENTS, 461
NIKAT ASSOCIATES, 309, 461
Noise, 3, 72, 181, 218
Non-metals (see also Plastics, Seals)
  Boron, 208, 209, 210, 230, 235
  Carbon, 37, 204, 208, 210, 258
  Glass, 63, 74, 258
  Phosphorus, 209, 225
  Silicon, 27, 39, 147, 208, 209, 211, 225, 230, 231, 275
  Sulphur, 25, 208, 211, 230
  Wood, 44
  Xylene, 225, 227
NOVASINA, 461
NTU — see Nephelometry

O

Odour, 22, 26, 72
Off-line — see In-line, On-line, Off-line
Oil (see also Water)
  Acidity (TAN), 25, 26, 115, 416
  Additives, 208, 209
    Anti-corrosion, 20, 27, 34
    Anti-foam, 20, 27, 209
    Anti-oxidation, 20, 27
    Anti-wear, ZDDP, 20, 25, 34, 35, 210
    Demulsibility, 26
    Effect on oil, 25, 33
    Elements, 208, 209
    Vapour Phase inhibitor, 20
    Viscosity Index improver, 20
  Alkalinity (TBN), 25, 26, 27, 301, 416
  Analysis (see also Services), 30, 217, 220, 224, 236, 301, 443ff, 456
  Change, 26, 98, 306
  Contaminant in water, 307
  Film thickness, 56, 58
  Flash point, 27
  Oxidation, Oxides, 27, 115, 284
  pH, 301
  Pour point, 27, 416
  Refinery, 309, 313

Oil (see also Water)–*contd.*
  Transfer, 65, 70, 71, 107, 140, 184
  Types:
    Mineral, 23, 210
    Oil-in-water (HFA), 9, 21, 22, 23, 307
    Phosphate ester (HFD), 4, 21, 23
    Water glycol (HFC), 12, 21, 23
    Water-in-oil (HFB), 21, 23
OILAB LUBRICATION, 138, 148, 255, 263, 388, **416**, 435, 461
Oilcheck [UCC International], 115, **155**, 417
OilView [CSI], 469
OLYMPUS, 437, 461
On-line — see In-line, On-line, Off-line
Optical, 109
  Absorption — see Light obscuration
  Diffraction — see Fraunhofer diffraction
  Image analyser — see Image analysis
  Opacity — see Nephelometry
  Reflectance — see Nephelometry
  Scatter — see Light scatter
  Time of transition — see Time of transition
OPTOKEM INSTRUMENTS, 461, 469
Optomax [Ai Cambridge], 441, 446
Origin of debris — see Debris
OTSUKA ELECTRONICS, 130, 134, **202**, 418, 461
OUTOKUMPU, 228, 230, 231, 233, 461
Ovoid shape — see Shape
OXFORD INSTRUMENTS, 228, 229, 230, 231, 462

P

PACIFIC SCIENTIFIC HIAC/ROYCO Div, 123, 125, **181**, 397, 398, 455, 462
Packaging, 28
Paint, 30, 42, 173

Pall Contamination Analysis [Pall], 443
PALL INDUSTRIAL HYDRAULICS, 443, 462
PARKER HANNIFIN, 138, 148, 419, 443, 462
PAR-TEC [Lasentec], 131, 168, 177, **402**
PARTECH (ELECTRONICS), 127, 462
Par-Test [Parker Hannifin], 443
Particle — *see* Calibration particles, Debris
Particle characterisation — *see* Shape
PARTICLE DATA, 116, **196**, 385, 462
PARTICLE MEASURING SYSTEMS, 125, 126, **184**, **187**, 422, 423, 463
PARTIKEL-MESSTECHNIK, 125, **185**, **424**, 463
Particle Quantifier — *see* PQ
Patch test, 138, 148ff, **414**, **419**
PDA [Dantec], 128, **382**
PDA 2000 [Rank Brothers], 129, **166**, **421**
PDPA [Aerometrics], 128, **362**
Pebble shape — *see* Shape
Peening, 28, 39
Performance, 3, 72, 327
Perimeter — *see* Shape
PERKIN ELMER, 213, 216, 218, 223, 227, 463
Phase/doppler scatter, 111, **128**, 362, 446
Phase-locked loop (PLL), 153
Photometric dispersion, 111, **129**, **166**, **421**
Photon correlation spectroscopy, 111, **130**, **200**, 368, 381, 406, 412, 418
Pi C 9000 [Mahle], 459
Piezoelectric, **164**
PILKINGTON GLASS, 63
Pipe and hose, 16, 80, 81, 82, 86, 93, 165, 317, 318
Pistons, 12, 15, 21, 37, 58, 159, 208, 209, 252, 275, 302

Pitting — *see* Bearings
Planned maintenance, 1, 325ff
Plasma — *see* Direct current plasma, Inductively coupled plasma
Plastics (Acrylic, Nylon, PTFE, etc), 21, 30, 35, 58, 95, 147
Platelet shape — *see* Shape
Plating, 209, 210
PLCA 520 [Horiba], 126, 187, **401**
PMS — *see* PARTICLE MEASURING SYSTEMS
PMT — *see* PARTIKEL-MESSTECHNIK
Polarization, 40, 218, 241, 246, 257
Polishing, 5, 38
Polymer, 39, 44, 53, 294, 399, 446
POLYTEC, 126, **188**, **425**, 463
Porosity — *see* Shape
Powder (see also Airborne particles), 170, 295
PQ [Swansea Tribology Centre], 93, 122, **155**, **420**
Predict [BP], 443
Predict [CMS] — *see* Tribo Predict
Predictive maintenance, 160, 308, 309, 325ff
Preventive maintenance, 1, 325ff
Printed circuit boards (cleaning), 140
Proactive maintenance, 160, 326
Projected diameter — *see* Size
PROTOS [Ai Cambridge], 441, 446
Prototron — *see* SPECTREX
PSDA [Horiba], 168, **399**
PUMP & PACKAGE, 87, 463
Pump wear, 9, 12, 23, 62, 63, 82, 209, 299, 310, 312, 317, 318, 319
Pure Alert [Micro-Pure], **164**
Purity Controller RC 1000 [Hydac], 125, 184, **426**

Q

QDM [Vickers Systems], 122, 151, 153, 251, 305, **427**
QUANTACHROME, 134, 168, **173**, 410, 445, 463, 464
Quantimet [Leica Cambridge], 242, 246, 441

Quarrying (see also Mining), 14, 40, 208

R

Radioactivation, Radio-isotope, 27, 107, **132**, 230, **252**, 306, 361, 433
Rail transport — *see* Trains
RANCO CONTROLS, 122, 151, 378, 464
RANK BROTHERS, 129, 166, 421, 464
Rayleigh scatter (definition), 126
RC 1000 — *see* Purity Controller
RDE — *see* Rotating disc electrode
Red oxide, 38, 39
Redwood Oilscan [SGS Redwood], 99, **313**, 443
Reflection — *see* Light
Refraction — *see* Light
Refractive index, 109, 127, 180, 266
Reliable, Reliability, 1ff, 27, 54, 147, 159, 160, 272, 284, 298, 312, 322, 325, 326
REO Ferrograph [VD Amos], 122, 263, **388**
Reservoir, sump, tank
  Debris entry, 32
  Design, 26, 29, 71, 76, 77, 85
  Monitoring, 9, 107, 317ff
  Oil sampling, 99, 140, 310
Resolution, 181, 243, 244
Reynold's number, 80, 134
RF oscillator, 151ff
RION, 125, 126, **185**, 189, 428, 429, 464
Road vehicles — *see* Automobiles
Robotics, 14, 150, 309
Rolling fatigue — *see* Wear
Roll shape — *see* Shape
Rotary Particle Depositor — *see* RPD
Rotating disc electrode (RDE, Rotrode), 104, 212, 218, **219**, 222, 223, 307
ROTH SCIENTIFIC, 395, 464
Rough, Roughness — *see* Shape
Royco — *see* HIAC/ROYCO
RPD [Swansea Tribology Centre], 122, 255, **263**, **430**

Running-in, 37, 38, 42, 79, 165, 287, 288, 304, 313
Russell sensor, 184, 372
Rust, 28, 38, 39, 42, 44, 147

S

Safeguard [Castrol (UK)], 444
Safety, 15, 63, 133, 305, 318, 323, 325
Salt crystal — *see* Sodium chloride, Sodium sulphate
Sampling
  Bottle, 17, 18, **95**, 108, 182, 183, 191, 240
  Cleaning, 18, **95**, 191
  Points, Valves, 17, 18, 19, **93**, 107, 144, 319ff
  Sample, 79, 108, 151
  Syringe, 99, 104, 107, 144, 187
  Technique, 81, 94, **98**, 144ff, 183, 296, 299, 300
Saturation — *see* Coincidence
Scanning electron microscope (SEM)
  Analysis, 38, 233, **242**, 259, 270
  Photography, 36, 233
  Examples, 38, 235, 243, 269
  Preparation, 68, 69, 234ff, 244, 259
Scanning laser beam — *see* Time of transition
Scatter — *see* Light scatter, Mie scatter, Phase/doppler scatter, Rayleigh scatter
Scientific Services Oil Analysis [British Rail Research], 444
Scopeman [Moritex], 248, 440
SDM 100, 162, 305
Seals, sealing (see also Non-metals), 21, 209
  Design, 9, 16, 32
  Ethylene Propylene, 21
  Fluorocarbon, 21
  Neoprene, 21, 209
  Nitrile, 21
  Rubber, 21, 30, 44
  Viton, 21
Sedigraph [Micromeritics Instrument], 134, 168, **172**, **431**

Sedimentation, 114, **134**, **171**, 203, 295, 364, 367, 369, 392, 400, 410, 431
Seeding — *see* ACFTD
SEESCAN, 442, 464
SEIKO INSTRUMENTS, 228, 231, 464
Seizing, 60
Sensitivity to contaminant, 61, 77, 78
Separator
   Oil/air, 76, 88
   Oil/debris, 88, **239**
SERADYN, 293, 465
Services (including laboratories), 183, 184, 224, 270, 275, 300, 305, 313, **443**, 456
Servo controls, 283
Severity of wear index, 262
SGS REDWOOD, 99, 257, 258, 259, 275, 276, **313**, 443, 465
Shape characterisation, 245ff, 436ff
   Aerodynamic factor, 134
   Aspect ratio, 33, 40, 44, 47, 48, 82, 84, 120, 245, 247
   Centroid, 48
   Chunk, 30, 39, **40**, 50, 51, 94, 249, 287
   Circularity, 48, 247
   Curl, **43**
   Dynamic, 267
   Fibre, 28, **44**, 246, 249
   Flake, 28, **42**, 300
   Form factor, 48, 49
   Index, 295
   Lumpiness, 48
   Ovoid, 40
   Pebble, 39, **40**
   Perimeter, 48
   Platelet, 41, **42**, 51, 52, 83
   Porosity, 48, 82, 109, 120, 247
   Roll, **43**
   Rough, Roughness, 48, 247
   Round, Roundness, 48
   Slab, **40**, 52
   Sliver, 30, **43**
   Sphericity, 37, 39, 44, 48, 120, 131
   Spiral, **43**, 51
   Strand, **44**
   Texture, 38, 246, 247
SHAPESPEARE, 438, 465
Sharpness, 249
Sheet metal mills, 104, 307
SHEFFIELD UNIVERSITY, 328, 465
SHELL, 34, 299
Ships — *see* Marine
Shock Pulse Monitoring, 4, 11
Sieve analysis, 47, **135**, 168, **173**, 295ff, 365, 386, 391
Silica, siliceous, 13, 40, 147, 251
Silting, Silting index, 30, 34, 61, 116, **136**, **158**, **162**, 250, 366
Size, sizing methods, **44**, **179**
   Brownian motion diameter, 47
   Element techniques, 211ff
   Equivalent spherical diameter, 46
   Feret's diameter, 45, 48
   Intensity, 147ff
   Largest diameter, 45
   Martin's diameter, 46
   Mesh — *see* Filter blockage
   Projected diameter, 46
   Range, 50, **141 (chart)**, 212, 221, 227, 247
   Shape of distribution, 142
   Sieve diameter, 47
   Stokes' diameter, 47
   Volume diameter, 47
SKF, 8, 56, 57, 58, 445, 465
Slab shape — *see* Shape
Slide comparison — *see* Conpar
Slivers — *see* Shape
Slurries, 20, 85, 132, 176, 234
Smell — *see* Odour
SOAP — *see* Spectrometric analysis
Sodium chloride, 216
Sodium sulphate, 24
Solcenic fluid, 22
Solenoid failure, 61
Solvent (see also Diluent), 95, 96, 101, 117, 139, 149, 187, 202, 210, 240, 301
Sonic analysis, 274
SONOFLOC, 91, 465
Sooty, 275
Sound — *see* Ultrasonic
Source of debris — *see* Origin
SOUTHAMPTON INSTITUTE OF HIGHER EDUCATION, 328, 465

Spalling — *see* Bearings
Spatial Filtering System [Monitek], 124
SPC-510 [Spectrex], 123, 193, **432**
SPECTREX, 123, **190**, 193, 432, 466
SPECTRO ANALYTICAL, 218, 219, 222, 223, 227, 444, 466
Spectroil [Spectro], 223
Spectrometric analysis (SOAP), 4, 52, 108, 204ff, 211, **212**, 224, 228, 254, 270, 302, 305
Speed, 3, 134, 317
Spheres (see also Latex spheres), **37**
    Examination, 38, 83, 242, 258, 259
    Examples, 33, 38, 243
    Meaning, 29, **37**, 51
Sphericity — *see* Shape
Spiral shape — *see* Shape
SPIRE, 132, 433, 466
Spi-Wear [Spire], 132, 252, **433**
Standards, 49, 65, **271**
    Comparison, 283, 287
    List of relevant standards, 295ff
    including:
      AS 4059, 277, **278**
      ASTM E11 (Sieve), 386
      ASTM f-52-65T, 34
      BS 410 (Sieve), 386
      BS 3406, 289
      BS 5295, 95
      Conpar, 277, **282**
      DEF STAN 05-42, 277, **281**
      ISO 3310 (Sieve), 386
      ISO 3722, 96
      ISO 4021, 98
      ISO 4405, 146
      ISO 4406, 61, 73, 96, 160, 277, **279**, 283, 289, 313
      NAS 1638, 277, **278**, 283, 286, 309
      SAE 749D, 277, **280**
      STANAG 3510, 286
Statistics, 46, 65, 125, 180, 199
STAVELEY NDT TECHNOLOGIES, 121, 151, **154**, 384, 466
Steady state monitoring, 3
Steel works, 14, 104, 308
STEPTECH INSTRUMENT SERVICES, 410, 466

Stokes' diameter — *see* Size
Stokes' law, 37, 47, 83, 84, 134, 364
Stokes–Einstein equation, 201
Storage, 28, 322
STORMAGE SCIENTIFIC (see also SEIKO), 466
Strand shape — *see* Shape
SULZER, 88, 466
Sump — *see* Reservoir
Surface layer activation (SLA) — *see* Radioactivation
Surface tension, 82
SWANSEA TRIBOLOGY CENTRE, 36, 122, 151, 155, 247, 255, 257, 328, 352, 420, 430, 467
Swarf, 41, 51
SYMPATEC, 124, 168, **170**, 434, 467
Syringe — *see* Sampling
System 20 — *see* CM20

T
Table of instruments, 141
TAN — *see* Oil
Tank — *see* Reservoir
TBN — *see* Oil
TCP, 20
TecAlert — *see* Continuous debris monitor
TEDECO — *see* VICKERS SYSTEMS
Temperature (see also Colour, Ferrography, Thermography)
    Effect on analysis, 5, 25, 43, 44, 139, 258, 317
    Monitoring, 3, 10, 11, 13, 72, 303
Terotechnology, 298, 316, **323**
Test points — *see* Sample points
Texture — *see* Shape
THERMAL CONTROL — *see* HOWDEN WADE
Thermography (see also Temperature), 4, 8, 327
Thin layer activation (TLA) — *see* Radioactivation
Thin layer sensor wear (wear of thin film), 139, **163**, 249, 394
Time of transition, 111, **131**, 176, **265**, 395, 402

483

Timken machine, 37, 38
Training, **326**, 458
Trains (see also BRITISH RAIL), 14, 301
Transmissions (see also gears), 248, 283, 287
Transport — *see* Aircraft, Automobiles, Marine, Trains
Trend, 15, 72, 227, 232, **271**, 285, 286, 443
Tribology — *see* Bearings, Lubrication, Oil, Wear
TRIBOMETRICS, 122, 151, **153**, 308, 435, 467
Tribo Predict [CMS], 444
Tunnel boring machine, 312, 313
Turbidity — *see* Nephelometry
Turbidity, definition of, 127
Turbine, 16, 162, 300, 305
Turbulence — *see* Flow

U
UAS — *see* UNITED AIR SPECIALISTS
UCC INTERNATIONAL, 115, 125, 151, 155, **185**, 375, 417, 467
Ultrasonic
  Cleaning (bath), 66
  Filter — *see* Filters
  Testing, 137, 191
Ultrasound, 3, 9, **137**, 164, 409
Underground — *see* Mining
UNITED AIR SPECIALISTS, 91, 138, 148, 371, 467
Universities — *see* name of town/city

V
Vacuum, 26, 34, 76, 101, 103, 115, 119, 146, 230
Valves, 24, 55, 60, 61, 62, 66, **93**, 144, 197, 209, 317
Van der Waal, 83
VARIAN, 213, 214, 216, 223, 224, 468
Varnish — *see* Lacquer
VD AMOS, 122, 255, 388, 468

Vehicle fleets — *see* Automobile
VG ELEMENTAL, 236, 468
Vibration (including pulsation)
  Effect on analysis, 131, 181, 184
  Monitoring, 3, 4, 10, 11, 15, 104, 142, 274, 304, 305, 309, 327
  Property, 7, 8, 16
VICKERS SYSTEMS, 88, 89, 90, 116, 122, 151, 153, 427, 468
Video microscopy, 120, 132, 244, 248, 267, 268, 395
VIDS [Ai Cambridge], 441, 446
Viscosity
  Effect on analysis, 33, 80, 119, 125, 140, 181, 183, 212, 217, 225, 226, 266
  Measurement, 199, 274, 301, 416, 443ff
  Property, 23, 25, 27, 29, 34
Visi-cross — *see* Water content
Visual appearance, 3, 51, **138**, **179**, 243, 370, 387, 414, 416, 419

W
Wall friction, 81
Water (see also HFA etc, Corrosion Distilled water, Oil)
  In oil (contaminant), 22, 27, 115, 165, 181, 197, 275, 284, 301
  In oil (measurement), 25, 274
    Crackle test, 25
    Dean & Stark, 25
    Infra-red, 25
    Karl Fischer, 25
    Microwave, 25
    Visi-cross, 25
  Monitoring, 5, 462
Wear (see also Bearings, Pump, Thin layer sensor wear), **51**, **56**
  Abrasive, 5, 6, 52, 57, 140
  Adhesive, 6
  Cutting, 51
  Combined rolling and sliding, 52
  Fatigue, 7, 8, 41, 52
  Fretting, 44, 52
  General wear, 12, 15, 52
  Normal wear, 51

Wear–*contd.*
  Rolling fatigue, 51
  Rubbing, 51
  Scoring, 5, 6, 302
  Scuffing, 5, 6, 7, 52
  Severe sliding, 37, 41, 52
WEARCHECK, 444
WEARCHECK
  LABORATORIES, 444, 468, 469
Wear Debris Monitor [Fulmer Systems], 139, **163**, 249, **394**
Wear Particle Analyzer [Tribometrics], 122, 151, **153**, 308, **435**
Wear Particle Atlas [British Coal], 36, 333
Wear Particle Atlas [Spectro], 36, 39, 242, 257, 261, 354
Weight — *see* Gravimetric
Welding, 29, 37, 38, 64, 66

WHATMAN LABSALES, 103, 240, 469
WYATT TECHNOLOGY, 127, 155, **157**, 383, 469

X
X-ray, 3, 134, 171, 172, 228ff, 233ff, 410
X-ray diffraction (XRD), **233**
X-ray fluorescence (XRF), **228**, 302, 305

Z
ZDDP — *see* Oil additives
Zeta potential, 111, 203

Numbers
2-D, **45**
3-D, 45, **46**, 440

AUTHORS AND REFERENCES INDEX

Chapter 15 is an extensive bibliography with details of individual papers and books. The index below provides the location of all the references in both Chapter 15 and the main text and, where appropriate, those in Appendix 1.

A
Albidewi, 247, 330
Allen, C., 342
Allen, T., 330
Alsop, 330
Anderson, D.P., 9, 104, 212, 222, 333, 345
Anderson, M., 246, 330
Asher, 253, 306, 330, 351, 362
Astridge, 153, 305, 331, 427

B
Baker, 283, 284, 331, 376
Bayvel, 331
Beerbower, 52, 331
Bell, 63
Belman, 331, 427
Berg, 332, 385
BFPA, 66, 70, 277, 332
Blatchley, 332, 433
Bogue, 152, 332, 378
Bott, 333, 379
Bowen, 9, 51, 52, 104, 333
Bowns, 199, 341
BP Oil, 33
Bruno, 333, 409
Buck, 211, 212, 341
Burton, 342

C
Campbell, 333, 389
Childs, 196, 334, 432
Clayton, 334, 432
Clevenger, 56
Clifton, 328, 334
Conlon, 330
Connelly, 136, 162, 196, 336
Coulter, 334
Cowan, 334, 403
Cox, 345
Cumming, 105, 214, 219, 311, 334
Czarnecki, 208, 335

D
Dadd, 335
Day, 31, 55, 290, 312, 335
Dickson, 335, 384
Dunn, 136, 336
Dunthorne, 91, 350
Dwyer, 136, 162, 196, 336

E
Eisentraut, 350
Elrick, 175, 340, 365
Endecotts, 47, 135, 336, 386
Evans, 336

F
Fairhurst, 353, 369
Faulkner, 152, 312, 336, 389
Faure, 5, 6, 336
Feran, 342
Fisher, 354
Fitch, E., 7, 84, 291, 337
Fitch, J., 337, 377
Fitzsimmons, 56
Flynn, 355
Foulds, 284
Frank, 335
Franzl, 353
Fuell, 337

G
Gandhi, 28, 348
Gaucher, 337, 409
Gilbert, 356
Glacier Metal, 56, 57, 337
Goldsmith, 103, 241, 338, 376
Graham, 338, 365
Gregory, 166, 167, 338, 428
Gulla, 208, 213, 222, 338
Guttenberger, 353
Guy, 338

H

Hammond, 152, 345
Hanna, 338, 365
Hanseler, 338, 402
Harrison, 351
Hart, 333, 379
Harvey, 8, 284, 317, 339
Heimann, 208
Henry, 339
Hermann, 339
Heron, 159, 160, 339, 366
Herraty, 8, 284, 317, 339
Hipkin, 208, 339
HMSO, 54
Holmes, 284, 304, 339, 394
Horowitz, 175, 340, 365
Hughes, 159, 160, 339, 366
Huller, 49, 340
Hunt, I., 453
Hunt, T., 9, 12, 98, 158, 196, 199, 289, 309, 340, 341, 350, 355, 390, 456
Huo, 262, 308, 356

I

Ionnides, 23, 56, 58, 341

J

Jacobson, 23, 56, 58, 341
Jantzen, 105, 211, 212, 341, 345
Jarvis, 34
Jin, 37, 38, 39, 342
Joffe, 342
John, preface v, 342
Johnson, 306, 342
Jones, A., 331
Jones, M., 9, 104, 342
Jones, W., 308, 342, 435

K

Karasikov, 343, 395
Kasten, 196, 343
Kibble, 12
Kim, 343
Kind, 343
Kohlhaas, 26, 27, 286, 343
Krauss, 343, 395
Kreikebaum, 157, 343, 373
Krulish, 334
Kuhnell, 351
Kwon, 343, 420, 430

L

Lantos, 22, 24, 59, 60, 84, 344, 447
Larkin, 308, 342, 435
Lasentec, 307, 344
Lewis, D., 302, 303, 304, 341, 384, 444
Lewis, R., 344, 435
Lines, 295, 345
Lloyd, O., 152, 305, 345
Lloyd, P., 345
Loipführer, 335
Lukas, 212, 222, 345
Luxmoore, 247, 330

M

MacIsaac, 152, 312, 389
Macpherson, 57, 350
Maier, 105, 345
Mann, 350
Mathieson, 300, 352, 384
McCadden, 208, 213, 222, 338
McFadyen, 346, 406
Miller, B, 200, 346
Miller, C., 155, 346
Mills, 152, 346, 355, 378, 459
Minns, 346
Montague, 351
Morley, 301, 346
Mucklow, 108, 347

N

Neale, 309, 315, 347
Nelson, 166, 167, 338, 421
Newell, 31, 347
Newhouse, 354
Nicholls, 309, 324, 347

O

Oh, 343

P

Perkins, 217, 347
Peterson, 348
Pipe, 10, 348
Prakash, 28, 348

Preikschat, 338, 402
Price, 247, 330, 348, 420, 430
Provost, 348

Q
Qiao, 262, 308, 356
Quantachrome, 349

R
Ramos, 349, 407
Ranco Controls, 310, 349, 378
Rao, 328, 349
Raubenheimer, 299, 349
Raw, 158, 196, 350, 389
Rhone, 350
Rice, 341
Rideal, 350
Riley, 350
Roylance, 247(2), 330, 348, 430
Rumberger, 155, 346

S
Saba, 212, 350
Sandoval, 356
Santilli, 272, 350, 394
Sasaki, 91, 350
Sayles, 57, 350
Schröeder, 105, 345
Schwabe, 253, 306, 351, 362
Seaman, 351
Seinsche, 335
Seow, 351
Shapespeare, 46, 134, 351
Silebi, 407
Sioshansi, 332, 433
Smith, G.J., 336
Smith, G.S., 256, 257
Sommer, 351, 397
Spair, 127, 352
Stafford-Allen, 302, 352
Stecki, 246, 330
Stewart, 10, 346, 352
Summerfield, 300, 352, 384
Swank, 196, 352

T
Tauber, 352, 427
Thomas, G., 353, 379
Thomas, J., 353, 369
Thompson, 229, 353
Tilley, 85, 86, 341, 353
Tofield, 330
Tweedale, 6, 353
Tumbrink, 55, 290, 335

U
Uedelhoven, 247, 353
Underwood, 354, 432

V
Vogel, 342
von Bernuth, 354, 391, 392, 393

W
Walsh, 229, 326, 327, 353, 354
Wang, C., 37, 38, 39, 342
Wang, G., 247, 330
Weichart, 49, 340
Wenman, 334, 403
West, 294, 355, 372, 374
Westcott, 51, 333
Whittington, 153, 355
Wicks, 336
Wilkes, 289, 309, 355, 390
Wilkins, 330
Williams D., 355
Williams P., 263, 265, 355, 430
Wilmott, 4, 355
Winer, 348
Wyatt, 305, 356

X
Xu, 262, 308, 356

Y
Yarrow, 300, 356
Yurko, 222, 356

Z
Zie, 348